D0713800

GLOBAL ENERGY

GLOBAL ENERGY

Assessing the Future

JAE EDMONDS
JOHN M. REILLY
Institute for Energy Analysis
Oak Ridge Associated Universities

New York Oxford
OXFORD UNIVERSITY PRESS
1985

Oxford University Press,
Walton Street, Oxford OX2 6DP

London New York Toronto
Delhi Bombay Calcutta Madras Karachi
Kuala Lumpur Singapore Hong Kong Tokyo
Nairobi Dar Es Salaam Cape Town

and associate companies in
Beirut Berlin Ibadan Mexico City

Published in the United States by
Oxford University Press, New York

LIBRARY OF CONGRESS CATALOGING IN PUBLICATION DATA

Edmonds, Jae.
Global energy.

Includes index.
1. Power resources—Mathematical models. 2. Energy consumption—Mathematical models. I. Reilly, John M. (John Matthew), 1955- . II. Title.
TJ163.2.E32 1985 333.79 84-27190
ISBN 0-19-503522-4

DISCLAIMER

This book is based on work sponsored by the U.S. Department of Energy, an agency of the United States Government. Neither the United States Government nor any agency thereof, nor any of their employees, makes any warranty, express or implied, or assumes completeness, or usefulness of any information, apparatus, product, or process disclosed, or represents that its use would not infringe privately owned rights. Reference herein to any specific commercial product, process, or service by trade name, trademark, manufacturer, or otherwise, does not necessarily constitute or imply its endorsement, recommendation, or favoring by the United States Government or any agency thereof. The views and opinions of authors expressed herein do not necessarily state or reflect those of the United States Government or any agency thereof.

Printing (last digit): 9 8 7 6 5 4 3 2 1

Printed in the United States of America

Preface

This book documents a research project begun in 1979. At that point there had been no attempt to draw on the huge literature devoted to forecasting energy use and apply it to the CO_2 issue, even though it was widely recognized that the rate of increase of atmospheric carbon dioxide was heavily dependent on fossil energy-use scenarios.

The project was originally conceived to be rather modest—perhaps four or five man-years of effort over two years. While the promised products were delivered to the Department of Energy on schedule, it was clear that more effort should be devoted to this research. The project has continued through this writing (February 1985). Obviously, much more remains to be done. While most of the work reported here was largely finished by late 1981, the writing of this volume, particularly the results sections, has stretched through to the present, and there has been major substantive editing of some earlier chapters to eliminate weaknesses or update and expand coverage.

The work grew out of a long history of energy-economic research at the Institute for Energy Analysis (IEA) which dates from 1975. The first efforts of the research are documented in an IEA report, *Global Energy Production and Consumption in 2000.* In producing this document, it became clear that some redirection would be useful if the effort was to have a lasting and interactive role with the physical science work being done on the CO_2 issue. A basic requirement was the ability to explore in some way the uncertainties associated with any long-term forecast. This strongly suggested the need for a computer-based simulation model of the global energy system.

The unique element of this project is that an analysis of supply and demand factors preceded construction of the model; the model was built to be consistent with and to represent the data available rather than the other way round.

Thus, the chapters discussing important determinants of supply and demand are not limited to the parameters and variables explicitly contained in the model nor should they be considered an after-the-fact construction to justify the model structure. They are the reference documents developed for use in constructing the model and developing data sets. We hope that the review of these factors will prove useful and provide some insights for the reader, quite apart from the model or results of the modeling exercise.

We also hope that the reader who disagrees with our input assumptions, or who views the results with disbelief, or who finds himself asking additional questions, can find the model itself useful. The computer tape of the model is on file at the Carbon Dioxide Information Center at the Oak Ridge National Laboratory, Oak Ridge, Tennessee, and we have transferred the model to the U.S. Environmental Protection Agency. The Gas Research Institute and a group of scientists at the Massachusetts Institute of Technology's Energy Laboratory have also used it as an aid in answering specific questions. Toward this end we have developed a user's guide and software description. We gratefully acknowledge the financial support of the U.S. Department of Energy, Carbon Dioxide Research Division, and particularly Fred Koomanoff, office director, and Roger Dahlman for making money for this project available. Moreover, they have firmly supported efforts to document the model and test its sensitivities. Without this level of documentation, the model would be considerably less accessible to other users.

Finally, we hope the reader will find the key energy and CO_2 results useful. More frequently, such results form the bulk of a book of this type and are viewed as the principal products of the research. Given the tremendous uncertainty about the long-term future, we believe that it would be unfair to expect the reader's blind trust or complete agreement with every assumption. We hope that we have provided enough background for readers to judge for themselves both the usefulness and limitations of the results.

Washington, D.C. J. A. E.
February 1985 J. M. R.

Acknowledgments

A book of this size is hardly possible without the assistance and cooperation of a vast number of people. The type of help we received in putting this book together ranges from financial support to the actual writing of chapters or major sections of chapters, research assistance, typing, editing, managing the publication process, and finally the technical review of various parts of the manuscript. In addition, there are the many intellectual debts one accumulates over time in give-and-take discussions with colleagues that become so deeply embedded in one's own views that it is no longer possible to separate them.

We owe a large debt to Ed Allen, who gave the project a firm push forward before retiring, and to Chester Cooper who, as director of the Washington office of the Institute for Energy Analysis, encouraged and supported the effort through literally years of work and who is responsible for bringing together a group of people who made the whole effort fun. Beyond that, both Chet and Ed helped shape the project in a thousand subtle ways that are simply impossible to acknowledge fully in writing.

Several individuals made specific contributions. Rayola Dougher was the primary author of chapters 4 and 11, and contributed a major portion of chapter 9. She provided invaluable research assistance throughout the project, surveying the literature, developing data sets, and compiling tables. Ray also patiently read through the chapters, helping to provide a clear focus through her careful attention to writing style and manuscript organization. Richard Mack wrote chapter 13 and part of chapter 14.

Doan Phung provided an initial draft of chapter 12. Nathan Keyfitz and Barbara Wiget provided a paper on population forecasts and Ed Allen developed labor force estimates from these forecasts. These two papers formed the background for chapter 2. Together, these papers provided an excellent basis for chapters 2 and 12 but were heavily rewritten to fit the style and format of the volume. We gratefully acknowledge all of these contributions but accept full responsibility for any errors or misinterpretations that have crept in.

We would also like to acknowledge the support of Alvin Weinberg, the director of the Institute for Energy Analysis, both for the project and for continued CO_2 research in general. Our general thanks go out to the entire Institute staff, most notably Bill Clark, who could always be depended on to provide a new perspective, challenging ideas, considered guidance, and boundless enthusiasm; thanks also to Gregg Marland, David Reister, and Ralph Rotty.

Many individuals outside the Institute provided formal and informal reviews of portions of the work. These include Jesse Ausubel, Philander Claxton, Bob Kuenne, Lester Lave, Marvin Miller, Alfred Perry, Peter Poole, Jeff Pransky, David Pumphrey, David Rose, and many others too numerous to mention.

We would also like to acknowledge the help of a long list of interns who spent a semester or summer working on the project, including Mary Christ, Barbara Gaige, Michelle Hall, Elizabeth Henderson, David Isola, Peter Natiello, Bridget O'Reilly, Lynn Quincy, Joy Roper, David Schmookler, Shelley Spencer, and Stephen Tull.

Over the several years the project has run, we have benefited from a tremendous amount of technical assistance. Dan Wahl provided expert computer assistance. Barbara Disckind, Yvonne Dunton, Paula Keene, and Fay Kidd battled through horrible handwriting and innumerable edits and re-edits of the book to provide a final typewritten manuscript. Ruth B. Haas provided capable, professional, and friendly editing. Susan Kincade took it upon herself to make sure the final version of the manuscript was in order after we had reached the point where we felt we couldn't possibly read through it one more time. And David Armbruster managed our side of the publishing end of the project—we don't know what this involved and are glad we don't know.

We would also like to acknowledge the support Oak Ridge Associated Universities gave to the publication of this manuscript. Finally, none of this would have been possible were it not for the financial support of the U.S. Department of Energy, Carbon Dioxide Research Division. We are particularly grateful to the director of the office, Fred Koomanoff, and to Roger Dahlman, with whom we have interacted over the course of many years.

Contents

GLOBAL ENERGY

Overview 1

This book looks across two generations to the future of man's global energy system in an attempt to answer the questions: During the next century, how might the world produce and consume energy? In what amounts? How will these consumptions of these amounts affect the environment? In the short run energy production and consumption take place within the context of existing capital stocks, but in the long term major shifts in both sources and uses of energy are possible since even long-lived capital, like utility power plants, will be replaced. Moreover, recognition of global interdependence implies that countries and regions may import or export fuels but the ability to import depends on the availability of resources for export elsewhere; that is, the global energy system must balance.

There are two natural viewpoints from which energy problems can be viewed—energy production and energy use. Both yield important insights, yet full understanding of the energy system is possible only when both are integrated to produce an overall global energy balance. The organization of this book flows naturally out of this realization and falls into four major sections: chapters 2 to 5 discuss long-term global energy use; chapters 6 to 15 look at long-term global energy production. These are followed by the presentation of an approach for simulating long-term global energy balance and a discussion of the supply/demand forecasting results.

ORGANIZATION OF THE STUDY

In this analysis the world is divided into nine geographic regions based on energy resources and reserves, political boundaries, economic and technical compatibility, social similarities, and geographical proximity. They include four developing regions, two centrally planned regions, and three OECD regions (figure 1-1). (Table 1-1 contains a list of countries included in each region.)

This division is a compromise between the problems of gathering disaggregated information and the usefulness of recognizing geographical diversity in the sources and uses of energy. In addition to the data problems associated with greater global disaggregation, long-term forecasts must also recognize that beyond a point, the increased information gained from the disaggregation process is outweighed by the greater uncertainty surrounding forecasts for numerous independent units. For some purposes combining countries like India and South Korea into a common region represents too great an aggregation. For others, four developing regions may represent too great a disaggregation. On balance, however, we feel that this level of aggregation has proved useful, allowing us to

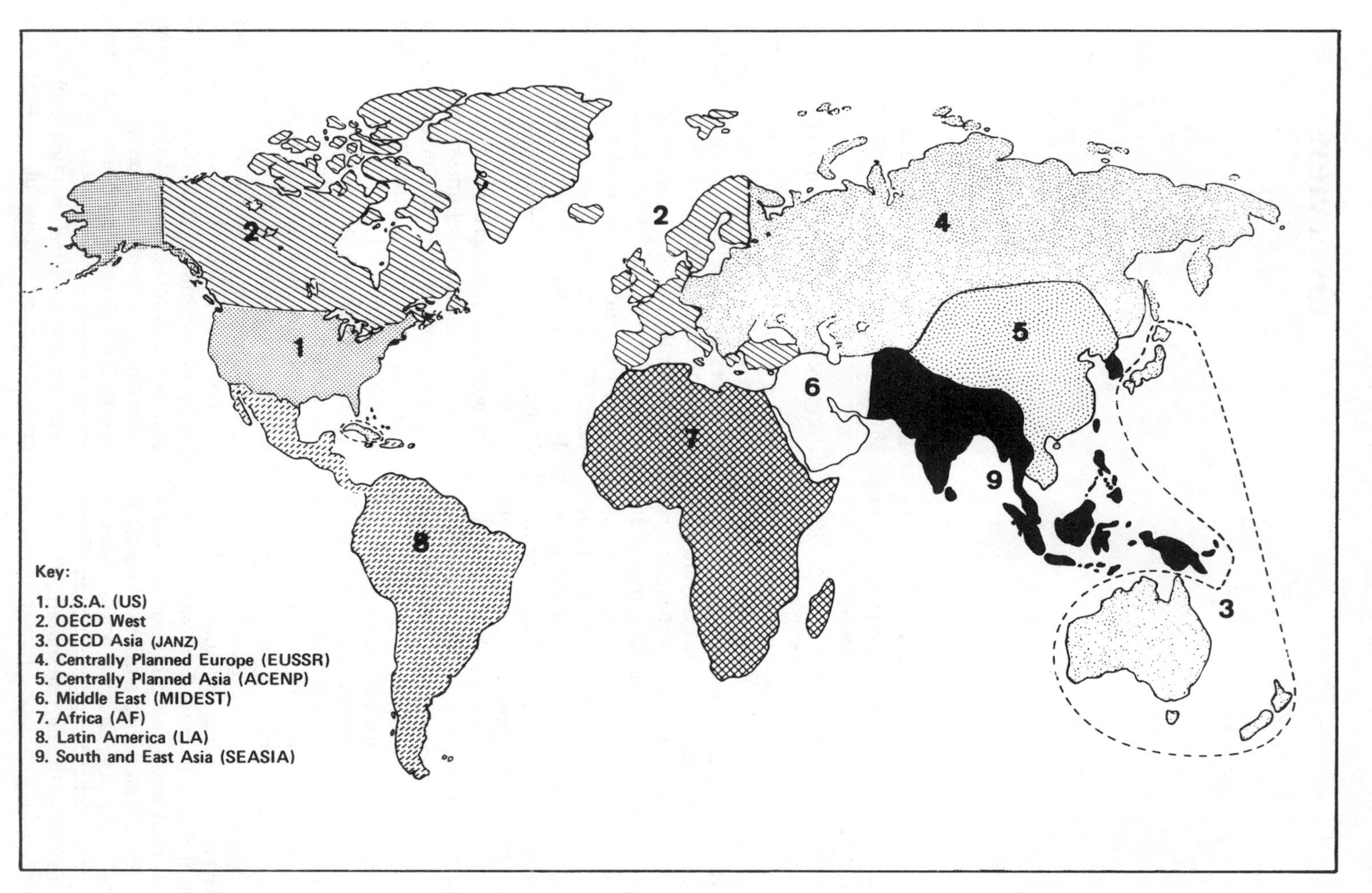

FIGURE 1-1. Geopolitical divisions used in the study.

TABLE 1-1. Regional Distribution of Countries in the ORAU/IEA Assessment Framework

Region 1 (US): United States	
United States	
Region 2 (OECD West): Western Europe and Canada	
Austria	Italy
Belgium	Luxembourg
Canada	Netherlands
Denmark	Norway
Finland	Portugal
France	Spain
Fed. Rep. of Germany	Sweden
Greece	Switzerland
Iceland	Turkey
Ireland	United Kingdom
Region 3 (JANZ): OECD Asia	
Japan	New Zealand
Australia	
Region 4 (EUSSR): Centrally Planned Europe	
Albania	Hungary
Bulgaria	Poland
Cuba	Rumania
Czechoslovakia	Soviet Union
East Germany	Yugoslavia
Region 5 (ACENP): Centrally Planned Asia	
China	Mongolia
Kampuchea	North Korea
Laos	Vietnam
Region 6 (MIDEST): Middle East	
Bahrain	Oman
Cyprus	Qatar
Iran	Saudia Arabia
Iraq	Syria
Israel	United Arab Emirates
Jordan	North Yemen
Kuwait	South Yemen
Lebanon	
Region 7 (AFR): Africa	
Algeria	Mali
Angola	Mauritania
Benin	Mauritius
Botswana	Morocco
Burundi	Mozambique
Cameroon	Namibia
Central African Empire	Niger
Chad	Nigeria
Region 7 (AFR): Africa, cont.	
Congo	Rwanda
Egypt	Senegal
Equitorial Guinea	Sierre Leone
Ethiopia	Somalia
Gabon	South Africa
Gambia	Sudan
Ghana	Swaziland
Guinea	Tanzania
Ivory Coast	Togo
Kenya	Tunisia
Lesotho	Uganda
Liberia	Upper Volta
Libya	Zaire
Malagasy Rep.	Zambia
Malawi	Zimbabwe-Rhodesia
Region 8 (LA): Latin America—Mexico, Central America, Caribbean, and South America	
Argentina	Jamaica
Bolivia	Mexico
Brazil	Netherlands Antilles
Chile	Nicaragua
Colombia	Panama
Costa Rica	Paraguay
Dominican Republic	Peru
Ecuador	Puerto Rico
El Salvador	Surinam
Guatemala	Trinidad and Tobago
Guyana	U.S. Virgin Islands
Haiti	Uruguay
Honduras	Venezuela
Region 9 (SEASIA): Non-Communist South, East and Southeast Asia	
Afghanistan	Nepal
Bangladesh	Pakistan
Brunei	Papua New Guinea
Burma	Philippines
Fiji	Singapore
Hong Kong	South Korea
India	Sri Lanka
Indonesia	Taiwan
Malaysia	Thailand

recognize varying rates of population, economic growth, and energy resource endowments without getting lost in excessive detail.

Four benchmark years were chosen for this analysis: 1975, 2000, 2025, 2050. The analytical tools for long-term energy assessments differ greatly from those used in short-term analyses.

As with geography, it often proves useful to aggregate energy by mode of delivery. There are four final use or secondary energy categories that we consider: liquids, gases, solids, and electricity. These four are, of course, aggregates of numerous refined products. Liquids, for example, consist of such disparate fuel types as motor gasoline, residual oil, and asphalts. Coal consists of anthracite, lignite, peat, and biomass. Electricity is the most homogeneous of the secondary energy types, but even here its form upon delivery varies with local conventions.

The primary energy types used to produce these four secondary fuels are even more varied. Six primary energy categories are used: liquids, gases, solids, nuclear, solar electric power, and hydroelectric power. Within these categories, nine subcategories or subaggregates are identified (see table 1-2).

Each of these subcategories in turn is disaggregated further. Thus, for example, liquids are made up of conventional oil, and a category of unconventional oil which contains enhanced recovery, shale oil, and tar sands. The contributions of each of these energy categories is discussed generally in the following section and in detail in chapters 6 to 15.

TABLE 1-2. Primary Fuel Categories in the Model

- 1.0 Liquids
  - 1.1 Conventional Oil
  - 1.2 Enhanced recovery, shale oil, and tar sands
- 2.0 Gases
  - 2.1 Conventional Gas
  - 2.2 Non-conventional Gas
- 3.0 Solids
  - 3.1 Coal
  - 3.2 Biomass
- 4.0 Resource-constrained renewables
  - 4.1 Hydro, geothermal
- 5.0 Nuclear
  - 5.1 Conventional reactors and breeder reactors
- 6.0 Solar
  - 6.1 Solar electric (other solar is associated with conservation), wind power, and tidal power, OTEC, fusion, and other advanced renewable technologies

ENERGY SOURCES

Energy resources defy easy, meaningful comparisons. Resource amounts can be estimated in conventional units—barrels, cubic feet, or tons—converted to joule or Btu equivalents and totaled to give world energy resources, but at once one confronts the problem of comparing energy sources which yield a fixed flow each year with those that represent a fixed stock that will be exhausted over time. It also stretches the limits of plausibility to argue that one can sum over disparate resource grades (e.g., low-cost oil versus high-cost coal liquefaction) and arrive at a total that is meaningful. In addition, there is an essential difference across energy carriers—liquids, solids, gases, and electricity—in terms of associated costs and quantities of useful energy in converting the energy to its final form (heat or power). While coal is cheap (dollars per joule) and abundant, the costs of producing private transport service (automobile transportation) from solid fuel are very high. For power requirements, electricity costs (dollars per joule of electricity) cannot be compared with liquids, gases, and solids because the transformation from heat to power (and the associated losses due to efficiency factors) is already accounted for in electricity whereas it is not accounted for in other energy carriers.

In view of these basically unresolvable problems, figure 1-2 summarizes the full spectrum of current and potential sources of energy. The figure categorizes energy sources along dimensions of abundance and cost. Resources found in the upper left are abundant and low cost. As one moves to the right, one finds resources that are available at greater cost; as one moves down, one finds resources that are constrained in terms of ultimate contributions to global energy supplies. Positions are approximate and judgmental. While discrete boxes and categories are drawn, these should be interpreted only as indicative of the resource category. In many cases a particular resource can be graded, yielding a

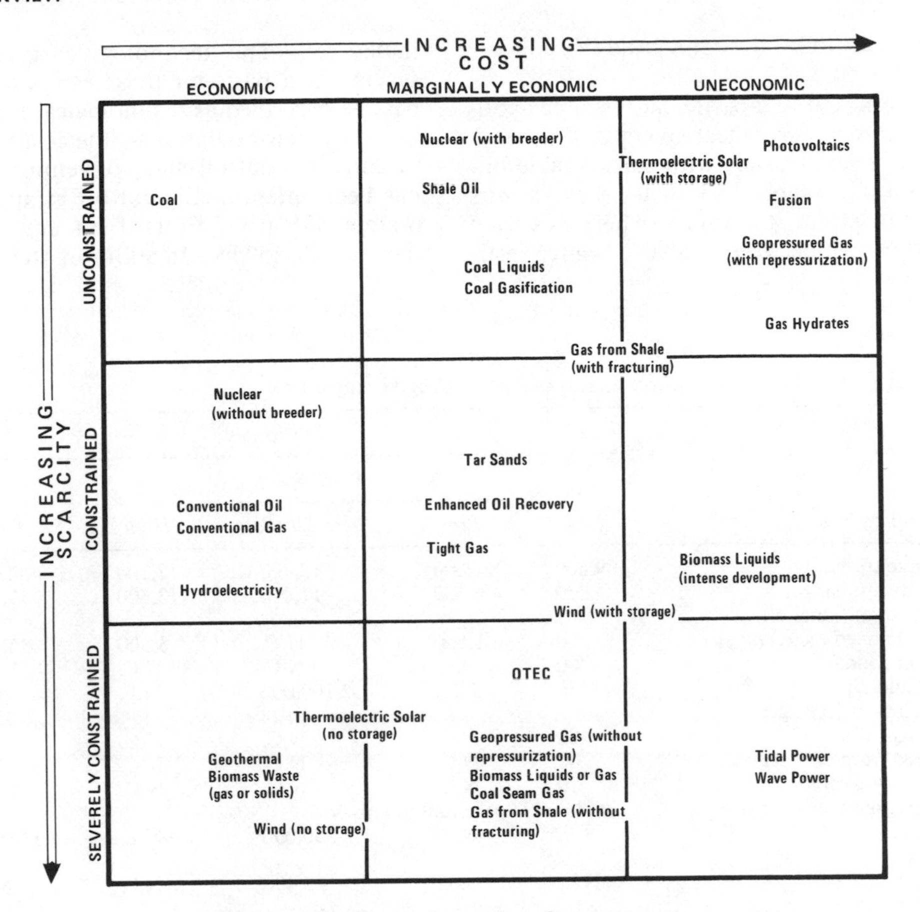

FIGURE 1-2. Cost and scarcity of energy resources.

considerable range of prices over which supplies will be forthcoming, or conversely, a range of amounts of the resource which will be available. However, in examining energy resources, one finds that cost variation within a particular resource tends to be less important than variation in cost across technologies. In part, and perhaps in large part, this conclusion represents a tautology; the examination of energy sources has been structured as an examination of different technologies. An alternative structure would have been to examine amounts of liquids available at different costs. One would have found large variation in costs across different grades (technologies) of primary liquid resources. Yet, such a structure, while perhaps a natural approach to examining final demands for energy, is forced when it is used to examine supplies of primary energy.

Complicating estimates of costs and amounts of energy supplies is uncertainty. There are three general types of uncertainty in resource estimates: (1) geological uncertainty—uncertainty as to the size of the stock or flow of the resource of particular resource grade; (2) technological uncertainty—uncertainty as to the amount of the existing resource that can be recovered; and (3) economic uncertainty—uncertainty as to the cost per unit of recovering a particular grade of the resource.

GEOLOGICAL UNCERTAINTY

The amount of a particular resource available under a given technology and price regime is known only with considerable uncertainty. Table 1-3 lists the various resources and the range (where possible) within which recoverable resource estimates fall. The technology/cost regime serving as a basis for these estimates can best be described as technological feasibility with a few exceptions—the cost regime in conventional oil and conventional gas has been intentionally cut off at approximately \$35 per barrel of oil equivalent (boe) (1979 dollars). In addition, the entire

TABLE 1-3. Global Resource Estimates and Range (exajoules)

	Full Range				
		Core Range			
Resource Type	*Low*	*Low*	*Best Estimate*	*High*	*High*
Conventional oil	9,800	11,600	13,400	12,300	34,300
Conventional gas	6,300	6,300	11,400	13,500	17,300
Unconventional oil					
Enhanced recovery[a]	0	1,500	N.C.	5,500	23,100
Tar sands	700	N.C.	N.C.	N.C.	7,500+
Shale oil	N.E.	N.E.	2,100,000	N.E.	N.E.
Unconventional gas					
Tight gas	400	N.C.	N.C.	N.C.	900+
Gas from coal seams	200	N.C.	N.C.	N.C.	900+
Geopressured brine	10	N.C.	N.C.	N.C.	5,300+
Hydrate zones	Exploitable potential unknown.				
Coal	N.E.	N.E.	330,000	N.E.	1,320,000+
Nuclear					
LWR	N.E.	600[b]	3,400[c]	N.E.	N.E.
HWR (e.g., CANDU)	835[d]	N.E.	9,200[e]	N.E.	N.E.
Breeder	No likely constraints.				
Biomass[f]	50/yr	N.C.	N.C.	N.C.	200/yr
Hydroelectricity	N.E.	N.E.	40/yr	N.E.	N.E.
Solar	Global insolation far exceeds any possible global energy requirements.				
Wind	Availability poses no likely constraint on development.				
Geothermal/tidal/wave	Severely constrained.				
Fusion	No resource constraints.				

N.E.: No estimates—relatively few (one) estimates available. Should be interpreted as order of magnitude estimate.

N.C.: No consensus—available estimates are approximately uniformly distributed over the given range.

+: Indicates possibility of much higher upper end—available estimates based largely on resources of one or a few countries with minimum extrapolation of occurrences to other land areas or none.

[a]Extraordinary recovery from conventional deposits. Based on cost rather than technology description usually associated with EOR. Includes deep offshore and polar oil. See chapter 8.

[b]Assumes 5.5×10^6 tons of U_3O_8 and 6,010 tons U_3O_8 (without recycle) required per 21 GWyr(e) produced.

[c]Assumes 24.5×10^6 tons of U_3O_8 and 4,750 tons U_3O_8 (with recycle) required per 21 GWyr(e) produced.

[d]Assumes 5.5×10^6 tons of U_3O_8 and 4,380 tons U_3O_8 (without recycle) required per 21 GWyr(e) produced.

[e]Assumes 24.5×10^6 tons of U_3O_8 and 1,770 tons U_3O_8 required per 21 GWyr(3) produced.

[f]Low estimate, only waste available; high estimate requires biomass farms on land not used for food and fiber requirements up to a total (energy plus other uses) of 40 percent of arable land under cultivation. Assumes efficiency of conversion is one third.

Estimates greater than 200 EJ rounded to nearest 100 EJ; estimates less than 200 EJ rounded to nearest 10 EJ.

question of geological uncertainty is moot for such superabundant energy resources as coal, shale oil, and solar; for these resources, cost is the crucial dimension. Similarly, fusion energy and nuclear breeder technology face no resource constraints.

The core range of conventional oil estimates is fairly small—the upper end is only 130 percent of the lower end. Conventional gas estimates are somewhat more uncertain, with the core upper end value approximately twice the lower end value. Such a level of uncertainty is relatively low in terms of other uncertainties connected with projections as far out as 2050. For other resource estimates the range of values is much greater.

COST AND TECHNOLOGICAL UNCERTAINTY

Technological and economic uncertainty can actually be viewed as two sides of the same coin. Technological recoverability has an economic dimension—unrecoverable resources often are potentially recoverable but only under very costly or somewhat perverse conditions. For example, energy has been generated by fusion technology but the amount generated is less than the amount of electricity required to contain the fusion reaction. Yet, such seemingly perverse situations—more energy used than produced—may still result in a sustainable, competitive industry if low-grade (low-cost) energy is used to produce a premium energy carrier. Despite these problems with defining technological uncertainty, the terminology is maintained because demonstrating technological feasibility represents a real-world process. It can best be thought of as a precursor to estimating costs of the technology and as an indication that ultimate costs are within some range which makes production on a pilot plant level possible, at least as a research project.

Figure 1-3 indicates the cost ranges for various sources of energy. Fusion energy and tidal and wave power are not represented because of the technological uncertainty connected with these sources. The cost ranges given for coal, conventional oil and gas, conventional hydro, and biomass feedstocks represent primarily different resource grades. Conventional thermal electric and nuclear power plant cost ranges indicate cost variation due primarily to scale, holding the cost of the fuel input constant. The cost ranges indicate cost uncertainty owing to technological immaturity. Cost estimate ranges where the high end value is two to three times the low end value must be considered relatively certain in terms of new technologies. Less certain estimates are available for many technologies, with the high end value five times or more the low value.

DETERMINANTS OF ENERGY USE

With the exception of energy prices, the central determinants of global energy use are greatly different from the determinants of global energy sources. We focus on five key factors: demographics, labor productivity, income, energy prices, and energy productivity.

Demographics

Demographics affect the level of energy demand both through the number of people who will be consuming energy, and through the number of laborers who will be using additional energy, producing goods, and using transportation outside of the household. A base population projection was prepared by Nathan Keyfitz of Harvard University. There is great uncertainty surrounding world population forecasts, with estimates of late twenty-first century global population varying between 8 and 12 billion. We find that the most compelling arguments support a lower number (8.2 billion by 2050 and 8.5 billion by 2075), with global population approaching stationarity. Both the developed and centrally planned economies are likely to reach stationarity before the less-devel-

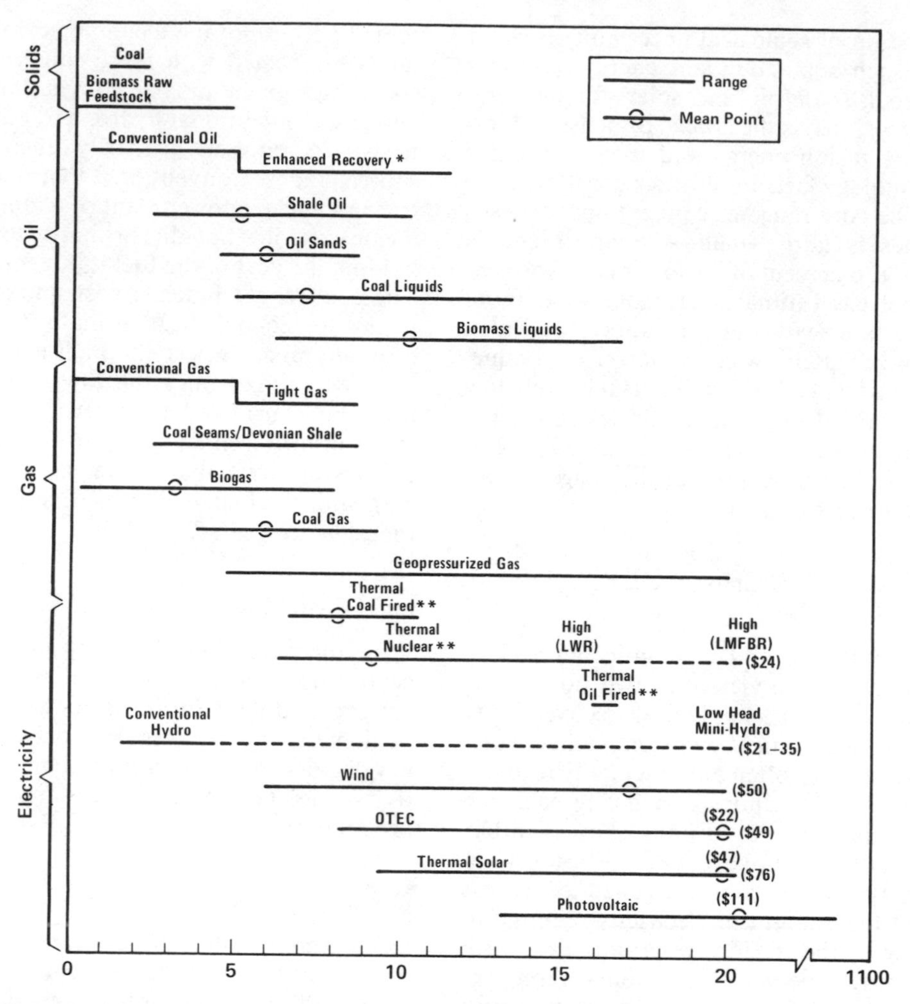

FIGURE 1-3. Range of cost estimates for fuels ($1979/gigajoule).

*By definition, enhanced recovery includes oil recovered at cost greater than conventional oil.

**Based on the following fuel costs ($1979/gigajoule): $1.39 for coal, $3.06 for nuclear, and $10.75 for oil. Costs are at the busbar.

oped economies, particularly India, Indonesia, Brazil, and Pakistan. Population growth by region over the period between 1975 and 2050 is shown in figure 1-4. As can be seen, the developing regions of the world which are home to less than half of the world's population in 1975 grow to over 60 percent by 2050. The developed and centrally planned regions continually shrink in population share, though their actual numbers continue to expand slowly as they move toward stationarity.

Population of labor force age is an important determinant of energy growth because it indirectly affects the determination of gross national product (GNP) and economic activity. The employed labor force affects energy use in the residential/commercial, transportation, and the industrial sectors.

This demographic cohort grows more rapidly than the global population, and the shift in its global distribution is pronounced. In 1975 there were 2.55 billion

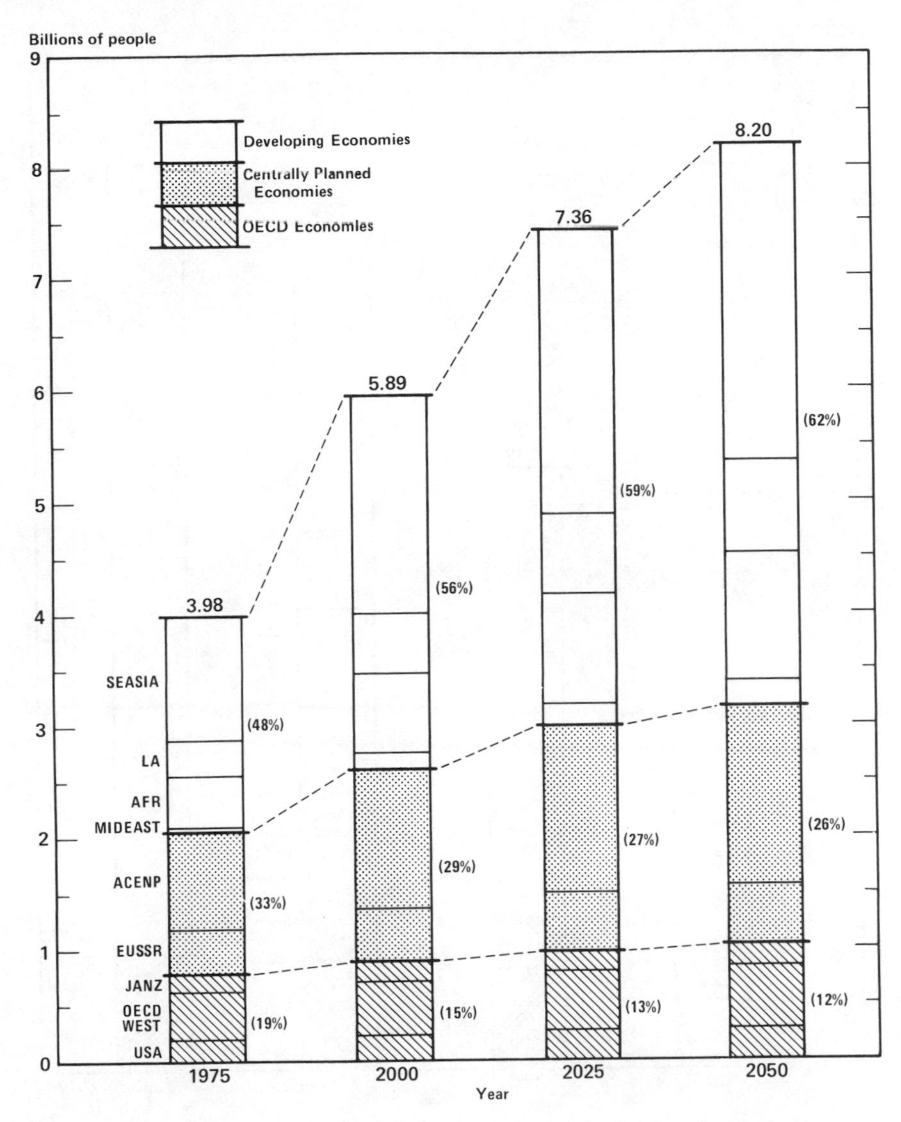

FIGURE 1-4. Global populations by region, 1975–2000 (billions).

persons of labor force age in the world (see figure 1-5), representing 64 percent of the global population. More than half of these were living in the OECD and centrally planned economies. By 2050 this number is expected to reach 6.52 billion, though most of the growth occurs by 2025. The growth in labor force age population is more rapid than the general population as a result of the trend toward stationarity, which gradually lowers the portion of very young persons in the population. By 2025, the global population of labor force age has reached 80 percent of the total population, up from 64 percent in 1975. In fact, the 2050 global average is higher than the U.S. average in 1975 (75 percent). Changing social institutions, which will encourage more women to enter the labor force, are expected to allow the labor force to grow at least as rapidly as the labor force age population.

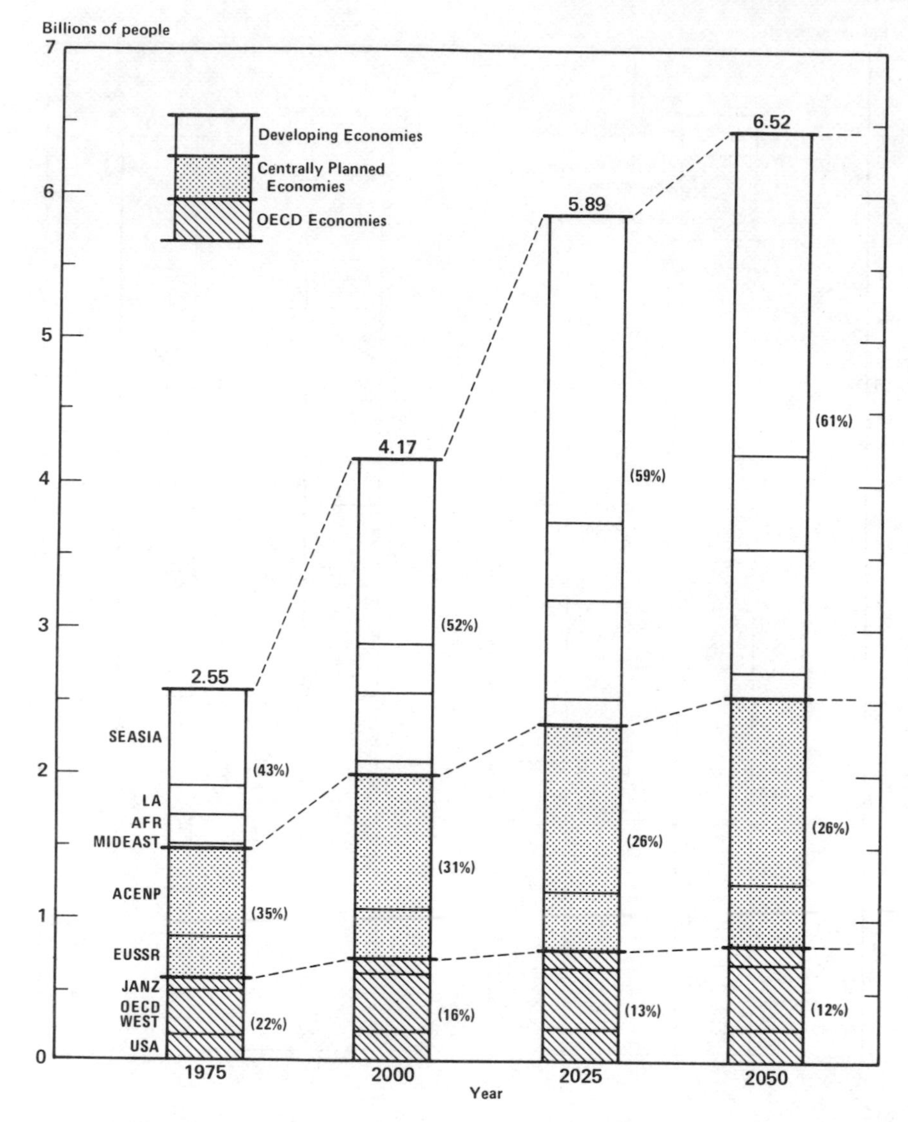

FIGURE 1-5. Global populations of labor force age, 1975–2050 (billions).

Labor Productivity

Labor productivity, or more accurately, the rate of productivity growth, has been a source of current controversy. It is also a major determinant of energy use. Moreover, the influence is indirect. Labor productivity interacts with the labor force to help determine the level of energy demand through its influence on gross national product. By definition, the product of labor productivity with the labor force is the GNP, which in turn is indicative of levels of income and output. Income is a major determinant of energy demand by households, while output is a major determinant of energy demand by producers. Higher rates of productivity growth are therefore associated with higher growth rates for energy use. In this study we have looked at the long-term statistics on labor productivity growth in order to couch future projections in the appropriate historical context.

There has been a great deal of pessimism concerning recent trends in labor productivity. OECD economies achieved relatively poor records of productivity growth during the late 1970s, and there are many who contend that, on the basis of this experience, future long-term economic growth, and hence energy demand, will likely be low.

However, it is useful to examine labor productivity assumptions in the light of longer term historical analysis. Since the year 2050 lies 75 years away from the initial benchmark date, 1975, we felt that the historical record should be examined at least as far back as 1900 wherever possible. Historical series of such length are available in only a few countries. Many modern countries cannot even trace their present geographical boundaries, let alone energy data series, further back than World War II. Gross national product itself is a post-1930s concept, so that data for earlier years, where they exist at all, are an after-the-fact reconstruction.

All of these problems notwithstanding, we have been able to develop long-term historical statistics for three economies—the United States, the United Kingdom, and Japan—and have examined Simon Kuznetz' work on several others. The results of these exercises indicate that long-term rates of labor productivity growth exhibit more stability than short-term rates. We can document only one instance in which labor productivity was actually lower after 25 years than at the beginning of the period (the United Kingdom from 1901 to 1931). We note that low and negative rates of labor productivity growth have been balanced by growth periods and that for developed economies productivity improvement has been the long-term norm. For the United States, productivity grew at an average annual rate of 1.5 percent between 1870 and 1980. For the United Kingdom, the rate was 1.2 percent between 1861 and 1971. And for Japan, the rate was 3.0 between 1886 and 1976.

As a consequence of our investigations, we feel comfortable with assumptions of continued growth in labor productivity. We are also comfortable with optimistic forecasts of labor productivity growth in the developing economies. The Japanese example indicates that rapid labor productivity growth is sustainable over long periods, even where there are times of devastation, such as that experienced in World War II. While the ebullient expansion in global productivity of the 1960s may not be a long-term norm, productions of flat productivity based on a few dismal quarters during the 1970s are almost certainly overly pessimistic.

Income

There is a very important relationship between income and energy demand, and since income and GNP are directly linked through accounting identities, there is an important relationship between GNP and energy. At one time there was a popular belief that energy and GNP resulted in equal percentage increases in energy use. A monumental amount of research has been undertaken in the period since 1973, and a much more sophisticated understanding of the relationship between income and energy use has resulted. Income has been shown to be an important determinant of energy use, but not the only one. Further, the influence that income exerts varies between developed and developing economies. The actual measured income elasticity of energy (the percentage change in energy use per 1 percent increase in income, other factors being constant), even for a developed economy like the United States, is subject to uncertainty. Similarly, there is further variance when one looks at individual fuels and end-use sectors.

In contrast, we observe energy-income elasticities for developing economies; prior to 1973, the average elasticity value for 16 developing countries was 1.5. For the developed economies, econometric studies indicate an average income elasticity of near unity. Higher energy income elasticities in the developing economies have been traced to the development process itself. First, developing economies are shifting away from traditional energy sources and toward marketed energy sources; second, the economies' structure is changing to reflect increased reliance on the more energy-intense modern sector. We also note that the effect of the 1973 increases in energy prices was to shift the energy-income elasticity downward.

Finally, we note that investigators have shown that energy (and particularly the energy price) has an effect on GNP. This feedback seems weak, however.

Energy Prices

The link between energy prices and energy demand is an important one. Prices are the signals which markets give to producers and consumers. Prices influence both the level and composition of energy demand.

The influence of energy prices on energy use is measured numerically in much the same way that income's influence on energy use was measured, using an energy price elasticity. The energy price elasticity measures the percentage change in energy use for each percent change in energy price, other things remaining constant. Since increased prices stimulate conservation and decrease energy use, elasticities are always negative. There have been numerous econometric studies investigating energy price elasticities. In most sectors (transportation is an exception here) interfuel substitution is easier than general energy conservation. However, for both fuel and aggregate energy elasticities, values range from near zero to over −1.0. Measured fuel price elasticities range as high as −2.0 to −5.0.

Energy Productivity

Energy productivity, like labor productivity, refers to the level of output that can be obtained for each unit of energy input. Both numbers are clearly variable since energy productivity may be raised by substituting labor on a project or vice versa. In fact, we have already observed that to some extent the development process is associated with just such a substitution of energy and capital for labor. However, pure technological change, an increase in output without an increase in any inputs, is also possible. The reduction in one or more inputs for a given level of output without a concommitant increase in other inputs is possible through technological innovation—better machines and better management. In fact, once an economy has achieved maturity, the effects of technological improvement have led to increases in both labor and energy productivity.

For the United Kingdom, we date this process from around 1880, while it occurs in the United States somewhat later, beginning around 1920. In each of these economies energy productivity (GNP divided by total energy consumption) reaches a minimum and begins to increase. In the United Kingdom this increase averages about 0.8 percent per year, while in the United States it averages about 1.0 percent per year (though the rate decelerates over the period 1920–75). Aggregate energy productivity measures are not the only measures which show improvement. The manufacturing sectors of the United States,[1] United Kingdom,[2] West Germany,[3] and France,[4] have all been investigated and all display continuous improvements in energy productivity over the postwar period, even when there was no energy price incentive spurring the process on. We have termed this nonprice-driven improvement in energy productivity technological change.

Uncertainty

Each of the factors influencing energy use that we have discussed is associated with uncertainty. For example, demographic estimates of the world population vary between about 8 and 12 billion persons. The aggregate long-term price elasticity of demand for energy ranges between −0.1 and

−1.1, an order of magnitude difference. The uncertainty arises out of the historical record (for example, fertility rates have been notoriously volatile) and out of limits on our ability to measure quantities (for example, long-term price elasticities have been illusive, because there are no data of a quality and time period that would allow us to unambiguously determine elasticities). In addition, any of these measures are or may be variable over time, changing systematically with some characteristics of the energy-using capital stock. Finally, there is a great deal of uncertainty surrounding the relationship between past behavior of these critical elements and their future behavior. Will the developed economies' rates of labor productivity growth diminish or remain near historic levels? Will fertility rates in developing economies approach replacement level rapidly or will they continue at present levels? The uncertainty ranges for most variables may fall far short of ideal, but in many cases it is possible to get a sense of the relative quality of various estimates. Our goal has been not only to examine the existing range of values for important measures of the energy system but also to obtain some idea of the quality of the range.

ENERGY BALANCE: A SYNTHETIC APPROACH

The structure we have chosen for analysis consists of three parts, which we refer to as the supply, demand, and energy balance modules. The first two modules determine the supply of and the demand for each of six major primary energy categories in each of nine global regions. The energy balance module assures global equilibrium in each global fuel market. (Primary electricity is assumed to be untraded, thus supply and demand balances in each region.)

Energy Demand

Energy demand for each of the six major fuel types is developed for each of the nine regions separately. Five major exogenous inputs determine energy demand: population, economic activity, technological change, energy prices, and energy taxes and tariffs.

An estimate of GNP for each region is used as a proxy for both the overall level of economic activity and as an index of income. While the level of GNP is an input to the system, it is derived from demographic projections of the labor-age population and an assessment of likely labor force participation rates and levels of labor productivity. These are discussed in chapters 2 to 4.

The exogenous energy productivity measure is a time-dependent index of energy productivity given constant energy prices and real incomes. In the past, technological progress has had an important influence on energy use in the manufacturing sector of advanced economies. The inclusion of energy productivity allows the development of scenarios that incorporate assumptions of either continued improvements or technological stagnation.

The final major factor influencing demand is energy prices. Each region has a unique set of energy prices which are derived from world prices (determined in the energy balance component of the model) and region-specific taxes and tariffs. The model can be modified to accommodate nontrading regions for any fuel or set of fuels. In the base case it is assumed that no trade is carried on between regions in solar, nuclear, or hydroelectric power, but all regions trade fossil fuels.

The energy demand module performs two functions. It establishes the demand for energy and its services, and it maintains a set of energy flow accounts for each region. Oil and gas are transformed into secondary liquids and gases, which are used either directly in end-use sectors or indirectly as electricity. The solid primary fuels, coal and biomass, can be used either in their solid forms, or may be transformed into secondary liquids and gases or electricity. Hydro, nuclear, and solar electric are accounted directly as electricity. Nonelectric solar is included with conservation technologies as a reduction in the demand for marketed fuels.

The four secondary fuels are consumed to produce energy services. In the three OECD regions (regions 1, 2, and 3 in figure 1-1), energy is consumed by three end-use sectors: residential/commercial, industrial, and transport. In the remaining regions, final energy is consumed by a single aggregate sector.

The demand for energy services in each region's end-use sector(s) is determined by the cost of providing these services, and the levels of income and population. The mix of secondary fuels used to provide these services is determined by the relative costs of providing these services using each alternative fuel. The demand for fuels to provide electric power is then determined by the relative costs of production, as are the share of oil and gas transformed from coal and biomass. Primary energy demands are calculated from technological parameters.[5]

Energy Supply

Three generic types of energy supply categories are distinguished: resource-constrained conventional energy, resource-constrained renewable energy, and unconstrained energy resources. There are eight different supply modes which are aggregated to form the six major primary energy types. Oil and gas are each the sum of conventional and unconventional components. The distribution of the eight different supply modes across these categories is given in table 1-4.

The regional supply of resource-constrained conventional and renewable energy modes depends on resource constraints and behavioral assumptions. The supply of unconstrained technologies depends on the technology description (embodied in a conventional supply schedule description) and the price of energy.

Energy Balance

The supply and demand modules each generate energy supply and demand estimates based on exogenous input assumptions and energy prices. If energy supply and demand match when summed across all trading regions in each group for each fuel, then the global energy system balances. Such a result is unlikely at any arbitrary set of energy prices. The energy balance component of the model is a set of rules for choosing energy prices which, on successive attempts, brings supply and demand nearer a system-wide balance. Successive energy price vectors are chosen until energy markets balance within a prespecified bound. Figure 1-6 displays the interactions necessary to achieve a global energy balance. The methodology is discussed in detail in chapters 16 to 18. The model is specifically designed to compute carbon dioxide emissions from fossil fuel use.

RESULTS OF THE ANALYSIS

The analyses of key factors influencing global energy production and use and the construction of a framework of analysis are

TABLE 1-4. Distribution of Energy Technologies across Supply Categories

Resource-Constrained Conventional Energy	*Resource-Constrained Renewable Energy*	*Unconstrained Energy Resources*
Conventional oil	Hydro	Unconventional oil
Conventional gas	Biomass	Unconventional gas
		Solids
		Solar
		Nuclear

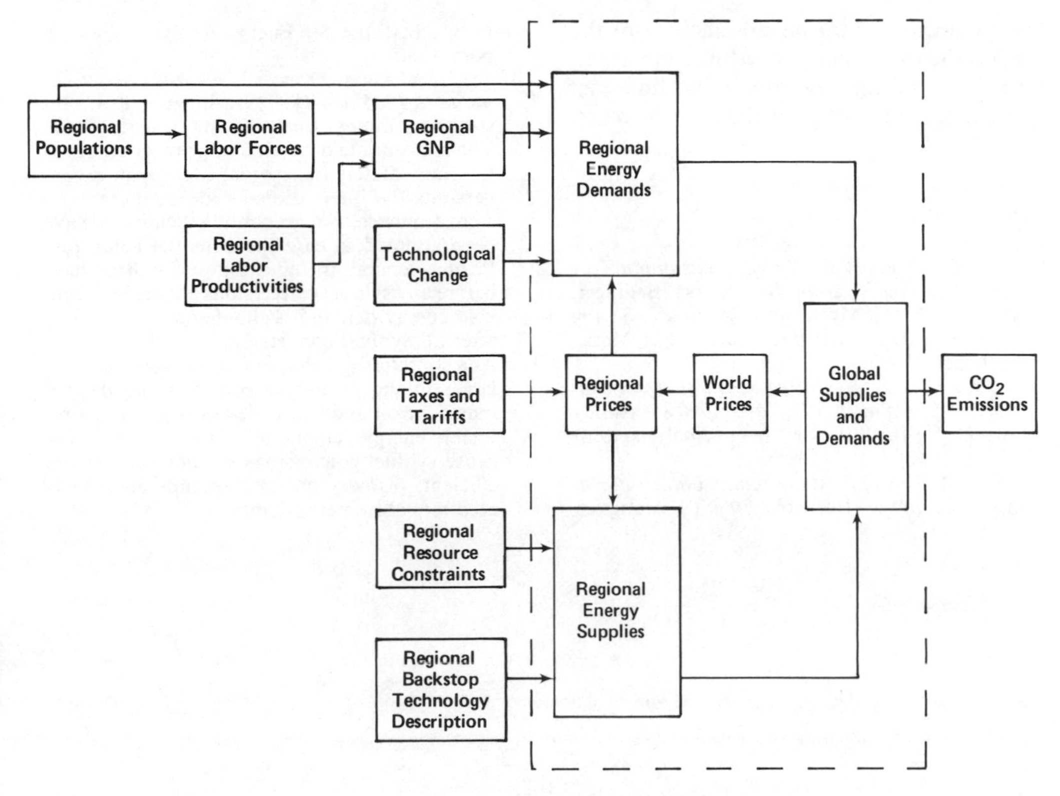

FIGURE 1-6. An energy-economic model of carbon dioxide emissions.

prerequisite to forecasting long-term energy use. The major lessons of this exercise go beyond the mere numbers presented in the base case scenario and its numerous alternatives. Our analysis has shown that man's global energy system is evolutionary, diverse, and uncertain. Because of diversity in energy-using technologies, the long-lived nature of capital stocks, institutional and cultural variations, and various other economic factors, energy systems emerge only slowly over time. Coexistence of competing energy technologies is the norm. Finally we observe that the ultimate level and configuration of energy production and consumption is highly uncertain over the long term. Therefore scenarios must be viewed as benchmarks, conditional statements about the way the global energy system might evolve. Over a 75-year time horizon, not only are key inputs such as resource constraints, technological and environmental costs, economic growth, energy-economy interactions and economic policy potential sources of uncertainty, but so too is the entire energy production, transformation, delivery, and end-use system.

The remainder of this book examines the issues and questions raised in the previous pages in considerably more detail. The primary focus is on the trends and forces shaping the level of energy use and the mix of fuels used to meet aggregate energy demands. The carbon dioxide issue as it relates to the level of emissions of CO_2 from fossil fuels provides a particular focus for these projections. The level of energy use and the mix of fuels used to meet these energy needs are initial dimensions of the carbon dioxide issue, providing a key input into determining how much and how fast climate may change over the next 75 years.

NOTES

1. See J. G. Myers et al., *Energy Consumption for Manufacturing* (Cambridge, Mass.: Ballinger, 1974), and J. G. Myers, and Nakamura, *Saving Energy in Manufacturing* (Cambridge, Mass.: Ballinger, 1978).
2. See J. A. Edmonds, *United Kingdom: Industrial Energy Demand in 1985 and 1990* (Washington, D. C., Institute for Energy Analysis, draft report, 1980).
3. See J. M. Reilly, *West Germany: Industrial Energy Demand in 1985 and 1990* (Washington, D. C., Institute for Energy Analysis, draft report, 1980).
4. See E. A. Allen, *France: Industrial Energy Demand in 1985 and 1990* (Washington, D. C., Institute for Energy Analysis, draft report, 1980).
5. The appropriate definition of primary energy is not immediately transparent when synfuel conversions and international trade are present. As a consequence, two accounting definitions have been adopted to ease the potential confusion. Primary energy includes the demand for fuels before any synfuel conversions are made. Refinable energy demand is the demand for energy after all synfuel conversions have been made. As a consequence, the demand for coal is always larger in the primary accounts, while the demand for oil and gas is always larger in the refinable energy demand accounts. Similarly, because synfuel conversions are not 100 percent efficient, primary energy demands always exceed refinable energy demands.

Demographic Impacts on Energy Consumption 2

POPULATION DYNAMICS AND FORECASTING PROBLEMS

Demographic changes are a key to understanding mid- (20–25 years) and long-term energy consumption patterns.[1] Population growth and its composition are key determinants of the labor force which, combined with assumptions of labor productivity, determines gross national product (GNP). GNP can be viewed both as a measure of economic activity and as an indicator of income levels. Economic activity in turn is a key determinant of energy demand in the industrial sector, while per capita income is an important factor in energy demand in the residential and transport sectors. These general relationships are illustrated in figure 2-1.

This chapter begins with a review of various long-term global population projections and methodologies for making future population projections. It continues with the development of detailed long-term population projections by age and sex. These estimates allow the projection of future labor force numbers by region.

POPULATION PROJECTION: FORECAST METHODOLOGIES

Population projections are derived from estimates of birth and death rates. The problem in making accurate future estimates of population has been in understanding the relationships that drive changes in these rates. While Malthus proposed that population growth would continue unabated until exhaustion of agricultural resources and starvation resulted in negative growth, modern demographic theory has come to accept the concept of a transition whereby societies move from high death and birth rates to a period of low death and birth rates.[2] Such a transition is inferred to have taken place in the past in the developed countries and is presently occurring in the developing countries. Although this insight into population dynamics is important and is critical in determining population age structures, future population size remains highly uncertain. The absolute and relative timing of the death and birth rate transitions and the speed at which they occur has not indicated a constant relationship between the determinants.

The failure of demographers working in the 1920s to anticipate the speed with which the death rate would fall in the developing countries (and the failure of the birth rate to closely track the death rate) led to considerable underestimation of future population size. During the 1960s and early 1970s, failure to anticipate the speed with which the birth rate would fall led population estimates to err on the high side. The recent trend has been to lower fu-

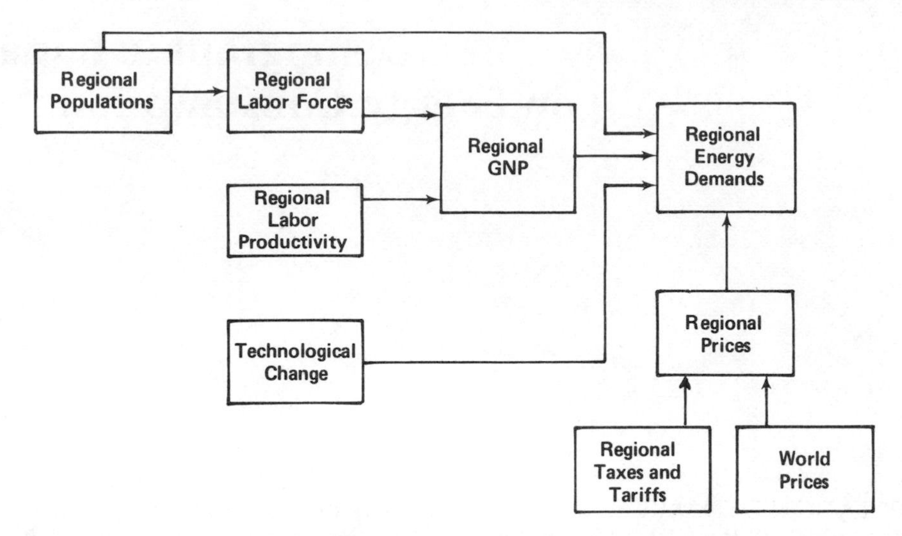

FIGURE 2-1. Key components of ORAU/IEA long-term, global, energy demand assessment module.

ture estimates of population to account for the rapid decrease in birth rates in recent years.[3]

The paradigmatic view of population dynamics is that of preindustrial society with high birth rates but also with high death rates, so that the population is stationary. As development occurs, medical care and health maintenance improves, resulting in a reduced death rate. The low death rate–high birth rate combination leads to a rapidly increasing population. For various and disputed reasons, birth rates eventually fall so that the post-transition society again finds itself with a stationary population.[4]

Apart from problems of identifying the timing and speed of such demographic transitions, producing an operational projection model based on this transition paradigm requires further specification of what is meant by stationarity. In the popular definition, death and birth rates balance to produce a stationary population level.[5] This is a razor's edge solution to stationarity in underlying birth and death rates. Assuming, for the moment, that stationarity is reached in birth and death rates, the more general characterizations of population stationarity are in terms of the rate of growth. The rate could be either negative or positive, with the stationary *level* being the special case where the stationary *rate* is zero.

Arguments for the special case are made on the basis of the absurdity of any other solution in the long run. Any positive rate yields an exponentially increasing population which must eventually face planetary constraints; any negative rate eventually leads to extinction of the species. However, these are broad bounds. Chou et al., Kahn et al., and Linneman et al., have variously reported conventional agriculture as capable of producing 10 to 40 times current output and thereby comfortably supporting similar increases in population, providing food distribution is more equitable.[6] Moreover, these estimates do not take into account such possible technologies as sea farming and hydroponics.

The current methodology for very long-term projections is to "predict" a date when the birth rate reaches the replacement level for a given country, interpolating birth rates for the years between the present and the replacement date. This leads to an eventual steady-state population, a generation or two after the replacement date, as the age structure adjusts.

In thinking about the relative long term (2050–2075), and the possibility for divergence from some best guess, it is necessary to keep in mind that the estimates are not cast in stone. In particular, one should consider the possibility that birth rates could fall rapidly and far enough to yield a nearer term peak in the population size, with declining populations thereafter. On the other hand, one must reasonably include the possibility that current trends in birth rates may not lead to a drop to replacement levels in the next 75 to 100 years. Rather, population may continue to grow, albeit at reduced rates from those now being experienced. While such specifications would eventually lead to untenable results, within fairly broad limits different-sized populations can be supported.

Briefly then, a stationary population level is not a natural result of biological limits, economic progress, or other time-dependent processes. Second, while the earth has a maximum carrying capacity, that capacity is not known with any certainty, but is unlikely to be reached in the next 75 to 100 years. As a result, the maximum carrying capacity offers little guidance to population projections over that time frame. Third, criteria for establishing an optimum population level (as opposed to the maximum carrying capacity) have not been identified. If national population control measures are truly as effective as they now appear to be,[7] it is not clear whether countries may set birth rate goals below replacement levels, resulting in eventual declines in population from some peak. Moreover, it is unclear how the optimum population level might change with changes in resource availabilities, technology, and political and other factors.

The concept of "optimum" population does not usually appear in the forecast literature. However, it is generally clear, and will not surprise readers, that nearly all observers see slower growth in population as "good news" and believe a stationary population must be achieved—the sooner the better. These observations imply a predominant view that additional population leads to an overall decrease in societal welfare. While the "optimum population" concept is almost certainly not a working concept, it is useful in judging the likelihood of future population scenarios. Its value is that it forces one to question whether an eventual projected stationary population, limited through population programs, is "good enough." Might society be better off at some lower population level or might technological advances in managing agriculture, resources, and environmental problems make population control less critical at some future date? At the same time, one might ask whether forces which lead to reductions in the fertility rate might overshoot the replacement level, as they have in several developed countries. Rather than a temporary phenomenon, as implicitly assumed under the stationary hypothesis, this level becomes permanent. In this case one might find more population programs aimed at increasing the fertility rate, as in the Soviet Union today.

A COMPARISON OF POPULATION FORECASTS

As already noted, long-term forecasts generally rely on the concept of stationarity to project ultimate population levels. Shorter term projections (to the year 2000) more often rely on extrapolation of current trends, combined with information concerning population programs in individual countries. Relatively few projections of global population beyond 2000 have been made. This section begins by comparing several year 2000 population estimates and then compares longer term estimates.

Table 2-1 summarizes five separate population forecasts for the year 2000, with four of the five forecasts providing high, medium, and low variants. Of particular interest is the comparison between the two United Nation's (UN) forecasts since they give an idea of how population forecasts

TABLE 2-1. Year 2000 Population Projections

Region		U.S. Census Bureau, 1979			World Bank, 1978	UN, 1978			UN, 1980			Community and Family Study Center, 1978		
		High	*Med.*	*Low*		*High*	*Med.*	*Low*	*High*	*Med.*	*Low*	*High*	*Med.*	*Low*
World	1975	4,134	4,090	4,043	4,033	3,981	3,968	3,941	4,066	4,066	4,066	4,017	4,017	4,017
	2000	6,797	6,350	5,921	6,054	6,638	6,254	5,840	6,337	6,119	5,837	5,975	5,883	5,736
More developed West[a]	1975	708	708	708	711	711	708	707	709	709	709	710	710	710
	2000	842	809	781	822	897	842	806	831	812	783	810	806	800
USSR	1975	384	384	384	384	385	385	385	383	383	383	384	384	384
Eastern Europe	2000	480	460	442	473	482	466	451	474	461	449	458	457	451
China	1975	978	935	889	897	844	839	829	928	928	928	897	897	897
	2000	1,468	1,329	1,176	1,192	1,209	1,148	1,073	1,294	1,257	1,226	1,136	1,131	1,120
Developing regions[b]	1975	2,064	2,063	2,062	2,041	2,041	2,036	2,020	2,046	2,046	2,046	2,026	2,026	2,026
	2000	4,007	3,752	3,522	3,567	4,050	3,798	3,510	3,738	3,589	3,378	3,571	3,489	3,365

Sources: U.S. Bureau of the Census, *Illustrative Projections of World Populations to the 21st Century,* Special Studies Service, No. 79, 1979; World Bank, Population and Human Resources Division, Development Economics Department, prepared by K. C. Zachariah and My Phi Vu, (Washington, D.C.); *Stationary Population,* April 1978, pp. 6–11; United Nations, *World Population Prospects as Assessed in 1973,* ST/ESA/SER.A/60, 1977; Community and Family Study Center, University of Chicago, *Projected Population of the World, Regions and Nations for the Year 2000,* February 1978; United Nations, *World Population Prospects as Assessed in 1980,* ST/ESA/SER.A/78, 1981; (As reported in N. Keyfitz, et al; "Global Population 1975–2075," Institute for Energy Analysis, Oak Ridge, Tenn., Research Memorandum (ORAU/IEA-83-6(M), 1983.

[a]Includes Australia, New Zealand, Japan, the United States, Canada, and all of Europe, except Eastern Europe.

[b]Includes all countries not included elsewhere.

have changed as a result of the unexpected success of population programs in the developing countries. The total 2000 population forecast for the world is only about 2 percent less in the 1980 assessment than in the 1973 assessment. However, the actual decline in the growth of population is more pronounced. This results from an upward revision of nearly 100 million people for the 1975 population of China between the two studies. If one uses the base population for 1975 as estimated in 1980 by the UN and applies the population growth rate of the 1973 study, the difference in the 2000 population projections is 5 percent in the medium variant. Differences in base population estimates for other regions are much less pronounced.

Comparing across all studies, one finds fairly close agreement among the UN 1980 and the U.S. Census Bureau estimates except for the developing countries, where the Census Bureau is somewhat higher. The UN 1973 study tends to be high, which is due to underestimation of the impact of family planning programs in developing countries. The World Bank's single case is also very similar to the medium variant estimates of the UN, particularly if China's 1975 population is comparably based. The Community and Family Study Center (CFSC) tends to be lower than the others and shows less difference between the extreme variants. In fact, the high CFSC variant is very similar to the low variant of the Census Bureau.

Among these estimates, only the World Bank figures are based explicitly on a model which predicts a date at which replacement fertility levels are reached in each country and declines in the fertility rate are interpolated from the present level. The other studies are based on various extrapolation models. However, to some extent, replacement-level fertility rates are a target. For example, the UN notes that for countries near or below the replacement level, "it is assumed that fertility will converge to a replacement level starting from the end of this century."[8]

LONGER TERM PROJECTIONS—A COMPARISON

Projections of global population beyond 2000 include those of Frejka, the World Bank, Keyfitz and others, and the UN in its most recent projection. With the exception of the UN projections, these studies use the years-to-replacement-fertility-level method as a basis for future projection. The result of this assumption is an eventual stationary population level which is implicitly maintained indefinitely once it is reached.

Table 2-2 compares longer term and ultimate stationary population levels. The UN series is not based on the stationarity projection methodology and the projections extend only to 2025.

The Frejka estimates are the earliest. They are also of a somewhat different nature. The lowest projection supposed that fertility dropped to replacement levels immediately (1973) and all population increase resulted from the lagged response of population growth due to changes in age structure. The high case supposed replacement fertility levels were reached in 70 years. Middle cases included replacement at 10, 20, and 50 years. Frejka's analysis was aimed at supposing various population transitions and asking what were the consequences of such assumptions. In contrast, the other projections, and the UN projections in particular, are based on an examination of current fertility rates, past history, family planning efforts, and economic factors in each country and therefore rank as an informed best guess about the future population size rather than a supposition. Each of these forecasts falls well within the wide bounds set by Frejka. Among these, the Keyfitz estimates tend to be low while the World Bank's estimates are high. The difference between Keyfitz and the World Bank is over 1 billion in 2025 and nearly 2 billion in the stationary population level. The UN estimates show a difference of nearly 2 billion between the high and low estimates in 2025.

TABLE 2-2. Long-Term and Stationary Population Levels of the World (billions)

Year and Region	*Frejka, 1973*[a] *Low*	*High*	*World Bank, 1982*	*UN, 1980* *Low*	*Med.*	*High*	*Keyfitz, 1981*
Year 2025							
World	5.2	10.1	8.6[b]	7.2	8.2	9.1	7.4
Developed	1.2	1.5	1.4[b]	1.3	1.4	1.5	1.5
Developing	4.0	8.5	7.2[b]	5.9	6.8	7.6	5.9
Year 2050							
World	5.5	13.4	9.8[c]	N.R.	N.R.	N.R.	8.2
Developed	1.3	1.7	1.4[c]	N.R.	N.R.	N.R.	1.5
Developing	4.2	11.7	8.4[c]	N.R.	N.R.	N.R.	6.7
Ultimate stationary							
World	5.7	15.1	10.4	N.A.	N.A.	N.A.	8.5[d]
Developed	1.3	1.8	1.4	N.A.	N.A.	N.A.	1.6[d]
Developing	4.4	13.3	9.0	N.A.	N.A.	N.A.	6.9[d]

Sources: T. Frejka, *The Future of Population Growth: Alternative Paths to Equilbrium,* (Wiley, New York, 1973); My T. Vu and Ann Elwan, "Short-Term Population Projection, 1980–2000 and Long-Term Projection, 2000 to Stationary Stage by Age and Sex for All Countries of the World," World Bank, Washington, D.C., unpublished, July, 1982. United Nations, *World Population Prospects as Assessed in 1980,* Population Studies, No. 78 (New York, 1981); N. Keyfitz, et al., "Global Population: 1975–2075," Institute for Energy Analysis, Oak Ridge, Tenn., Research Memorandum (ORAU/IEA-83-6(M), 1983).

N.R. = None reported.

N.A. = Not applicable.

[a]Extreme scenarios; five standard scenarios were produced by Frejka.

[b]Year 2030.

[c]Year 2055.

[d]Year 2075, which for practical purposes is the stationary population.

Uncertainty

While the various projections give some spread in terms of future forecasts, these spreads do not appear to be good indicators of the uncertainty connected with forecasts. The UN methodology could indicate degree of uncertainty, but the UN does not view the estimates in this way; it does not subjectively attach a probability to the band generated by extreme scenarios. Rather, each scenario is a forecast conditioned on achieving or failing to achieve, in various degrees, population planning goals, together with uncertainty in underlying demographic transition relationships. Attaching probabilities to these outcomes would imply some probabilities also attached to the success of population planning programs and to advances in underlying economic determinants of the demographic transition. The UN has been unwilling to make such judgments. By implication, the medium variant is usually taken to be a best-guess scenario but it is not specified whether the extreme variants might be interpreted as having a 5 or 95 percent probability.

Further, if one judges the various best guesses as observations yielding information on the projected population variance, one is faced with the problem that the distribution of best guesses will necessarily have a smaller variance than the population. This property is almost certainly at work in the 2025 estimates where both the World Bank's and Kefitz's best guess are within the extremes of the UN. This leads the stationary population high-low differences in these two estimates to be no greater than the high-low differences of the UN in 2025. This result is in direct conflict with the usual observation that uncertainty increases as one extrapolates further into the future.

Could one use Frejka's range as a measure of uncertainty? Like the UN, no probabilities are attached to the bands resulting from extreme assumptions. Moreover, as noted earlier, the assumption of eventual stationarity in population may tend to constrain the problem. While dropping to a replacement-level fertility immediately might be considered unreasonably extreme, a population as low or lower than 5.7 billion by 2075 may be somewhat more likely; fertility rates may stay below replacement in the developed countries and fall gradually but continue to fall to some level below replacement in the developing countries. Corresponding arguments can be made on the high side.

Keyfitz, et al. comment briefly on the uncertainty of future estimates. They note that:

> Most observers are confident that replacement for most countries will be reached before the end of the first quarter of the 21st century. That would make the ultimate world total population between 8 and 12 billion and puts the average annual rate of increase between 0.7 and 1.1. Can we take it that there is a two-thirds probability that the rates of increase of 0.7 and 1.1 percent would straddle the population performance during the 100 years after 1975? If the past work on individual countries is any guide, this is too narrow. Workers in the field now seem to agree unanimously on a range narrower than the narrowest found in the past for any large number of forecasts that have been checked against subsequent performance.[9]

A BASE SET OF DEMOGRAPHIC PROJECTIONS

This section adopts a base set of population projections and develops labor force estimates from the population estimates. As noted at the beginning of this chapter, labor force growth is an important determinant of economic activity, and as a result is also an important determinant of energy use.

Table 2-3 summarizes a set of population projections which have been adopted as a base case. These projections are aggregated from country projections into the nine regions adopted for this study.

Labor force estimates are developed in a two-step process. First, the labor force age population is calculated for males and females from the detailed demographic projections. Then labor force participation rates are applied to age cohorts to generate male and female components of the labor force, which are summed to form the total labor force.

Definitions of Work Age Population

There are two major international conventions used in defining work force age pop-

TABLE 2-3. World Population Projections, 1975–2075 (thousands)

	1975	*2000*	*2025*	*2050*	*2075*
US	213,925	254,437	281,812	288,138	292,301
OECD West	405,029	476,377	527,881	552,805	561,844
JANZ	127,961	153,719	163,714	166,932	168,599
USSR/E.EUR	394,582	471,756	516,078	532,945	541,362
CHINA/Et Al.	910,964	1,248,451	1,498,939	1,612,743	1,647,157
MIDEAST	81,371	146,685	199,208	231,651	240,820
AFRICA	399,370	697,000	942,749	1,101,464	1,150,477
LATIN AMER	312,631	540,233	718,473	822,432	848,621
SEASIA	1,129,957	1,903,646	2,514,947	2,888,408	2,994,843
World totals	3,975,790	5,892,304	7,363,811	8,197,518	8,446,024

Source: N. Keyfitz, et al., *"Global Population, 1975–2075."*

ulation. One includes all persons who are 15 years of age or older.[10] The second convention limits labor force age eligibles to those who are between 15 and 64 years; this age group is sometimes referred to as of prime labor force age.

Within the United States, the trend has been toward earlier retirement from the labor force, which has been made financially possible by the expansion of Social Security and private pension plans. In the short run, this trend is expected to continue. For 1990, Flaim and Fullerton have estimated projected labor force participation at 18 percent of males 65 years and older and 7 percent of women in the same age group.[11]

However, trends to under-65 retirement may soon be reversed. In the United States, the mandatory retirement age has been raised to 70. This change reflects not only increased longevity, but also the growing ability of persons over age 65 to perform required physical and mental tasks. Improved health in this age group is a global phenomenon. According to current estimates, the proportion of the total population 65 years of age and older will increase from the current 10 percent to over 25 percent in the next century.[12] Since our interest in this study runs to 2050, we expect longer working years to become general. Therefore, we have elected to follow the definition of labor force eligibles which includes all persons 15 years of age or more, with no cutoff at 65. Table 2-4 shows estimates of world populations of labor force age for 1975, 2000, 2025, and 2050.

Global Trends in Work Age Population

As table 2-4 shows, the world's labor force age population is projected to increase from 2.5 billion in 1975 to 6.5 billion by

TABLE 2-4. Populations of Labor Force Age, 15 Years and Over (thousands)

	1975			*2000*		
Region	*Females*	*Males*	*Total*	*Females*	*Males*	*Total*
1. US	83,015	76,767	159,782	102,041	96,777	198,818
2. OECD West	155,363	144,959	300,322	185,477	180,693	366,170
3. JANZ	49,181	46,729	95,910	61,933	59,637	121,570
4. EUSSR	158,746	135,450	294,196	189,732	178,326	368,058
5. ACENP	293,685	306,841	600,526	460,301	468,745	929,046
6. MIDEAST	22,212	22,700	44,912	46,979	47,956	94,936
7. AFR	113,035	109,752	222,787	225,568	223,095	448,663
8. LA	91,179	89,647	180,826	182,907	181,785	364,692
9. SEASIA	316,647	329,879	646,526	627,000	646,993	1,273,988
World totals	1,283,063	1,262,723	2,545,786	2,081,935	2,084,006	4,165,938

	2025			*2050*		
	Females	*Males*	*Total*	*Females*	*Males*	*Total*
1. US	115,408	110,151	225,559	118,316	113,416	231,732
2. OECD West	211,400	206,841	418,242	224,631	218,821	443,452
3. JANZ	66,726	64,507	131,233	68,124	66,127	134,251
4. EUSSR	209,294	201,957	411,251	217,597	210,307	427,903
5. ACENP	587,279	586,540	1,173,815	647,810	640,109	1,287,916
6. MIDEAST	74,185	75,291	149,475	91,696	91,836	183,532
7. AFR	351,163	351,435	702,598	435,302	433,272	868,575
8. LA	274,598	273,761	548,360	329,426	325,873	655,299
9. SEASIA	1,060,392	1,073,652	2,134,039	1,144,380	1,145,947	2,290,320
World totals	2,950,443	2,944,133	5,894,570	3,277,281	3,245,704	6,522,976

Note: Due to rounding, columns may not add exactly to totals.

TABLE 2-5. The Average Annual Growth of the World's Population 15 Years and Older for Selected Years 1975–2050 (percent)

Region	*1975–2000*	*2000–25*	*2025–50*	*1975–2050*
1. US	0.88	0.51	0.11	0.50
2. OECD West	0.80	0.53	0.23	0.52
3. JANZ	0.95	0.31	0.09	0.51
4. EUSSR	0.90	0.44	0.16	0.50
5. ACENP	1.76	0.94	0.37	1.02
6. MIDEAST	3.04	1.83	0.82	1.89
7. AFR	2.84	1.81	0.85	1.83
8. LA	2.85	1.64	0.72	1.73
9. SEASIA	2.75	1.63	0.74	1.70
World totals	1.99	1.24	0.56	1.26

2050. Among the country groups, the less developed countries of the Mideast, Africa, Latin America and South and East Asia (regions 6, 7, 8, 9) account for a large and growing share of the world's labor force age population. In the year 2050 these regions are projected to account for 61 percent of the world's population of 15-year olds and older, compared with 43 percent in 1975. The largest decline in the world's share of labor force age population occurs within the industrialized OECD countries (regions 1, 2, 3), which falls from 21.8 percent in 1975 to 12.4 percent in 2050. The Communist countries of regions 4 and 5 also decline during this period, from 35.2 percent to 26.3 percent.

These changes in the world's labor force age population are reflected in the average annual growth rates shown in table 2-5. The labor force age population of all of the country groups is projected to grow at slower rates in the future. The world's average annual growth rate of 1.99 percent for the period 1975–2000 is projected to decline to 1.24 percent during the years 2000–25 and then fall to 0.56 percent during 2025–50. The world average for the period is 1.26 percent.

During the period 1975–2025, the total world population is projected to grow at an average annual rate of 0.97 percent. With the labor force age population growing at an average annual rate of 1.26 percent, the world's population is projected to age in the future. As table 2-6 shows, the proportion of the world's population 15 years old and older is projected to increase from approximately 64 percent in 1975 to about 80 percent in 2050. The most pronounced shift occurs within the less-developed countries of regions, 6, 7, 8, and 9 where the share of the labor force age population grows from roughly 57 percent in 1975 to approximately 79 percent in 2050. Communist Europe increases from 74.6 in 1975 to 80.3 percent in 2050, and Communist Asia increases from 65.9 to 79.9 percent during this same period. The least amount of change occurs in the OECD countries of regions 1, 2, and 3 whose population of 15 year olds and older increases from roughly 75 percent in 1975 to about 80 percent in 2050.

TABLE 2-6. World's Labor Force Age Population as a Share of Total Population, 1975–2050

Region	*1975*	*2000*	*2025*	*2050*
1. US	74.7	78.1	80.0	80.4
2. OECD West	74.1	76.9	79.2	80.2
3. JANZ	75.0	79.1	80.2	80.4
4. EUSSR	74.6	78.0	79.7	80.3
5. ACENP	65.9	74.4	78.3	79.9
6. MIDEAST	55.2	64.7	75.0	79.2
7. AFR	55.8	64.4	74.5	78.9
8. LA	57.8	67.5	76.3	79.7
9. SEASIA	57.2	66.9	75.8	79.3
World totals	64.0	70.7	77.0	79.6

Labor Force Benchmarks

The benchmark growth forecasts of work age population need to be adjusted by appropriate labor force participation rates to obtain regional labor force benchmarks. The labor force participation rate is the fraction of the work age population which is either employed or actively seeking work. Because labor force participation rates vary from region to region as well as over time, each of the nine regions has been analyzed separately.

OECD Countries

It is assumed that male labor force participation rates for developed OECD nations will remain relatively static between 1975 and 2000, declining perhaps 2 percent over the 25-year period in response to more liberal retirement policies. Female labor force participation rates, on the other hand, are assumed to expand from a little over 42 percent of age-eligible females to about 52 percent by the year 2000 (see table 2-7). While total participation rates will be moving up slowly, the growth of the labor force will slow as a consequence of fewer teenagers reaching labor force age.

It is anticipated that by 2000 in the OECD countries (regions 1, 2, 3) as a whole, 76 percent of the male age-eligibles will be in the labor force, and 52 percent of the females. For the United States, labor force projections made by the Social Security Board have been adopted. For the rest of the OECD, it is assumed that participation rates are 75 percent for males and 50 percent for females beyond 2000.

USSR/Eastern Europe

In the USSR, the growth of population is tapering off, as it is in most industrial countries, and this trend is expected to continue until "a steady state" is reached shortly after 2050. Participation in the labor force has been running relatively

TABLE 2-7. OECD Countries of North America, Europe, and Asia: Current and Projected Labor Force and Participation Rates, 1975–2050

	Labor Force (10^3)			*Participation Rates (%)*		
Year and Region	*Male*	*Female*	*Total*	*Male*	*Female*	*Total*
1975						
1. US	59,187	38,602	97,789	77.1	46.5	61.2
2. OECD West	110,352	61,095	171,447	76.1	39.3	57.1
3. JANZ	37,638	22,157	59,795	80.5	45.1	62.3
Total	207,177	121,854	329,031	77.2	42.4	59.2
2000						
1. US	73,551	63,265	136,816	76.0	62.0	68.8
2. OECD West	133,765	88,679	222,444	74.0	47.8	60.7
3. JANZ	47,555	29,572	77,127	79.7	47.7	63.4
Total	254,871	181,516	436,387	75.6	51.9	63.6
2025						
1. US	82,613	57,704	140,317	75.0	50.0	62.2
2. OECD West	155,132	105,702	260,834	75.0	50.0	62.4
3. JANZ	48,380	33,363	81,743	75.0	50.0	62.3
Total	286,125	196,769	482,894	75.0	50.0	62.3
2050						
1. US	85,062	59,158	144,220	75.0	50.0	62.2
2. OECD West	164,116	112,317	276,433	75.0	50.0	62.3
3. JANZ	49,595	34,063	83,658	75.0	50.0	62.3
Total	298,773	205,538	504,311	75.0	50.0	62.3

TABLE 2-8. The Soviet Union, Eastern Europe, and Other Centrally Planned Economies: Current and Projected Labor Force and Participation Rates, 1975–2050

Year and Country	*Labor Force (10^3)*			*Participation Rates (%)*		
	Male	*Female*	*Total*	*Male*	*Female*	*Total*
1975						
Soviet Union	70,179	67,395	137,574	82.6	64.6	72.7
Eastern Europe	29,980	26,127	56,107	76.7	61.2	68.7
OCPE	8,762	7,201	15,963	76.7	61.2	68.8
Total	108,921	100,723	209,644	80.4	63.4	71.3
2000						
Soviet Union	92,945	81,905	174,850	80.4	65.5	72.7
Eastern Europe	33,724	31,284	65,008	72.9	65.0	68.9
OCPE	11,980	10,769	22,749	72.9	65.0	68.9
Total	138,649	123,958	262,607	77.8	65.3	71.3
2025						
Soviet Union	105,320	89,396	194,716	80.0	65.0	72.3
Eastern Europe	36,146	33,197	69,343	71.9	64.3	68.0
OCPE	14,404	12,945	27,349	71.9	64.3	68.1
Total	155,870	135,538	291,408	77.2	64.8	70.9
2050						
Soviet Union	109,684	92,787	202,471	80.0	65.0	72.3
Eastern Europe	36,526	33,619	70,145	70.9	63.6	67.2
OCPE	15,361	13,962	29,323	70.9	63.6	67.2
Total	161,571	140,368	301,939	76.8	64.5	70.6

Note: OCPE = Other centrally planned economies.

high by western standards, amounting to approximately 72 percent of age-eligibles in the 1970–80 period. Continuing this rate to 2050 results in the labor force size shown in table 2-8.

Labor force growth in Eastern Europe and other centrally planned economies (OCPE) is adversely affected by the relatively large number of persons 65 years of age and older. In Eastern Europe, this number is proportionately larger than in the USSR. Participation rates are expected to peak in the 1975–85 period, and to decline moderately thereafter to 2000. At the same time, participation rates are expected to become much more uniform than they were in the earlier postwar years. Eastern Europe will be a region dominated by relatively mature economies, with slow demographic and economic change. Hence, we would not expect to see participation rates vary considerably from the present level, which is about 69 percent of age-eligibles (see table 2-8).

Developing Country Regions and China

No estimates of labor force participation rates in the developing country regions and China have been made. Defining labor force in the developing countries poses considerable problems.

In the traditional sector of developing countries, the separation of household labor force tasks of a commercial nature (e.g., harvesting a marketed crop) from other duties (e.g., child care and food preparation) is not done on a systematic basis. Estimates of labor force participation in low-income agricultural countries may be very high, reflecting more intensive seasonal use of labor rather than year-round employment. Since there are major differences in reporting participaton, there are significant problems in comparing across countries. Moreover, there are problems of hidden unemployment and underemployment which make a change in the size of the participating labor force. difficult to interpret.[13]

For this reason, we adopt the growth of the labor force age population as a proxy for the labor force. This implies constant participation rates. It is an improvement over population because it captures large responses of the labor force to population growth changes. Using such a proxy for labor force makes productivity changes "crude" rates of productivity increase. The crude rate would implicitly incorporate changes in the participation rate and such factors as unemployment and underemployment. If one was to postulate any changes in these underlying factors, the changes would almost certainly be related to economic development as measured by per capita income growth. Thus, whatever the expected rate of productivity improvement was for the developing countries, the crude rate would tend to be lower, reflecting increasing participation and reduction of unemployment and underemployment. The crude rate would also tend to accelerate relative to more refined measures as the trends in participation rates saturated and unemployment and underemployment rates declined to developed country levels.[14]

Using this convention, the labor force age population projections reported in table 2-4 provide direct estimates of labor force growth for regions 5–9. Apart from the difficulty of making useful observations about the future changes in labor force participation, one finds, as detailed in the next chapter, that available long-term historical data and data on the developing countries are generally not disaggregated sufficiently to allow much more than "crude" productivity growth estimates for historical periods. Thus, to the extent that these historical estimates form a benchmark for future projection, it is appropriate that the future labor force estimates are "crude" estimates.

CONCLUSION

The base demographic characteristics developed in this chapter provide the foundations for developing GNP growth projections. As an indicator of economic activity and income levels, GNP is an important determinant of energy use. Several studies have presented estimates of future population size. The disagreement among studies is, in general, relatively minor.

Several questions arise in regard to using existing population estimates directly as input in energy studies. The first question is: Have the studies been directed at answering the question of what is likely to happen or the question of what should happen? The answer appears to be that the studies are some unspecified combined answer to the two questions. Whether or not it is a difficult or, more appropriately, an occupationally hazardous task to offer a best-guess projection, a best-guess projection is what is needed for our purposes. It was for this reason that Keyfitz was asked to specifically answer the question: What is the most likely future population size and structure?

A second question is: How good an understanding do we have of uncertainty concerning future population forecasts? The answer is, a relatively poor understanding. While most studies have provided high, mid, and low cases, probabilities have not been attached to the various cases. Thus, there is no explicit statement of whether the subjectively "best guess" case is even contained within the range given, though one implies, perhaps incorrectly, that the mid-case is the best guess. But, even working with this assumption, should one understand the band implied by the extreme cases to capture 1 percent or 99 percent of possible outcomes? The issue of uncertainty in population forecasts is one which requires further examination.

Finally, one must ask a more fundamental question: What types of biases exist in currently used projection methodologies and how might these biases affect our best-guess estimate and an estimate of the uncertainty in future forecasts? From an examination of this question, a more difficult question follows: What are the goals of population programs? Zero population growth? A population level perhaps lower or higher than the current population level? Or might the target be dependent on underlying factors that change over time? Despite these questions, available population estimates and existing knowledge of the dynamics of demographic change provide considerable useful information concerning future population levels.

NOTES

1. See for example, Edward Allen, *Energy and Economic Growth in the United States* (Cambridge, Mass.: MIT Press, 1979). Lester Lave comments in Electric Power Research Institute (EPRI), *Proceeding of the Workshop on Modeling the Interrelationships between the Energy Sector and the General Economy* (Palo Alto, Calif.: EPRI, 1976), and Lester Lave, "What we have Learned from these Scenarios" *Modeling Energy-Economy Interactions,* Charles Hitch (Resources for the Future, Washington, D.C., 1977).
2. Thomas Robert Malthus, "An Essay on the Principle of Population," *Classics of Economics,* Charles W. Needy (Oak Park, Ill.: Moore Publishing, 1980), pp. 47–56.
3. See Nathan Keyfitz, et al., "Global Population, 1975–2075." Institute for Energy Analysis, Oak Ridge, Tenn., Research Memorandum (ORAU/IEA-83-6(M), 1983).
4. Economic demographers posit an income-leisure tradeoff against net benefits of child rearing. Other researchers have posited that children are desired as old age insurance and that if more survive to adulthood, the couple needs to have fewer children to assure caretakers for their old age. Still others posit that the provision of government-supplied old-age benefits is related to economic development and reduces the need for producing children who will serve as caretakers of the parents. These "natural" forces are contrasted with differing views as to the success of government-sponsored programs limiting population.
5. This concept of stationarity is used in T. Frejka, *The Future of Population Growth: Alternative Paths to Equilibrium* (New York: Wiley, 1973); in K. C. Zachariah and My Thi Vu, "Stationary Populations," World Bank, Population and Human Resources Division, April 1978, and in Nathan Keyfitz, et al. "Global Population, 1975–2075," Institute for Energy Analysis Oak Ridge, Tenn., Research Memorandum (ORAU/IEA-83-6(M), 1983). In contrast, U.S. Bureau of the Census, *Current Population Reports,* Series P-25, No. 704, "Projections of the Population of the United States: 1977 to 2050" (U.S. Government Printing Office, Washington, D.C., 1977), uses an extrapolation method.
6. Marylin Chou, David P. Harmon, Jr., Herman Kahn, and Sylvan H. Wittwer, *World Food Prospects and Agricultural Potential* (New York: Praeger, 1977). H. Kahn, W. Brown, and L. Martel, *The Next 200 Years: A Scenario for America and the World* (Croton-on-Hudson, New York: Hudson Institute, 1976). H. Linneman, J. DeHoogh, M. A. Keyzer, and A. D. J. VanHenst, *MORIA, Model of International Relations in Agriculture* (Amsterdam: Elsevier, 1979).
7. See Keyfitz et al., "Global Population, 1975–2075."
8. United Nations, *World Population Prospects as Assessed in 1980,* Population Studies, No. 78, New York, 1981, p. 3.
9. Nathan Keyfitz, et al., "Global Population, 1975–2075."
10. The United States often uses 16 years of age or older.
11. Paul O. Flaim and Howard N. Fullerton, Jr., "Labor Force Projection to 1980: Three Possible Paths," *Monthly Labor Review,* December 1978, pp. 25–35. Data are for the high labor force growth path.
12. Nathan Keyfitz, *Estimates of Population, 1975–2075, World and Major Countries* (Cambridge, Mass.: Harvard University Press, 1980), p. 24.
13. For a general discussion of some of these problems, see, for example, Gerald M. Meir, *Leading Issues on Economic Development* (New York: Oxford University Press, 1976), pp. 167–248.
14. Structural relationships between employment and other factors might be specified and one might then fit a reduced form directly to economic development. Such a hypothesis might be examined in the vein of H. Chenery and M. Syrquin, *Patterns of Development, 1950–1970* (London: Oxford University Press, 1975).

Productivity and Income 3

Gross national product (GNP) is the measure of the market value of a nation's total output of final goods and services during a particular time period (usually a year).[1] It is also generally accepted as a measure of a nation's economic activity and income.

The level of GNP for any given year can be defined as the product of two general factors, labor force and productivity. Productivity (*PRO*), as it is used in this study, is defined to be the average output of GNP per unit of labor (*L*).

$$PRO = GNP/L \tag{3-1}$$

Rearranging these terms yields the common identity.

$$GNP = L \cdot PRO \tag{3-2}$$

Chapter 2 discussed the factors affecting labor force formation. This chapter briefly examines factors affecting rates of change in productivity and then compares both observed historical rates of productivity change and projected rates of productivity change embodied in various projections of GNP.

As noted in chapter 2, available historical data and knowledge of future trends in such things as unemployment and underemployment over the long term are limited. Thus, we have chosen to work with a crude productivity measure. These crude productivity measures are GNP divided by labor as in table 3-1 where labor is defined as size of the labor force or of the population of labor force age. More disaggregated productivity measures define labor as the employed labor force or as hours of work, and may be developed by sector or industry.

Numerous factors influence the crude productivity measure as defined. These can be broken down into two groups. One set of factors influences the quality level embodied in each laborer and the other includes nonlabor inputs, with production process changes in any of these factors contributing to changes in crude productivity and transformed into changes in GNP. The problem of understanding the impact on productivity changes in these underlying factors is the problem of understanding the forces of economic growth.

There is a large body of literature that examines the impacts of changes in these factors on economic growth.[2] While this research has contributed to an understanding of the growth processes, it has yielded neither predictive success nor prescriptive solutions. At present, various forecasters and futurists are concerned both with the possibility of little or no productivity growth and the continued poverty such an outcome would imply, as well as the displacement of workers resulting from massive automation and its implications for greatly increased labor productivity.[3]

TABLE 3-1. Factors Influencing Crude Labor Productivity

Labor quality
Sectoral distribution of labor force
Rate of unemployment
Hours worked per laborer
Sexual composition of the labor force
Age composition of the labor force
Education
Experience
Health
Social, institutional, and psychological factors
Nonlabor inputs
Capital stock
Economies of scale
Technology state
Terms of trade
Other effects

HISTORICAL TRENDS IN PRODUCTIVITY

While future rates of productivity improvement are neither dictated by nor limited to historical rates, the latter form useful benchmarks for projections. In examining long-term futures, we maintain that it is useful to step outside the experience of the past 6 months or 10 years. There is little reason to believe that so short a period can be representative of the performance that one might expect over periods of 50 to 100 years. Unfortunately, long-term records are scarce.

In this section, long-term measures of crude productivity are constructed for three economies—the United States, the United Kingdom, and Japan. Shorter term global records are also examined.

The discussion of productivity trends in the United States, the United Kingdom, and Japan is supplemented by a more general assessment of the experiences of other developed western economies. These economies are unique in the sense that they are members of a relatively select group of countries which have developed successfully. For the centrally planned economies and less-developed economies (LDCs), data are less complete. An attempt has been made to compare basic productivity trends among nine major world regions over the period 1960–75. In all instances, the measure of productivity under discussion is crude productivity as defined in the previous section.

Long-Term Trends in Productivity

Three Countries

Table 3-2 presents data on the rates of crude productivity change for three countries where long-term data were available; the United States, the United Kingdom, and Japan. Included are average annual rates of change over 10-year, 70-year, and to-date periods.

Before discussing the trends in table 3-2, a few comments on data validity are necessary. For the United States, the estimates of GNP prior to 1909 were prepared by John W. Kendrick and Simon Kuznets.[4] The values from 1909 to 1980 are official estimates prepared by the U.S. Department of Commerce. The margin of error for the figures prior to 1919 is considered to be fairly wide. Since 1919, however, the reported values of GNP are thought to be very reliable. The estimates of labor force are considered to be reliable over the entire period.[5] Similar caveats apply to the United Kingdom and Japanese data. The reconstruction of historical data series offers considerable room for uncertainty. Nevertheless, the data are at least indicative of changes occurring in earlier periods and offer the only information available on very long-term trends.

Not surprisingly, one finds considerably more fluctuation in the 10-year averages. In the United States, the 10-year averages range from −0.8 percent per year recorded during the 1880s to 3.9 percent annually recorded during the 1940s. In the United Kingdom, the range is from −1.6 in the 1911–21 period to 3.2 during the 1960s. In Japan the rate ranged from a low of 0.9 during the 1920s to 8.7 during the 1960s. These ranges are somewhat distorted for the United Kingdom and Japan owing to the lack of data for the World War II pe-

TABLE 3-2. Long-Term Trends in Crude Productivity: Three Countries (annual percentage rates)

	United States: 1870–1980[a]			*United Kingdom: 1861–1971*[b]			*Japan: 1886–1976*		
	Rate of Productivity Growth:			*Rate of Productivity Growth:*			*Rate of Productivity Growth:*		
Year	*Previous 10 years*	*Previous 70 Years*	*From Date to 1980*	*Previous 10 Years*	*Previous 70 Years*	*From Date to 1980*	*Previous 10 Years*	*Previous 70 Years*	*From Date to 1976*
1981	0.7	1.6	N.A.	2.5	1.3	N.A.	7.2[b]	3.3[b]	N.A.
1971	2.4	1.7	0.7	3.2	0.9	2.5	8.7	3.0	4.5
1961	1.1	1.6	1.5	2.0	0.8	2.8	6.3	2.1	7.3
1951	3.9	1.4	1.4	N.A.	0.7	2.6	N.A.	N.A.	6.9
1941	1.3	1.2	2.0	N.A.	N.A.	N.A.	N.A.	N.A.	N.A.
1931	1.1	N.A.	1.9	0.6	0.8	2.0	0.9	N.A.	4.0
1921	0.7	N.A.	1.7	−1.6	N.A.	1.8	3.0	N.A.	3.4
1911	1.8	N.A.	1.6	−0.1	N.A.	1.3	1.9	N.A.	3.3
1901	1.6	N.A.	1.6	2.1	N.A.	1.1	1.9	N.A.	3.1
1891	−0.8	N.A.	1.6	1.2	N.A.	1.2	N.A.	N.A.	3.0
1881	2.8	N.A.	1.4	1.4	N.A.	1.2	N.A.	N.A.	3.0
1871	N.A.	N.A.	1.5	2.4	N.A.	1.2	N.A.	N.A.	N.A.
1861	N.A.	N.A.	N.A.	N.A.	N.A.	1.3	N.A.	N.A.	N.A.

Sources:

United States: GNP for 1870–1970 comes from tables F1-5 in U.S. Bureau of the Census, *Historical Statistics of the United States: Colonial Times to 1970,* Bicentennial Edition, Part 1 (Washington, D.C.: U.S. Government Printing Office, 1975), p. 224. Labor force for 1870–1970 comes from table D11-25 in *Historical Statistics of the United States,* p. 127. GNP and labor force for 1980 come from U.S. Department of Commerce, *Survey of Current Business,* July 1981.

United Kingdom: GNP for 1861–1951 comes from tables 90 and 91 in Phyllis Deane and W. A. Cole, *British Economic Growth, 1688–1959: Trends and Structure* (Cambridge, England: Cambridge University Press, 1962). GNP for 1961 and 1971 comes from Central Statistical Office, *Economic Trends: Annual Supplement, 1980 Edition* (London: Her Majesty's Stationery Office, 1979), pp. 9, 12, and 36. Labor force for 1861–1941 comes from tables, "Labour Force I, B-D" in Brian Mitchell and Phyllis Deane, *Abstract of British Historical Statistics* (Cambridge, England: Cambridge University Press, 1962), pp. 60–63. Labor force for 1951, 1961, and 1971 comes from CSO, *Economic Trends,* P. 97. For 1980, data are from International Monetary Fund, *International Financial Statistics, Yearbook 1982* (Washington, D.C.: 1982).

Japan: GNP for 1886–1971 comes from tables A3 and A4 in Kazushi Ohkawa, Miyohei Shihohara, and Larry Meissner, eds., *Patterns of Japanese Economic Development: A Quantitative Appraisal* (New Haven, Conn.: Yale University Press, 1979), pp. 256–60, GNP for 1976 comes from World Bank, *World Tables,* 2nd ed. (Baltimore, Md.: Johns Hopkins University Press, 1980), pp. 250–51. Labor force for 1886–1966 comes from table A53 in Ohkawa, et al., p. 392. Labor force for 1971, 1976 come from oral communication with the economic-commercial division of the Embassy of Japan. GNP for 1971 uses table A4 from Ohkawa by converting 1965 constant yen on the assumption that national income and GNP grew proportionally between 1950 and 1952. GNP for 1931 uses the average of the values found in tables A3 and A4 of Ohkawa, et al.

Note: Crude productivity is measured as output per member of the labor force, except for the United Kingdom where for 1861–1951 Mitchell and Deane estimates have been scaled to conform to the labor force standards used in the 1980 CSO tables. Thus, for the United Kingdom as opposed to the United States and Japan, values for the "employed labor force" for the United Kingdom are used, and are not strictly comparable to the labor force used elsewhere.

N.A. = Not available or not applicable.

[a]All data for the United States are from 1860, 1870, 1880, etc., rather than from 1861, 1871, 1881, etc.

[b]Data for 1976 rather than 1981; growth rates are computed from 1966 and 1976 for the ten-year period rate.

riod. However, the average over the war period (1931–57) for the United Kingdom was a 1.3 percent per year improvement. For Japan (1936–51), the average annual change in productivity was −0.7 percent.[6]

The surprising result is that while the most recent decade has been a relatively poor one for productivity growth compared with the 1960s, it has not been exceedingly bad compared with most past performances. In fact, for both the United Kingdom and Japan, the 1970s showed the second or third best performance of the ten or so decades for which data were available. Recently diminished productivity growth rates, particularly in the United States, have stimulated great interest among economists. Many studies have attempted to define the causes of the recent sharp drop in productivity growth with respect to the relatively rapid growth experienced since World War II.[7] The United States still has a very high absolute productivity level, and some economists have interpreted the low 1970–80 rate as a short-term fluctuation with a negligible long-term effect. However, as Denison explains, the continued slowdown since 1974 has been puzzling and it is now thought that this contemporary slowdown may signal a turning point in long-term growth trends.[8]

It should be noted, in counterpoint, that this downturn is by no means unique. Productivity growth rates were equally low in the period 1910–20 and actually were negative for 1880–90. By itself, then, the poor productivity growth of the past decade is not without precedent.

An interesting observation is the suggestion of a leader phenomenon in the data.[9] The United Kingdom, the early leader in development, fell upon a period of dismal growth from about 1900 until after world War II. This allowed other countries, and in particular the United States, to surpass the United Kingdom in terms of per capita production and income. Under this hypothesis, the United States, now long the leader, may be entering a period of slower growth, allowing a latecomer like Japan to spurt ahead. One observes that the 1970s showed the United States to be particularly hard hit by the widespread slowdown in productivity increase and that even during the 1950s and 60s the rates of increase were considerably below those for Japan and the United Kingdom. Such a hypothesis would be consistent with the underlying need for the front runner to undertake basic research to achieve continued growth. Research and product and process development costs draw on limited investment resources and draw funds away from investment in more capital. In a follower country, nearly all investment can go directly to investment in known technologies developed by the leader country. Interestingly enough, the productivity growth in the United Kingdom over its leader phase (the late part of it since data are unavailable for earlier years), which we count from 1861 to 1901, is very similar to the rate experienced during the United States over its leader phase (1920 to 1970); the rates are 1.8 percent per year versus 1.9 percent per year. The ranges of rates during these periods are also very similar, with the exception of the United States during the war-driven 1940s.

While the data suggest the possibility of a leader phenomenon, the switchover, which has occurred only once in modern history, cannot be viewed as a necessarily generalizable theory; nor does it offer particular predictive power as to how long a leader position can be maintained or what country is likely to take over as a leader country falters. Olson has examined in much more detail the forces leading to the rise and fall of economic powers. He views the decline as determined by the entrenchment of interest groups, which occurs over long periods of political and economic stability and thus tends to occur in an economically and politically powerful nation.[10]

70-Year Records

The overlapping 70-year rates range from 1.2 to 1.7 percent per year in the United

States, from 0.7 to 1.3 percent per year in the United Kingdom, and from 2.1 to 3.3 percent per year in Japan. In comparing the countries, one must observe the differences that exist. Japan differs from the United States and the United Kingdom in that it has been in a "catch-up" or "transitional" phase through most of the postwar period. Its phenomenal success has been generally noted.

The average annual rate of Japan's productivity growth from 1951 to 1976 is 6.9 percent. This corresponds with an 8.6 percent average annual rate of growth of real GNP over the same period. Denison and Chung suggest that much of this postwar jump is due to "above-average contribution from changes in employment, hours of work, and the distribution by age and sex of total hours worked; . . . a greater contribution from capital; . . . application of new knowledge to production; and . . . reallocation of resources away from agriculture and from nonagricultural self-employment."[11] In addition, economies of scale are important in Japan's growth, as are improvements in education and the reduction of trade barriers. The spread of technologies was fostered by the very high savings rates achieved in the country and the levels of investment the high savings rates implied.[12]

Denison and Chung define the "transitional component" of economic growth as the amount of growth made possible solely by virtue of a nation arriving late on the scene of economic development. They note that the transitional component of the Japanese growth rate is exceptionally big, but that probably all countries have a transitional component.[13]

Other Long-Term Data

Kuznets has constructed a long-term data series on economic growth for several western economies. His data have been adapted to be comparable to the crude productivity estimates used in this study, as noted in table 3-3. The rates of productivity growth resulting from the Kuznets data are not consistent with those presented in table 3-2. Reasons for this inconsistency include the facts that the statistical sources are different, and that the margin of error introduced by several crude assumptions in the adaptation of Kuznets' data may be large. The estimates provide an additional indication of rates of growth experienced over the long-term course of development in Eurpose and North America.

It is impossible to calculate productivity growth over very long periods for large regions of the world. In much of the developing world, the process of modern economic development and associated increases in per capita GNP did not begin until the postwar period. It is usually inferred that, in these countries, per capita income levels were relatively unchanged over long periods prior to the onset of the modern development process. To the extent that this inference is correct, productivity growth was zero by implication. Cippolla has provided an interesting survey of preindustrial advance in Europe (1000–1700 A.D.), challenging the idea that no productivity growth occurred.[14] However, even the most rapidly advancing industries showed rates of improvement of 1.2 percent per year, while agriculture showed little if any improvements in most cases. It is not clear how these results transfer, if at all, to the preindustrial experiences of other world regions.

Short-Term Productivity Trends

While it is not possible to develop long-term productivity series on a global basis, data do exist for constructing a shorter term series. This section presents global productivity series for the 1960 to 1975 period. Summary data are presented in table 3-4 for the nine world regions. Other regional aggregations have also been reported to facilitate comparison with other studies. The OECD region is an aggregate of regions 1, 2, and 3, while the LDC region is an aggregate of regions 6, 7, 8, and

TABLE 3-3. Long-Term Percent Growth of Crude Productivity for Selected European Economies: 1870 Through 1913 (I) and 1913 Through 1960 (II)

Country	Growth in Output per Man-Hour per Year (%)[a] I	Growth in Output per Man-Hour per Year (%)[a] II	Decline in Hours per Laborer per Year (%)[c]	Growth in Crude Productivity per Year (%)[c] I	Growth in Crude Productivity per Year (%)[c] II
1. United Kingdom/Great Britain[d]	1.52	1.80	0.27	1.25	1.53
2. France	1.81	2.09	0.44	1.36	1.64
3. Belgium	2.03	1.67	0.34	1.68	1.32
4. W. Germany (1871)[e]	2.07	1.97	0.38	1.68	1.58
5. Netherlands (1900)[e]	1.09	1.69	0.58	0.50	1.10
6. Switzerland (1890)[e]	1.56	2.39	0.46	1.09	1.92
7. Denmark	2.56	1.79	0.54	2.01	1.24
8. Norway	1.78	2.71	0.33	1.44	2.37
9. Sweden	2.67	2.29	0.37	2.29	1.91
10. Italy (1890)[e]	1.87	2.35	0.87	0.98	1.46
11. United States (1871)	2.38	2.40	0.49	1.88	1.90
12. Canada	2.07	2.22	0.52	1.54	1.69

Source: All values calculated from tables 6.6 (pp. 352–53) and 2.6 (p. 73) in Simon Kuznets, *Modern Economic Growth: Rate, Structure, and Spread* (New Haven, Conn.: Yale University Press, 1976).

[a]Growth in output per man-hour per year is calculated using the data in table 6.6 (Kuznets) and the following algorithm:

$$\text{Growth in output per man-hour per year (\%)} = \left\{\left[\frac{\text{\% growth in product/man-hour/decade}}{100} + 1\right]^{1/10} - 1\right\} * 100\%$$

[b]Rate of decline in hours per laborer per year is calculated using the data in table 2.6 (Kuznets) and the following algorithm:

$$\text{Rate of decline in hours per laborer per year (\%)} = \left[\left(1 - \frac{\text{\% decline over total period}}{100}\right)^{(1/\text{length of period})} - 1\right] * 100\%$$

The decline in hours per laborer per year is assumed to be constant over periods I and II.

[c]Growth in crude productivity per year is calculated using the following algorithm:

$$\text{Growth in crude productivity per year (\%)} = \left\{\left(\frac{\text{\% rate of growth in output per year}}{100}\right) - \left(\frac{\text{\% decline in hr/laborer/yr}}{100}\right) - \left[\left(\frac{\text{\% growth in output/yr}}{100}\right) * \left(\frac{\text{\% decline in hr/laborer/yr}}{100}\right)\right]\right\} * 100\%$$

[d]Data for the United Kingdom and Great Britain are used interchangeably on the assumption that their respective growth rates are essentially the same over these periods.

[e]The date in parentheses is the initial date of the first period (I), if different from 1870.

9. Regions 4 and 5 are aggregated as the centrally planned economies (CPEs); non-CPEs are aggregated as the world outside of the Communist area (WOCA). Before discussing trends in productivity, it is useful to point out an arithmetic property of aggregating across regions. It is arithmetically possible that the average rate for an aggregate region falls beyond the extremes for individual regions (countries) making up the aggregate region. This occurs in the case of the OECD, LDCs, and MIDEST aggregated as WOCA for the period 1970 to 1975; the average is 1.79 percent per year while the lowest rate for an individual region is the 1.95 percent per year in the OECD. This occurs because the labor force is growing rapidly in the LDCs, where the average productivity rate is much lower. Thus, at the end of the period, the low productivity in the LDCs implicitly has a much higher weight in the average than at the start of the period.

In examining table 3-4, several points are worth noting. First, overall rates of productivity growth are quite high for the

TABLE 3-4. Global Productivity Summary: 1960–75

	Productivity (10^3 1975 U.S. $/laborer)			Productivity Growth (%/yr)		
Regions	*1960*	*1970*	*1975*	*1960–70*	*1970–75*	*1960–75*
US	12.17	15.48	16.06	2.44	0.73	1.87
OECD West	6.08	9.53	10.72	4.60	2.38	3.85
JANZ	3.70	7.64	9.11	7.52	3.58	6.19
E USSR	2.82	4.23	4.91	4.14	3.03	3.77
A CENP	0.23	0.34	0.41	3.99	3.82	3.93
MIDEST	3.28	5.26	7.32	4.84	6.83	5.50
AFR	0.68	0.92	1.07	3.07	3.07	3.07
LA	1.93	2.71	3.21	3.45	3.44	3.45
SEASIA	0.35	0.48	0.55	3.21	2.76	3.06
Aggregated Regions						
World	2.29	3.30	3.66	3.72	2.09	3.18
CPEs	1.13	1.63	1.89	3.73	3.00	3.49
WOCA[a]	3.03	4.30	4.70	3.56	1.79	2.97
OECD	7.27	10.84	11.94	4.08	1.95	3.36
LDCs	0.64	0.89	1.05	3.35	3.36	3.36
MIDEST	3.28	5.26	7.32	4.84	6.83	5.50

Sources: GNP values calculated using data in *World Bank Atlas, 1977,* pp. 14–23. All labor force values are found in International Labour Office, *1950–2000 Labour Force,* vols. I–IV (Geneva: ILO Publications, 1977), tables 1 and 4. Labor force values for Taiwan are calculated using the information in World Bank, *World Development Report, 1979* and *1980,* and the ILO, *1950–2000 Labour Force, Asia.*

[a]World outside Communist areas.

world and all regions during the 1960–70 period. The Mideast had remarkable success even before OPEC price actions. The world average productivity growth rate of 3.72 was truly remarkable. It exceeded the productivity growth rate of any prewar Japanese decade (see table 3-2). Moreover, growth was fairly well balanced in that productivity growth increases by the LDCs were only three-fourths of 1 percent lower than OECD growth rates, and if the Mideast is included within the group, the discrepancy becomes minor (3.7 for the greater LDC group versus 4.1 for the OECD). In comparison with the western experience, the less developed CPEs experienced somewhat slower productivity growth than the more developed economies.

During the 1970–75 period, world productivity growth rates fell markedly. The OECD region's productivity growth rate was halved. Similarly, growth in productivity was curtailed in both the LDCs and CPEs. Only the Mideast succeeded in increasing its productivity, and this was largely the result of greatly improved terms of trade in oil, rather than a consequence of more conventional productivity enhancements. However, the general slowdown in productivity growth among the world's regions was not shared equally. It was most pronounced in the OECD, less pronounced in the CPEs, and negligible in the LDCs. In fact, the rate of productivity growth was virtually unchanged in Africa and Latin America. Similarly, the decline in productivity growth rates for the CPEs was not shared equally.

While these groupings are somewhat different than those in table 3-4, making comparisons of actual rates subject to the arithmetic problems noted earlier, the general conclusions made earlier remain unchanged.

Basically, productivity growth slowed during the 1970s, but the impact was disproportionally felt in the industrial countries of the OECD.

Much has been made of this world productivity slowdown, and increased energy costs are frequently cited as a major cause of this downturn. The analysis is, however, more complicated, and recent research into U.S. productivity has revealed that energy played only a minor role.[15] More important, especially in terms of our analysis, is the global recession which followed the OPEC price increase. This may also explain somewhat the relatively small drop in LDCs and CPEs; these regions responded with a lag to the multiplier effect of reduced OECD demand for LDC exports and to a lesser extent CPEs.

In some cases, LDC regions experienced productivity growth because major segments of the region are oil producers. This is clearly the case in Africa, where the exclusion of Algeria, Libya, and Nigeria results in a decline in the productivity growth rate for the remainder of the region. However, even without the oil economies, African productivity growth remains near the world average. In the preceding 10 years, productivity growth for this region had been far below the world average (2.4 percent for nonoil Africa versus 3.6 percent for the world in 1960–70 and 1.9 percent versus 2.1 percent for 1970–75).

This is clearly not the case in Latin America. In Latin America, both Venezuela and Mexico saw productivity growth fall, with Venezuela showing the sharpest drop. Brazil, on the other hand, increased its productivity growth rate from 3.5 percent (1960–70) to 6.2 percent (1970–75). Similarly, the decline in South and East Asian productivity is due almost exclusively to the dismal performance of India, where productivity growth fell from an annual rate of 2.4 percent (1960–70) to 0.6 percent (1970–75). The remainder of the region held its growth rate steady, with impressive increases registered in such nonoil economies as South Korea, the Philippines, and Malaysia.

Part of the reason that LDC and CPE productivity growth rates remained relatively high over the 1970–75 period is that both groups of economies were increasing the level and share of GNP allocated to investment. For example, the low-income economies increased their average share of GNP allocated to investment from 14 to 21 percent between 1960 and 1977. Similarly, the middle income group of economies raised their investment share from 21 to 25 percent. These increases have come almost exclusively as the result of increased rates of domestic savings.[16] The Soviet Union also increased its share of GNP allocated to investment from 21 to 30 percent between 1960 and 1977.[17] This approaches the level of investment in Japan, which was stable over most of the period and held at about 32 percent in 1978.[18] These high rates of investment allowed the formation of capital stocks and facilitated the transfer of technology. In contrast, the industrialized economies held their rate of investment relatively constant at 22 percent.[19]

To update the "statistics" for the world, which run only to 1975 in the historical section, the comparative growth in (gross domestic product) in the 1960s and 1970s is given in table 3-5 for various aggregations of countries.

A COMPARISON OF PROJECTED GNP GROWTH RATES

Table 3-6 reports future GNP growth rates for five regions and two aggregate world regions as projected in nine major studies.

A quick examination of the figures in this chart reveals several relationships and trends. Through 2000, projected GNP growth rates for the United States tend to fall between 2.3 and 3.5 percent per year. Projections for the non-U.S. OECD regions are similar to the United States but tend to be 0.1 to 0.2 percentage points higher. In both cases the OECD study is considerably more optimistic about future rates of economic growth.

For studies offering post-2000 projections, growth rates taper by 0.5 to 1.5 percentage points. This tapering phenomena is observed for the CPE and LDC regions

TABLE 3-5. Comparative GDP Growth, 1960s and 1970s (percent per year)

Country Group	GDP Growth		Approximate Crude Productivity[a] Growth	
	1960–70	1970–80	1960–70	1970–80
Low-income economies[b]	4.4	4.6	2.8	2.8
India and China	4.5	4.9	3.0	3.2
Others	4.4	3.5	2.1	1.3
Middle income economies[c]	5.9	5.6	3.8	3.1
Oil exporters	6.2	5.5	3.7	3.2
Oil importers	5.8	5.6	3.8	3.1
Industrial market economies	5.2	3.2	4.1	2.0
High-income oil exporters[d]	N.A.	5.3	N.A.	2.9

Source: World Bank, *World Development Report 1982* (New York: Oxford University Press, 1982), pp. 112–113.

N.A. = Not available

[a]Calculated by subtracting crude labor force growth rates from GDP growth rates assuming 1970–75 labor growth was characteristic of the period 1970–80. Labor force is from Kefitz, et al. "Global Population: 1975–2075," Institute for Energy Analysis, Oak Ridge, Tenn., 1983.

[b]Below $410/capita income in 1980, in 1980 U.S. dollars.

[c]Between $410/capita and $4,500/capita in 1980, in 1980 U.S. dollars.

[d]Libya, Saudia Arabia, Kuwait, United Arab Emirates.

TABLE 3-6. Comparison of GNP Growth Among Major Energy Studies

	EXXON 1979–2000	OECD 1975–2000[a]	U.S. DOE 1979–1995	HK 1985–2005	RFF 1972–2000	RFF 2000–2025	WAES (Low) 1972–2000	WAES (High) 1972–2000
U.S.	2.7[c]	5.3[d]	2.5	3.0	2.7	2.3	2.6	3.6
Other OECD	2.7	5.1	3.1	3.1[f]	3.3	2.4	2.6	4.0
CPE	2.9	3.9[c]	N.G.	N.G.	4.3	3.2	N.G.	N.G
Non-oil LDC's	4.8	6.2[h]	5.6	6.4[i]	5.4[j]	5.3	4.0	5.2
Oil LDC's	5.4	4.5[k]	5.1	7.7[l]	m	m	5.8	7.6
WOCA	3.2	4.4	3.4	4.2	3.5	3.4	3.0	4.2
WORLD	3.1	4.3	N.A.	N.A.	3.7	3.3	N.A.	N.A.

Sources:

Lovins: Amory B. Lovins, L. Hunter Lovins, Florentin Krause, and Wilfrid Bach. *Energy Strategy for Low Climatic Risks.* Prepared for the German Federal Environmental Agency (San Francisco, Calif.: International Project for Soft Energy Paths, June 1981).

OECD: Organisation for Economic Co-operation and Development. *Interfutures, Facing the Future (Paris, 1979).*

RFF: Ronald G. Ridker and William D. Watson, *To Choose a Future* (Baltimore, Md.: Johns Hopkins University Press, for Resources for the Future, 1980).

WEC: The Full Report to the Conservation Commission of the World Energy Conference. *World Energy Demand* (New York: IPC Science and Technology Press, 1978).

DOE: U.S. Department of Energy. *1980 Annual Report to Congress.* Report # DOE/EIA-0173(80)/3 (Washington, D.C.: U.S. Government Printing Office, 1981).

H-K: Hendrik S. Houthakker, and Michael Kennedy. "Long-Range Energy Prospects." *Journal of Energy and Development,* Autumn 1978.

WAES: Carroll L. Wilson, Project Director. *Energy: Global Prospects 1985–2000.* Report of the Workshop on Alternative Energy Strategies (New York: McGraw-Hill, 1977).

EXXON: Exxon Corporation. *World Energy Outlook,* December 1980.

IIASA: Wolf Hafele, Project Leader. *Energy in Finite World: A Global Systems Analysis.* Report by the Energy Program Group of the International Institute for Applied Systems Analysis (Cambridge, Mass.: Ballinger, 1981).

N.A. = Not applicable.

N.G. = Not given.

as well. While not all the studies provide estimates of labor force, such a time trend in GNP is consistent with an underlying slowing in population growth and, therefore, eventually slower rates of increase in labor forces.

A look at the centrally planned economies reveals a growth range between 3 and 4.5 percent. These figures also drop quite sharply after the year 2000 with average drops of approximately 1 to 1.2 percentage points.

Growth of GNP in the developing countries is projected to be uniformly higher than in other groups by anywhere from 1 to 3 percentage points. Within this group, the oil-exporting countries are projected to grow faster—by 0.5 to 1 percentage points—than non-oil LDCs. The exception is the OECD and U.S. Department of Energy studies, which show slower growth for oil LDCs.

Looking at the figures for the world outside of the centrally planned economies (WOCA), one notes a trend in GNP growth rates that ranges between 3 and 4 percent. Predictions for reductions in growth rates after 2000 range from between 1.2 and 1.3 percentage points. Projections for the world as a whole parallel these WOCA trends, with growth rates of 3–4 percent and drops after 2000 ranging from 0.5 to 1.3 percentage points.

The more rapid growth in developing economies can be explained by differences in both productivity and labor force growth assumptions. More rapid labor force growth can be used to account for up to 1.5 percent per year more rapid growth in GNP. Greater differentials must, in general, be explained by more rapid growth in productivity.

Future Rates of Productivity Growth: An Assessment

The model used to generate future estimates of energy use and described in chap-

Lovins[b]			*WEC*[c]				*IIASA*			
			(Low)		*(High)*		*(Low)*		*(High)*	
1975–2000	*2000–2030*	*2030–2080*	*1972–2000*	*2000–2020*	*1972–2000*	*2000–2020*	*1975–2000*	*2000–2030*	*1975–2000*	*2000–2030*
2.4	1.0	0.0	2.5[d]	2.5	3.5	3.1	2.4[d]	1.0	3.7	2.2
2.5	1.3	0.0	3.0	3.2	4.0	3.2	2.5	1.3	3.8	2.2
3.4[g]	2.2	0.0	3.5	3.0	4.7	3.7	3.4	2.2	4.4	3.4
4.1	2.8	1.1	3.8	3.5	5.0	4.6	4.1	2.8	5.3	3.5
5.0	2.4	1.1	4.8	3.8	6.8	5.5	5.0	2.4	6.4	4.0
2.9	1.6	0.44	3.0	3.1	4.1	3.6	2.9	1.6	4.1	2.6
3.1	1.8	0.38	3.1	2.9	4.2	3.7	3.1	1.8	4.2	2.8

[a]Using OECD's D (3) scenario reported in table 21 in the source.

[b]Using IIASA's 1979 "low" scenario table 14-4 in the source.

[c]Using ILASA's "high" scenario as base GNP (1975).

[d]U.S. = North America

[e]Includes Japan, W. Europe, other non-EEC, Australia, and New Zealand.

[f]Europe only.

[g]CPE does not include China here.

[h]Includes sub-Saharan Africa, SE Asia, and Latin America.

[i]Latin America, Asia, and Pacific.

[j]Latin America, other non-Communist Asia, Africa.

[k]Includes S. Asia, N. Africa and W. Asia.

[l]Middle East and Africa.

[m]Not separated from non-oil LDCs. See note j.

ter 16 provides for an endogenous GNP impact resulting from changes in energy prices. As a result, from the standpoint of the modeling exercise, thinking about appropriate rates of future productivity growth requires a separation of productivity growth in the absence of increases in energy prices from the impact of rising energy prices on economic growth and hence productivity.

Beyond supplying input numbers for a model, the exercise is useful in that it forces one to address directly the issue of how large an impact on GNP one can expect from a given change in energy prices. This latter issue is one which can and has been explored separately. This section outlines plausible rates of productivity growth for various regions of the world. A discussion of the energy impact on GNP appears in chapter 4.

As we have argued elsewhere, long-term forecasting requires an explicit attempt to discount current experiences when forming assumptions about the future since a year or two of observations are more likely to be aberrant than indicative of long-term trends.

While the 1950s and 60s exhibited unprecedented rates of productivity improvement compared with past performances, productivity growth rates of the 1970s were unprecendented relative to all decades except the 1960s. Given this background, what might constitute a best-guess scenario for future rates of productivity increase in the absence of energy price increases? Table 3-7 contains our best-guess estimates.

There are, of course, very large uncertainties connected with these estimates. Some observers would refuse to offer such an estimate, and would rather construct scenarios of higher and lower rates of growth which would result under different assumptions about the ability of the world or countries to control inflation; the likelihood of increasing or decreasing trade barriers, different assumptions about the tradeoff between growth and equity or growth and the environment and other factors. To argue that one has a best unconditional guess implies some assessment of these factors. Rather than attempt to assess such factors, we will argue that a best guess for long periods of time is roughly the long-term historical experience unless there appears to be a strong reason to suspect change.

The expected U.S. rate is somewhat slower than long-term historical rates. It would appear that the United States suffered the largest drop in productivity, in percentage terms, during the 1970s. The difficulties of maintaining rapid growth as the technological leader and low savings and investment rates appear to signal a significant long-term change in the United States. OECD West growth is set at the higher levels achieved by the United States over long periods. This is a considerable slowing from the 1960s and even the

TABLE 3-7. Best-Guess Estimates of Future Rates of Productivity Increase (percent per year)

	1975–2000	*2000–2025*	*2025–2050*
US	1.5	1.5	1.5
OECD West	1.7	1.7	1.7
JANZ	3.0	2.5	2.0
EUSSR	2.5	2.5	2.5
ACENP	3.0	3.0	3.0
MIDEST	3.0	3.0	3.0
AFR	2.7	2.7	2.7
LA	3.0	3.0	3.0
SEASIA	2.6	2.6	2.6

1970s. The argument for such a trend is that the postwar years represented a period when Europe was able to grow rapidly as it put in place technologies developed in the United States. As the gap between Europe and the United States is narrowed, European countries will be unable to take advantage of new technologies which represent quantum leaps in productivity since the larger part of the capital stock in place already represents high productivity, recent vintage equipment.

A similar argument holds for JANZ. Japan is the major economy. Even at the slowed rates of productivity increase, per capita income will exceed U.S. per capita income by 2025. Traditionally high savings and investment should maintain growth rates above other OECD regions for some time, however. The Eastern block countries are expected to continue to make relatively rapid progress, but will not catch up to the western industrialized countries by 2050.

In all developing country regions, productivity improvement is expected to be rapid. The availability of highly productive technologies from the industrialized countries makes possible rapid increases even with relatively limited investments. In most cases, and particularly in the larger economies, savings rates have increased through the 1960s and the 1970s. On average, savings rates in the developing countries exceed the average rate for the developed countries.[20] Investment rates have run even higher as a result of borrowing and private direct investment. To some extent, borrowing was used to counteract the growth impacts of the global recessions of the 1970s and this has been cause for concern. Bankers may be retrenching somewhat in making loans but even without net inflows of new capital, savings rates are high enough to support rapid economic growth. Moreover, there is little reason to believe that retrenchment in lending is a permanent phenomenon. As economies catch up to the currently high debt levels, private banks are likely to reenter, although perhaps with somewhat more caution.

Rates of improvement in these regions of 3.0 percent per year are generally assumed to be likely. This represents somewhat slower rates than experienced in the 1960s and early 1970s, in the belief that this represented a particularly expansive period which over the longer run will be averaged in with periods of poorer performance. The rate is similar to 70-year averages achieved by Japan which, we note, include World War II.

Africa and South and East Asia are expected to grow somewhat slower, in keeping with their historical experiences. Both regions have many or large economies which have yet to demonstrate the ability to sustain the very rapid economic growth achieved in Latin America or Japan. In addition, the South and East Asia region is a particularly broad group of countries that includes rapidly growing economies like Korea, Taiwan, and Singapore and poorer performers like India and Bangladesh. The poorest countries will continue to have rapid population and labor force growth while the richer countries will soon be seeing declining rates of labor force growth. This will lead to an aggregation problem similar to that presently existing between the OECD and LDCs. The growing weight of lower productivity economies like India and Indonesia will tend to depress the average productivity growth rate even if all the countries grow at equally rapid rates.

Uncertainty

As already noted, levels of uncertainty surrounding estimates are high. On the low side, much has been made about growth versus resource limits and the environment in the developed countries. Nevertheless, growth continues to make income redistribution policies more palatable. Thus, it seems unlikely that developed economies would choose a zero-growth society. The environmental tradeoff has become less a tradeoff between growth and no growth and more a tradeoff between where growth will occur. Provision of environmental quality is becoming a market activity which adds to the GNP and requires scarce inputs, including energy. Here we point out again that the GNP measure is a measure of economic activity and income but not necessarily societal welfare. Even with a growing GNP, welfare may be constant or fall if larger shares of GNP are needed to *maintain* the environment or if welfare losses from environmental degradation exceed welfare gain from increased GNP.

In the developing countries, political instability appears to be the largest obstacle to growth. If equity questions are not addressed or the social fabric is threatened through too rapid change, as in Iran, instability could result and slow growth. The possibility of financial collapse and inability of countries to borrow offers a threat to growth as well. In a single country, these events are likely to happen in combination, but it is less clear that instability in different countries is correlated with time. Historical rates of productivity increase include historical levels of instability; further depression of productivity increases for large regions would mean postulating more widespread instability.

On the high side, one can observe that the overall trend in productivity improvement has been for accelerating rates. Based simply on the historical trends, one might predict that the 70s and early 80s represent a decadal slowdown in a accelerating trend. If this is true, one would expect productivity growth to rebound to levels above the 1960s. One might further justify such a projection by observing that as more and more countries (Europe and Japan) catch up to the United States, they will increasingly participate in basic research. With more effort devoted to research, one might expect the global level of technological change to accelerate. In addition, the savings rates in the developing countries may continue rising even beyond presently high levels and, with higher incomes, political stability in these regions may increase.

SUMMARY

While productivity research has gone far in illuminating the factors affecting the rate of growth and has in some cases succeeded in quantitatively attributing historical growth to the various underlying causes, predictive power remains poor. While technological change and increases in human skill levels are major determinants of growth, the abilty to predict these factors is poor or nonexistent. Wide cross-sectional differences in saving rates, between Japan and the United States, for example, have not been explained other than as a cultural phenomenon. As such, the reference point for future growth projections must be representative of historical experience, adjusted to the extent that one can identify fundamental changes in underlying factors. In using historical rates as a reference, one must keep in mind, first, that historical periods in general include considerable political instability, economic mismanagement, and various crises and shortages. Thus, postulating instability, economic mismanagement, or various crises in the future should not affect the historical reference rate unless one is postulating a change in the intensity or frequency of occurence of these events.

Our base rate of global productivity expansion represents a slowing from the rapid rates of the 1960s but relatively rapid expansion relative to longer term historical rates. Continued rapid expansion is premised on the fact that nearly all major countries have become integrated into the world economy and will be able to take advantage of existing and newly available productivity-increasing technologies. This projection assumes no additional energy price increases. These impacts are endogenously determined in the model and are discussed in chapter 5.

NOTES

1. In some cases, the output of a nation is measured as gross domestic product (GDP) rather than GNP. GDP differs from GNP by net factor payments abroad. GDP measures only the output resulting from resources originating in the country of interest while GNP measures total final output, regardless of net imports and external debts. For reasons of consistency, this study will consider only GNP.
2. For those interested in pursuing this analysis, see E. F. Denison, *Accounting for Slower Economic Growth: The United States in the 1970s* (Washington, D.C.: Brookings Institution, 1970); or E. F. Denison and W. K. Chung, *How Japan's Economy Grew So Fast: The Sources of Postwar Expansion* (Washington, D.C.: Brookings Institution, 1976); or P. W. Jorgenson and E. Griliches, "The Measurement of Productivity," *Survey of Current Business,* 52, Part II, May 1972.
3. For examples of expressions of both views, see U.S. Senate, Committee on Labor and Human Resources, *Productivity in the American Economy, 1982* (Washington, D.C.: U.S. Government Printing Office, 1982).
4. U.S. Bureau of the Census, *Historical Statistics of the United States: Colonial Times to 1970,* Bicentennial ed., part 1 (Washington, D.C.: U.S. Government Printing Office, 1975), p. 216.
5. Ibid., pp. 121–123.
6. These figures are calculated from data given in the sources listed in table 3-2.
7. The Joint Economic Committee, Congress of the United States, presented an extensive report on recent productivity and its future implications in the *Midyear Review of the Economy: The Outlook for 1979* (Washington, D.C.: U.S. Government Printing Office, 1979). Also, see The Conference Board, "Productivity," in *Economic Road Maps,* nos. 1822–1823 (New York, January 1978); Frank A. Weil, "Managment's Drag on Productivity," in *Business Week* (December 3, 1979), p. 14; Jerry Flint, "Lagging Productivity: Planners are Stymied," in *New York Times* (January 27, 1979), pp. 1 and 28; and Denison, *Accounting for Slower Economic Growth,* among the many studies.
8. Edward F. Denison, "The Puzzling Drop in Productivity," in *The Brookings Bulletin* 15 (no. 2), pp. 10–11.
9. The importance of the date of entry into the modern world economy in determining the development rate of a nation and its relative position is discussed by Simon Kuznets in *Modern Economic Growth: Rate, Structure, and Spread* (New Haven, Conn.: Yale University Press, 1976), p. 291.
10. Mancur Olson, *The Rise and Decline of Nations* (New Haven, Conn.: Yale University Press, 1982).
11. Denison and Chung, *How Japan's Economy Grew,* p. 10.
12. The recent high savings rates are documented in J. A. Edmonds and E. L. Allen, "Japan: Estimates of Future Energy/GNP Relationships in Energy Use," Institute for Energy Analysis, Oak Ridge, Tenn., Research Memorandum (ORAU/IEA-79-21(M), 1979).
13. Denison and Chung, p. 116.
14. Carlo M. Cippolla, *Before the Industrial Revolution* (New York: W. W. Norton, 1976), pp. 115–138.
15. See, for example, U.S. Department of Energy, *Interrelationships of Energy and the Economy,* Office of Planning and Policy Analysis, Washington, D.C., and E. F. Denison, *Accounting for Slower Economic Growth* (Washington, D.C.: Brookings Institution, 1979).
16. See World Bank, *World Development Report,* p. 134.
17. See National Foreign Assessments Center, *Handbook of Statistics,* October 1978, Washington, D.C., p. 45.
18. See J. A. Edmonds and E. L. Allen, "Japan: Estimates of Future Energy/GNP Relationships in Energy Use," Institute for Energy Analysis, Oak Ridge, Tenn., Research Memorandum (ORAU/IEA-79-21(M), 1979), p. 16.
19. World Bank, *World Development Report,* p. 135.
20. Ibid.

Energy Consumption and Economic Growth 4

The relationship between energy consumption and economic growth can be expressed as a ratio which measures the amount of energy consumed per dollar of GNP (E/GNP ratio). This equation is a very simple and straightforward tool which measures a very complex and constantly changing relationship. The E/GNP elasticity is the percentage change in energy use over the percentage change in GNP. It is a unitless statistic which can be compared across countries. We have avoided comparing E/GNP *ratios* among countries. Care must be taken in drawing conclusions from such comparisons. Various studies have pointed to low E/GNP ratios in European countries as evidence that the United States, for example, should or could consume radically less energy. The explanations for such differences include pricing policies, import dependence, product mix, energy embodied in imports and exports, GNP accounting differences, real product (exchange rate) conversion problems, climate, and geography to name a few. With the causes for differences being both real and merely statistical and driven by both policy and geographical considerations there is little ground to place blame for high energy consumption, if one finds it necessary to place blame, on the basis of aggregate comparisons. As an aside, our purpose is not to lay blame but only to understand patterns of change. For this reason, most of the comparisons are made between E/GNP elasticities.

Changes in the E/GNP ratio and elasticity are a reflection of numerous underlying factors, such as the state of technology, the price of energy, environmental constraints, the level of activity in individual energy-using sectors, the composition of GNP, and demographic and sociological factors. It is beyond the scope of this chapter to provide an in-depth analysis of the influence of each of these factors on changes in E/GNP ratios. The purpose here is to examine the historical changes in E/GNP ratios for various countries and global regions, highlighting the major trends and likely factors influencing them, and compare these results with recent forecasts of future energy use and GNP growth. The examination of historical changes in E/GNP ratios covers three periods: a long-term period extending from the year 1700 to 1975 for the United Kingdom and from 1880 to 1975 for the United States; a mid-term period from the year 1925 to 1975 for several major industrialized countries and a few Communist countries; and a short-term period from 1950, 1955, or 1960 to 1975 for the rest of the world. The forecasts of future energy and economic growth encompass ten major studies of global and non-Communist energy and economic growth to the year 2000 and beyond published since 1970.

E/GNP—BACKGROUND

Analysts working in the 1960s and early 1970s characterized the E/GNP ratio as an unchanging relationship. Thus, every percentage increase in GNP was met by an equal percentage increase in energy consumption—the E/GNP elasticity was one. This rule was based on observations of highly aggregated data. In supplying some theoretical basis for the relationship, one could examine energy either as an input to production or as a product for which there was a demand. Clarifying the demand relationship, the consumer demand was for an energy service or product which embodied energy. The one-to-one aggregate relationship would be satisfied from the input side if one posited a fixed, Leontief production function and an unchanged product mix. From the demand side, the one to one relationship would be preserved if, with increases in income, consumers proportionally increased their demand for energy services, including embodied energy, and there was no ability to provide the same energy service with less energy input. Conceptualizing the problem as the Leontief production relation led many observers to fear that not only would energy use increase with GNP but if energy supply was curtailed, an equal drop in GNP could be expected. Apart from the validity of the Leontief production relation, the fear that GNP would be proportionately curtailed appears less real when the problem is viewed from the demand side; one could imagine changing consumption habits from more to less energy-intensive products and services if energy became less available.

As the 1970s progressed, and post-oil embargo data became available, it became clear that the one-to-one relationship was not a first law of energy use and economic growth. However, it is probably fair to say that the popular conception of the 1970s was that the period represented an unprecedented break with the past. The historical data examined in this chapter strongly question the validity of the one-to-one rule, even in the preoil embargo period. In fact, in examining the data it is difficult to understand how the one-to-one relation ever gained acceptance.

Before proceeding with an examination of the historical data, it is useful to discuss briefly a related strain of the literature on energy use and economic growth. The one-to-one relationship was viewed as holding in the case of the developed (industrialized) economies. The developing economies were viewed as countries with low E/GNP ratios, which, as development proceeded, would rise to the level of the developed countries. Thus, the early period of development would be characterized by a low E/GNP ratio, but with an elasticity greater than one. A continuing literature has examined the cross-country evidence in support of such a relationship. Empirical tests have been something less than convincing in establishing the relationship. A major obstacle is the set of problems and additional factors which influence energy use as put forth in the opening remarks in this chapter.

As a result of these problems in establishing firm relationships on an aggregate basis, the main body of recent energy literature has focused on disaggregated approaches and understanding relationships within individual countries. In fact, most interesting policy questions are best answered by such research. However, in examining long-term, global energy futures, the broad trends become important. In particular, we need an answer to the question, "How does energy use per unit of output change (independent of price and technological change) as development proceeds?"

HISTORICAL TRENDS IN E/GNP RATIOS

We begin the examination of historical trends and forecasts of aggregate behavior in the knowledge that the ability to separate causes underlying the trends in the historical data is poor. However, merely reporting these trends sheds some light, even if firm statistical causation cannot be established.

Very Long-Term Experiences

Data were available for the United States from 1880 to 1975 and for the United Kingdom from 1700 to 1975. Figure 4-1 displays the calculated E/GNP ratios for the two countries over these periods. In the case of the United Kingdom, the E/GNP ratio is clearly increasing up to 1880, with a falling trend thereafter. The United States shows a similar peaking, but around 1920. The pattern is somewhat less clear because there are no data for early years; however, it appears to parallel the United Kingdom experience.

Tables 4-1 and 4-2 provide the data on which figure 4-1 is based. In addition, discrete elasticities are calculated for 20- and 50-year periods. The 50-year elasticities reflect the general trend observed in figure 4-1; they are greater than 1 in early years and less than 1 in later years. The 20-year elasticities are interesting because they show considerable variation from one period to the next. From these data it is difficult to support a constant elasticity, much less an elasticity constant at one. However, the elasticities do support the overall rising E/GNP in early years and a falling E/GNP in later years. One might ask, is this a development phenomena or an historical accident? The somewhat later peaking of the United States gives some support to the development explanation since the United Kingdom is usually thought of as industrializing ahead of the United States. The later peaking is also supportive of the generality of the rise and fall of the E/GNP ratio as a development pattern in that it suggests that a major determinant of the energy use pattern is derivable from a factor unique to the country (its development history) rather than a common history of world events. However, this conclusion must be tentative since (1) there are re-

FIGURE 4-1. U.K. E/GNP ratios, 1700–1975, and U.S. E/GNP ratios, 1880–1975 (10^6 joules per U.S. 1975 dollars).

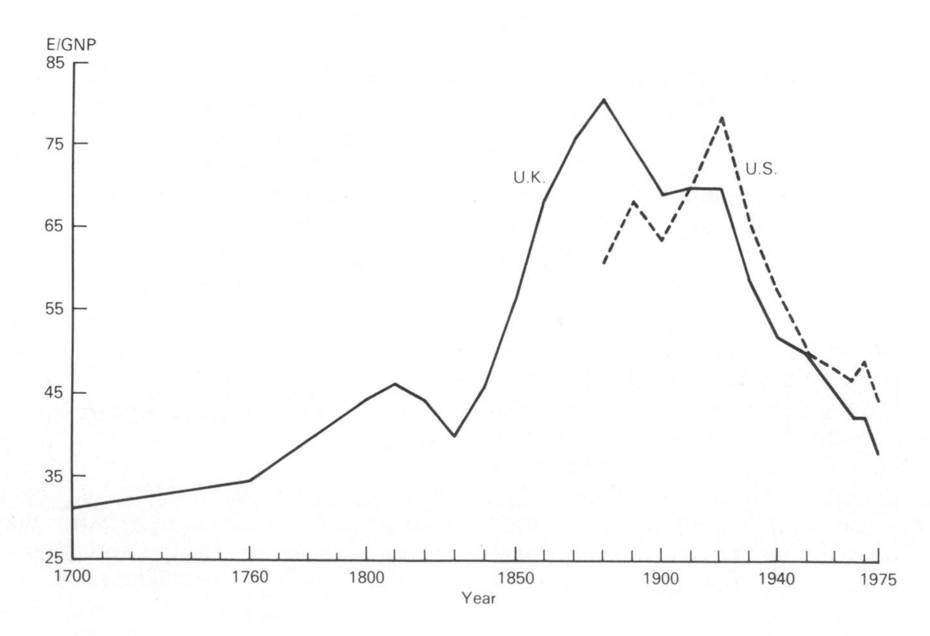

TABLE 4-1. United Kingdom E/GNP Ratios, 1700–1975

Year	Energy (10^{15} joules)	GNP (10^6 1975 US$)	E/GNP (10^6 joule/$)	E/GNP Elasticities	
				20-year periods	*50-year periods*
1975	8157	211,700	38.53	0.54 (15 years)	0.58 (25 years)
1970	8372	197,661	42.36		
1965	7577	176,509	42.93		
1960	6759	149,359	45.25	0.66 (22 years)	
1950	5761	114,725	50.22		0.50
1938	5018	94,687	53.00	0.05 (18 years)	
1930	4713	79,225	59.49		
1925	4660	73,659	63.26		
1920	4946	70,319	70.34	1.10	
1910	4812	68,278	70.48		
1900	4226	60,919	69.37	0.66	1.16
1890	3722	49,539	75.13		
1880	3194	39,705	80.44	1.40	
1870	2492	33,026	75.46		
1860	1846	26,965	68.46	1.85	
1850	1169	20,966	55.76		1.14
1840	814	17,626	46.18	1.05	
1830	558	13,977	39.92		
1820	437	9,772	44.72	1.01	
1810	349	7,545	46.26		
1800	275	6,185	44.46	1.43 (40 years)	1.43 (40 years)
1760	127	3,649	34.80	1.28 (60 years)	1.28 (60 years)
1700	77	2,474	31.12		

Sources: The energy figure for 1975 is from United Nations, *World Energy Supplies, 1973–1978,* ST/ESA/STAT/SER.J/22, New York, 1979. The GNP figure for 1975 is from the World Bank, *1977 Atlas,* Washington, D.C. Energy and GNP statistics before 1975 are from William S. Humphrey and Joe Stanislaw, "Economic Growth and Energy Consumption in the U.K., 1700–1975," *Energy Policy,* vol. 7, no. 1, March 1979, pp. 29–42. The figures from this volume were indexed to conform to the 1975 energy and GNP numbers.

Note: Elasticities calculated using the midpoint formulations.

markable similarities except for the peak date and (2) communication and transportation technology probably limited or slowed the transfer of event impacts between countries. Thus, the later peaking could be only the delayed reaction of the United States to the same event that caused the United Kingdom to peak earlier.

Even if rising energy intensity can be related to the development process in the United Kingdom and the United States, one might question whether this tendency necessarily exists in currently developing countries. We use the term "tendency" because it is fairly clear that in the United States and United Kingdom there were periods when the general tendency of rising E/GNP ratios in early years and a falling ratio in later years was discontinuous in rate of change or even in the direction of change. The most recent example is the period from 1965 to 1970. In the United States the elasticity was above one and in the United Kingdom it was near one. These periods register as small local peaks in figure 4-1 over 55 years of generally falling E/GNP ratios. During the 1960s we do know that the real price of energy, driven by low-cost oil discoveries in the Middle East, was generally falling. Thus, E/GNP elasticities of one which were observed over about 15 years in the United States and over 5 years in the United Kingdom may be nothing more than an anomaly generated by falling energy prices which encouraged energy use. The erratic behavior of the 20-year elasticities might be ex-

TABLE 4-2. U.S. E/GNP Ratios, 1880–1975

Year	Energy (10^{15} joules)	GNP (10^6 1975 US$)	E/GNP (10^6 joule/$)	E/GNP Elasticities	
				20-year periods	50-year periods
1975	68,020	1,519,890	44.75	0.85 (15 years)	0.79 (45 years)
1970	66,133	1,349,090	49.02		
1965	52,221	1,096,135	47.64		
1960	43,247	890,609	48.56	0.80	
1955	38,341	799,850	47.94		
1950	32,629	648,828	50.29		
1945	30,732	648,645	47.38		
1940	23,716	414,899	57.16	0.35	
1935	19,225	309,531	62.11		
1930	22,281	335,097	66.49		1.04
1925	21,062	327,610	64.29		
1920	20,092	255,660	78.59	1.31	
1915	16,695	227,355	73.43		
1910	15,568	219,320	70.98		
1905	12,417	175,857	70.61		
1900	9,010	140,430	64.16	1.09	
1895	7,200	114,316	62.98		
1890	6,590	96,238	68.48		
1880	4,700	77,428[a]	60.70		

Sources: Energy statistics for period 1950–1975 are from United Nations, *World Energy Supplies 1973–1978,* ST/ESA/STAT/SER.J/22, New York, 1979; and United Nations, *World Energy Supplies 1950–1974,* ST/ESA/STAT/SER.J/19, New York, 1976. Energy statistics before 1950 are from Sam H. Schurr and Bruce C. Netschert, *Energy in the American Economy, 1850–1975* (Baltimore, Md.: Johns Hopkins University Press for Resources for the Future, 1960). The figures in this volume were indexed to conform with the UN numbers. GNP statistics for the period 1960–75 are from the World Bank, *1977 Atlas,* Washington, D.C. GNP statistics before 1960 are from the U.S. Department of Commerce, *Historical Statistics of the United States,* Part 1. The statistics in this volume were indexed to conform with the World Bank figures.

Note: Elasticities are calculated using the midpoint formula.

[a]Decade average.

plained by similar types of events which persisted for a decade or so and then were eclipsed by technology or resource limits. An attempt to document changing energy price availability conditions over a longer period of history could aid in understanding the patterns exhibited, even if this information indicated only the sign and general order of magnitude of changes.

50-Year Observations

Energy and GNP data for eighteen countries were available for the 50-year period 1925 to 1975. Ten-year elasticities are given for these eighteen countries in table 4-3. The impact of falling real energy prices during the 1960s is clearly indicated in the fifteen market economies; eleven showed a higher elasticity in the 1960s than in the 1950s. The higher energy prices of the 1970s were reflected as twelve of the fifteen countries having lower elasticities in the 1970s than in the 1960s.

In addition, and in even a higher degree than in the long-term history of the United Kingdom and United States (though it is at least partly the result of the 10-rather than 20-year periods), the eighteen country history shows tremendous variability. Moreover, the 1970 to 1975 period, while, in most cases a clear break from the previous 10 and even 20 years, is less often unprecedented in terms of the 50-year history. A new low is reached in six cases and in one case a new high is reached.

It is also interesting to note that negative E/GNP elasticities have occurred over periods of 10 years and prior to the 1970s. In understanding the existence of negative

TABLE 4-3. E/GNP Elasticities for 18 Countries, 1925–75

Country	*1925–35*	*1938–50*	*1950–60*	*1960–70*	*1970–75*	*50-year period* *1925–75*
Market economies						
Australia	2.53	0.43	1.17	0.98	1.11	0.86
Belgium-Lux.	−0.02	−0.02	0.76	0.81	0.17	0.53
Canada	0.74	1.03	0.87	1.09	0.79	0.98
Denmark	1.25	0.85	1.23	1.48	−1.29	1.05
France	−3.74[a]	0.22	0.79	1.03	0.21	0.80
Germany	1.16	4.93[a]	0.68	0.89	0.17	0.45
Italy	1.69	−0.46	1.72	1.74	1.03	1.33
Japan	1.31	1.00[a]	0.96	1.11	0.60	1.03
Netherlands	1.83	1.19	0.82	1.35	1.54	1.15
Norway	0.87	0.71	1.48	1.44	0.43	1.05
South Africa	0.95	0.53	1.14	0.66	1.29	0.89
Sweden	1.74	0.76	1.79	1.51	−0.78	1.16
Switzerland	−0.64[a]	0.18	1.13	1.55	−0.02	0.94
United Kingdom	0.33	0.97	0.55	0.55	−0.62	0.44
United States	−0.96	0.62	0.89	1.02	0.24	0.86
Planned Economies						
Czechoslovakia	1.43[a]	2.29	0.98	1.25	0.50	1.13
Soviet Union	2.02	1.84	1.22	1.08	0.94	1.10
Yugoslavia	0.87	3.96	1.32	0.99	0.80	1.13

Sources: Energy statistics for 1950–75 from United Nations, *World Energy Supplies 1973–1978,* St/ESA/STAT/SER.J/22, New York, 1979; and United Nations, *World Energy Supplies 1950–1974,* ST/ESA/STAT/SER.J/19, New York, 1976. Energy statistics before 1950 from Joel Darmstadter, *Energy in the World Economy* (Baltimore, Md.: Johns Hopkins University Press, 1971). The statistics presented in this volume were indexed to conform to the UN numbers. GNP statistics for period 1960–75 from the World Bank, *1977 Atlas,* Washington, D.C.; and the World Bank, *World Development Report 1980,* Washington, D.C. GNP statistics before 1960 from the World Bank, *World Tables 1980* (Baltimore, Md.: Johns Hopkins University Press, 1980); Darmstadter, *Energy in the World Economy;* and the United Nations, *Statistical Yearbooks* for 1978, 1973, 1968, 1964, 1962, 1957 and 1953, New York. The statistics presented in these three volumes were indexed to conform to the numbers presented in the World Bank's *1977 Atlas.*

[a]Periods of declining GNP.

elasticities, it is important to remember that the E/GNP elasticity is an aggregate measure. It must be distinguished from an income elasticity (where a negative value is something of a perverse result) because the income elasticity is a partial concept, i.e., it is measured by holding all other things constant. Thus, the major implication of negative E/GNP elasticities is that other factors besides increased income (viewed from the demand side) or increased output (viewed from the energy as an input side) have had major impacts on energy use. Moreover, the view that such responsiveness had not been exhibited until the 1970s is incorrect.

A subtle trend that appears in the data is the increasingly parallel movement of countries over time. Between the 1925–38 and 1938–50 periods, eight of the eighteen countries showed rising elasticities, nine declined, and one was unchanged. Between 1938–50 and 1950–60, ten increased while nine fell. Between the next two periods eleven increased, one was unchanged, and seven fell. Finally, between the last two periods fifteen fell and only three rose. One should probably expect increasing parallel movements on the basis of the increasing integration of the global economies. The earliest periods suggest that deviations in the trend use per unit of output were dictated by country-specific factors, whereas in the 1960s and 1970s domestic changes in prices or technology were more likely to be mitigated by or transferred abroad via international trade.

Short-Term Elasticities

Table 4-4 presents 5-year elasticities over a period of 20 years for nine world regions and several aggregations of these regions. An important caveat is necessary before discussing trends in the data. The ACENP group shows a very high positive elasticity for the 1955–60 period, followed by a high negative elasticity in the following period. As with most data on China, there is reason to be skeptical of its accuracy.

The late 1950s in China were the years of the Great Leap Forward, with its accompanying backyard steel furnaces. It is possible, therefore, that the 1960 energy consumption figures reflect this experience, but they may also reflect inflated energy figures as other statistics for this period were subsequently found to be inflated. Another explanation may be revised GNP statistics which actually overcorrected for earlier overreporting of national output. In any case, there is good reason to suspect the accuracy of figures for the ACENP region for these years. As can be seen from table 4-4, the statistics for China have a fairly significant impact on groups in which they are included.

Table 4-4 also provides additional information on the trends in energy use over the course of development. Here we rely on

TABLE 4-4. E/GNP Elasticities by Region, 1955–75 (percent)

Region	*1955–60*	*1960–65*	*1965–70*	*1970–75*	*1955–75*
OECD	0.74	0.90	1.06	0.39	0.83
US	1.12	0.91	1.14	0.25	0.70
OECD West	0.58	0.98	1.06	0.46	0.82
JANZ	1.01	0.98	1.23	0.66	0.96
CPE	1.65	0.58	1.02	0.95	1.09
EUSSR	1.08	1.24	1.03	0.82	1.03
ACENP	1.73	−1.09	1.12	1.17	1.10
LDC	0.86	1.04	1.16	0.89	0.96
MIDEST	N.A.	1.10	1.83	0.70	1.15
AF	N.A.	0.98	0.79	1.26	1.01
LA	0.68	0.99	1.25	0.93	0.94
SEASIA	1.16	1.32	1.13	0.91	1.11
WOCA[a]	0.74	0.94	1.07	0.46	0.84
World	1.08	0.83	1.02	0.65	0.93

Sources: Energy statistics for 1950–75 from United Nations, *World Energy Supplies 1973–1978,* ST/ESA/STAT/SER.J/22, New York, 1979; and United Nations, *World Energy Supplies 1950–1974,* ST/ESA/STAT/SER.J/19, New York, 1976. Energy statistics before 1950 from Joel Darmstadter, *Energy in the World Economy* (Baltimore, Md.: Johns Hopkins University Press for Resources for the Future, 1971). The statistics presented in this volume were indexed to conform to the UN numbers. Taiwan's energy figures were derived from the Industry of Free China, *Taiwan Economic Statistics,* Taipei, 1979. GNP statistics for 1960–75 from the World Bank, *1977 Atlas,* Washington, D.C.; and the World Bank, *World Development Report 1980,* Washington, D.C. GNP statistics before 1960 from the World Bank, *World Tables 1980* (Baltimore, Md.: Johns Hopkins University Press, 1980); Joel Darmstadter, *Energy in the World Economy;* and the United Nations, *Statistical Yearbooks* for 1978, 1973, 1968, 1964, 1962, 1957, and 1953, New York. The statistics presented in these three volumes were indexed to conform to the numbers presented in the World Bank's *1977 Atlas.*

N.A. = Not available.

[a]WOCA = World outside Communist areas.

cross-section analysis. One observes that the LDCs as a group have a higher energy/GDP elasticity than the developed regions. Thus, the cross-section evidence supports the hypothesis that the process of industrialization and development tends to increase energy use per unit of output relative to developed economies. Again, we stress that this is high *relative* to the developed economies since the 1970–75 experience shows E/GNP ratios declining in all three of the aggregate groupings; CPE and LDC elasticities remain above that of the OECD group.

One also observes that the drop in the E/GNP elasticity between the periods 1965–70 and 1970–75 is most severe in the OECD region. This might be read as an indication that the price responsiveness of demand is greater in these countries. Such a conclusion may be unjustified, however. The 1973 oil price hikes included an embargo of oil to the developed countries. Thus, nonpecuniary costs may be partly responsible for drops in energy use. The embargo was not imposed in shipments to developing countries and the CPEs were not major importers; thus, neither of these groups were subject to the embargo costs.

A second consideration is the nature of energy pricing decisions in the developed countries. In most of these countries prices are set by government agencies, who have tended to be reluctant to pass through higher costs to consumers. As a result, the consumer energy prices were likely to have risen less than world prices and less than in the developed regions, with a full pass-through of higher costs delayed until sometime beyond 1975.

Due to these considerations, the smaller drop in the LDCs elasticity does not appear to be a significant indicator of less price responsiveness. In fact, our feeling is that the drop is relatively large given these considerations and suggests that there is probably little reason to believe that the price elasticity in the LDCs differs significantly from that in the OECD. (Price and income elasticities are independently considered in chapter 5.)

Historical Evidence on E/GNP Elasticities—Conclusions

Both the cross-section data, which include the developing and developed countries, and the time series data for the United Kingdom and the United States support the hypothesis that the early stages of development include a tendency for increasing energy intensity. However, this tendency does not appear to be particularly strong in the aggregate and, as the experience of the early 1970s indicates, can be offset by price impacts.

E/GNP GROWTH RATE ASSUMPTIONS IN LONG-TERM INTERNATIONAL ENERGY FORECASTS

The studies chosen for review here represent a broad range of views, including national and international groups, as well as private and public companies and institutions. All of them project energy demand and economic growth. With the exception of the forecast made by the Department of Energy (DOE) to the year 1995, all of the studies extend to the year 2000 and several forecast beyond this year. Of the ten studies surveyed, seven are international in scope and three encompass the world outside the Communist areas (WOCA). The E/GNP elasticities of each study were derived from calculations of forecast energy consumption and GNP. Thus, they include price, income, and other impacts on energy use.

As can be seen, most of the projected E/GNP elasticities fall within a fairly narrow range (from 0.4 to 0.80 for developed countries and from 0.9 to 1.3 for developing country regions). They tend to echo the historical experience, with developing

TABLE 4-5. Projected E/GNP Elasticities of Major Energy Studies

Regions	EXXON	IIASA		RFF		OECD	LOVINS [a]			Standard Oil	WEC		U.S. DOE	HK	WAES [b]
	1979–2000	1975–2000	2000–2030	1972–2000	2000–2025	1975–2000	1975–2000	2000–2030	2030–2080	1980–2000	1972–2000	2000–2020	1979–1995	1985–2005	1972–2000
U.S.	0.19	0.42[c]	0.74[c]	0.63	0.72	0.75[c]	−0.60[c]	−1.47[c]	N.A.[c]	0.64	0.60	0.51	0.45	1.45	0.77
Other OECD	0.37[d]	0.69[e]	0.76[e]	0.71	0.91	0.96[e]	−0.50[e]	−1.21[e]	N.A.[e]	0.64	0.70	0.39	0.43	1.24[f]	0.80
CPE	0.75	0.74	0.82	0.98	0.81	0.88	0.05[a]	−0.31[a]	N.A.[a]	1.12	0.84	0.97	N.G.	N.G.	N.G.
Nonoil LDCs	0.92	1.08[g]	1.05[g]	1.16[h]	0.94[h]	1.33[i]	0.40[j]	−0.16[j]	−0.13[j]	1.04	0.76	0.85	1.28	1.98[k]	1.05
Oil LDCs	1.02[e]	1.84[g]	1.02[e]			1.13[i]					0.93	0.76	0.68[l]	1.82[m]	0.86
WOCA	0.66	0.72	0.93	0.89	0.96	0.90	−0.18	−0.76	−0.92	0.76	0.79	0.56	0.54	1.70	0.85
WORLD	0.66	0.74	0.91	0.93	0.91	0.92	−0.11	−0.56	−1.80	0.87	0.81	0.72	N.G.	N.G.	N.G.

Sources:

Exxon: Exxon Corporation. *World Energy Outlook,* December 1980.

IIASA: Wolf, Hafele, Project Leader. *Energy in a Finite World: A Global Systems Analysis.* Report by the Energy Program Group of the International Institute for Applied Systems Analysis (Cambridge, Mass.: Ballinger, 1981).

Lovins: Amory B. Lovins, L. Hunter Lovins, Florentin Krause, and Wilfred Bach. *Energy Strategy for Low Climatic Risks.* Prepared for the German Federal Environmental Agency (San Francisco, Calif.: International Project for Soft Energy Paths, June 1981).

OECD: Organisation for Economic Co-operation and Development. *Interfutures, Facing the Future* (Paris, 1979).

RFF: Ronald G. Ridker, and William D. Watson, *To Choose a Future* (Baltimore, Md.: Johns Hopkins University Press for Resources for the Future, 1980).

Standard Oil: Standard Oil Company of California. *World Energy Outlook 1981–2000* (San Francisco, Calif. 1981).

WEC: The Full Report to the Conservation Commission of the World Energy Conference. *World Energy Demand* (New York: IPC Science and Technology Press, 1978).

DOE: U.S. Department of Energy. *1981 Annual Report to Congress.* Report # DOE/EIA-0173(81)/3 (Washington, D.C.: U.S. Government Printing Office, 1982).

HK: Hendrik S. Houthakker and Michael Kennedy, "Long-Range Energy Prospects," *Journal of Energy and Development,* Autumn 1978.

WAES: Carroll L. Wilson, Project Director. *Energy: Global Prospects 1985–2000.* Report of the Workshop on Alternative Energy Strategies (New York: McGraw-Hill, 1977).

Notes: For IIASA, OECD, WEC, and WAES, the calculated elasticities are based on an average of high and low scenarios.

WOCA = World outside Communist areas.

N.A. = Not applicable for these regions over the given periods. Lovins assumes no economic growth. As a result, any change in energy consumption yields an elasticity of $\pm$ infinity. In this case, energy consumption falls.

N.G. = Not given.

[a]Excludes China.

[b]Case C.1-high energy prices, high economic growth, vigorous conservation policies.

[c]Includes Canada.

[d]Including South Africa, and Israel, but excluding Turkey.

[e]Excludes Canada, includes South Africa and Israel.

[f]Europe only; the Asia Pacific region, including Japan, India, and others, is excluded from any of the grouping except WOCA.

[g]Middle East and North Africa region assume to be oil exporting, all other LDCs are included in nonoil group.

[h]Middle East or oil-exporting group was not broken out; all other LDCs are included in nonoil group.

[i]OPEC countries only.

[j]Includes China; oil-exporting group was not broken out.

[k]Latin America, see note f.

[l]OPEC plus fifteen other oil-exporting developing countries.

[m]Middle East and Africa.

country regions showing higher elasticities than developed country regions. The most recent studies (EXXON, IIASA, and U.S. DOE) tend to show lower elasticities than earlier studies. The outliers among the forecasts are the Lovins and Houthakker–Kennedy (HK) results, with Lovins projecting long-term negative elasticities and HK projecting elasticities well over 1.0. The Lovins estimates truly represent a break with the past; long-term, negative elasticities for large groups of countries or even single countries have not been experienced historically.

CONCLUSIONS

The results of this chapter provide some interesting views of the long-term evolution of the relation between GNP and energy use, both historically and as foreseen by various forecasters. Whether one believes the future will be like the past or radically different, it is the past which serves as the benchmark. Perhaps the result of most interest is that the past records are not nearly as uniform as often suggested by general comparisons of pre- and post-1973 experiences.

The Role of Price and Income in Determining Energy Demand 5

In 1980, an unusual event occurred. Total energy consumption in the United States fell 3.4 percent from its 1979 level, to 76 quadrillion Btu. Not only was this the largest single year's decline in post-war U.S. energy consumption, but in addition, it occurred in a year in which real gross national product remained constant. What makes this event unusual is the fact that prior to 1973, energy consumption and economic activity showed such a remarkably proportional relationship that many felt that economic growth without concomitant growth in energy consumption was impossible. But since 1972, the price of energy has increased by 150 percent, forcing the energy-GNP ratio down by 17 percent. These simple statistics point up the importance of two key factors, income and the price of energy.

While income and price are not the only factors which affect energy consumption, their importance by now lies completely beyond question. This conclusion is the result of an enormous research effort spanning a full decade and with roots extending much further back. This is not to deny the significance of such key variables as weather, demographics, or culture. But income and price are expected to play important roles in the determination of long-term future energy consumption patterns.

Much of the information concerning the relationship between energy demand, price, and income is expressed quantitatively in terms of constructs called income elasticities and price elasticities. It will be a useful diversion to review these concepts.

A PRIMER ON ELASTICITIES

Price Elasticities

The price elasticity of demand is denoted e_{ij}, and refers to the percentage change in the quantity demanded of good i in response to a 1 percent change in the price of good j. If i and j are the same good, for example, we might want to know how electricity demand will change as electricity prices change; then we have what is known as an "own-price elasticity." If i and j are different goods, for example, we might wish to know what effect an increase in gasoline price would have on automobile sales; then we have what is known as a "cross-elasticity."

Note that since the concept of elasticity is stated in terms of percentage changes in demand and price, it is independent of any units of measure. Thus, it is possible to compare directly energy studies conducted in West Germany and the United States, despite the fact that energy data were denominated in joules in the German study and in British thermal units (Btu) in the American study, and prices were deutschemarks in Germany and dollars in America. This convenient property makes elasticities the most useful measure available to evaluate demand responsiveness to price changes.

Own-elasticities

Own-elasticities are generally, though not always, the most interesting magnitudes to measure, and there are several benchmarks by which price responsiveness can be judged. If the elasticity is zero, then demand if said to be perfectly price inelastic; that is, price has no effect on demand. If the elasticity is -1, then a 1 percent increase in price will cause demand to fall by 1 percent. In this case, demand is termed "unitary elastic." If the price elasticity of demand is between zero and -1, then demand is said to be "inelastic," while an elasticity larger than 1 in absolute value is termed "elastic." Elasticities refer to demand curves, and some of these demand curves and matching elasticities are depicted in figure 5-1. The correspondence between price elasticities and economic terminology is shown below.

Classification of Goods by Own-elasticities

Elasticity	*Term*
0	Perfectly inelastic
$0 > e_{ii} > -1$	Inelastic
-1	Unitary elastic
$-1 > e_{ii} > -\infty$	Elastic
$-\infty$	Perfectly elastic

Only negative own-elasticities have been considered, and this is because economic theory suggests that as prices rise, demand falls, other things being held constant, so that if an own-elasticity is positive, the response can be said to be perverse.

Cross-elasticities

While own-elasticities should always be negative, cross-elasticities can take on any values at all. If a cross-elasticity is positive, then the two goods can be said to be substitutes. For example, electricity and natural gas are substitutes if, as the price of electricity increases, the demand for natural gas increases. if two goods are complements, then it follows that the cross-elasticity is negative. For example, as the price of gasoline increases, the demand for cars decreases. A summary of cross-elasticities and terms follows.

Classification of Goods by Cross-elasticities

Cross Elasticity	*Term*
$0 < e_{ij}$	Substitutes
$0 > e_{ij}$	Complements

Dynamic Elasticities

Energy products differ from most goods because demand elasticities come in both short-run (usually defined as one year) and long-run magnitudes. This is true for both

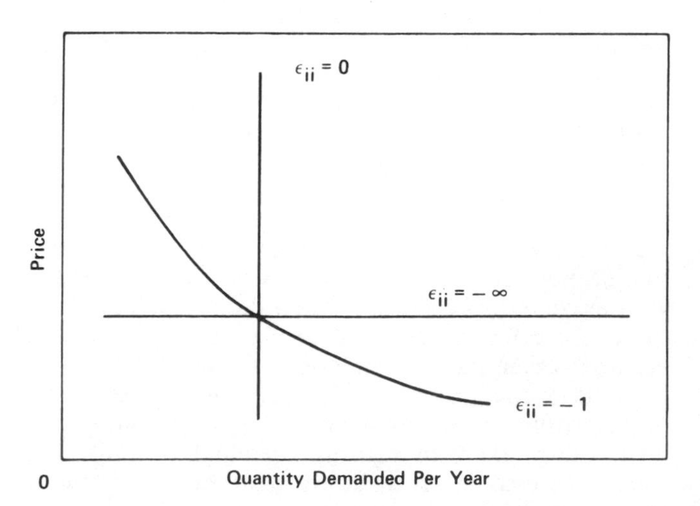

FIGURE 5-1. Own-price elasticities and the corresponding demand curves.

aggregate and disaggregate energy types. Energy is a good which even at the residential level is not consumed directly. It is rather the purest of intermediate goods, being used as an input to produce something which is only later consumed. For example, space heat is the product the consumer enjoys rather than fuel oil. The same holds true in industry, where the intermediacy of the product is even more apparent. Energy is used to produce steam that is used to make paper that the consumer purchases.

As a consequence of this indirection, energy is consumed in conjunction with other inputs, including capital stocks. For the consumer this means refrigerators and stoves; for the industrial user, machinery and blast furnaces. As long as a given capital stock remains in place, the possibility of substitution is limited, due to the technical features of the refrigerators, stoves, boilers, furnaces, and machines. Each piece of capital has an efficiency, or amount of insulation, which is given. In the short run then, substitution, even to different cheaper fuel types, is difficult, though not always impossible.

Appliances and machinery do not last forever though, and as the old capital stock is retired, it can be replaced by new stocks that are more energy efficient. These stocks may be smaller, better insulated, or have more energy-conserving characteristics. They may also facilitate the use of alternative fuels. Thus one suspects that long-run elasticities are larger in absolute magnitude than short-run elasticities. This suspicion is confirmed by numerous empirical studies.

Income Elasticities

Just as the effects of a change in price can be measured in terms of an elasticity, so too can the effects of a change in income. The income elasticity, here denoted n_i, measures the percentage change in the demand for good *i* resulting from a 1 percent change in income. In many ways income effects are simpler to discuss than price effects, since there is only one income elasticity associated with each good's demand.

The most conventional benchmarks for income elasticities center around the values zero and one. If the income elasticity exceeds one, then demand increases even more rapidly than income and the good is said to be superior, or in some references it is said to be a "luxury" good. In contrast, goods whose demand grows with income, but not more rapidly, are said to be normal goods. Goods whose income elasticities are less than one are sometimes categorized as "necessities." Any good whose demand actually decreases as income increases is said to be an "inferior" good.

Classification of Goods by Income Elasticities

Income Elasticity	*Term*
$n < 0$	Inferior good
$0 < n < 1$	Normal good
$1 < n$	Superior good

Income elasticities have the same time dimensional qualities as price elasticities. That is, income responsiveness of energy demand is constrained by the same capital stocks that constrain price effects.

For example, if incomes rise, one may not immediately buy a second car. But automobiles may be used more intensively in the process of increasing the general level of consumption. If the income increase remains in effect over an extended period, then the demand for automobiles may well rise, allowing still further increases in the demand for energy.

Production Elasticities

Only consumers have incomes which are allocated on a purely discretionary basis, so income elasticities of demand for energy apply strictly only to energy forms purchased directly by final consumers. As stated previously, much energy is consumed very indirectly by consumers; for example, coke burned in blast furnaces to make steel for cars. The demand by the industrial, commercial, and commercial transport sectors is influenced by the scale of operation. That is, if steel production increases, so will coal consumption. Thus, in addition to income elasticities, one finds "output" or "production" elasticities. These concepts are analytically very similar.

In many studies, however, the output elasticity is assumed to be one. This rather stringent assumption comes from the "neoclassical" theory of production in which it is assumed that a most efficient process always exists for making any good. Since this process is efficient, doubling output requires a simple repetition of the process and a doubling of all inputs. While the relative magnitudes of different inputs in the base process may change with prices, for any fixed set of prices the optimum process is simply scaled up or back with output changes. The assumption of unitary output elasticity is most appropriate in the long run. In the short run, it is usually abandoned and a short-term output elasticity is developed from statistical information.

GNP Feedback Elasticities

The elasticity concepts developed to this point have been conventional. They help measure the impact of exogenous variables such as income and price on energy demand. But energy is an important good, and as the oil embargo–recession of 1974–75 so aptly demonstrated, abrupt changes in the general availability of energy can in and of themselves affect the overall level of economic activity and income. Thus, the concept of a "GNP feedback" elasticity is a useful one.

In physical terms, the feedback elasticity is defined as the percentage change in GNP resulting from a 1 percent change in aggregate energy consumption. In monetary terms it is defined as the percentage change in GNP resulting from a 1 percent change in the price of energy. Most empirical work indicates that both of these magnitudes are inelastic and small.

Implications for 2050

Since the determinants of energy demand must be considered as far into the future as 2050, the long-run elasticities are clearly the most interesting conceptual magnitudes. In addition, the global nature of the CO_2 problem argues for an international assessment of elasticity variation.

MEASUREMENT OF ELASTICITIES

The measurement of price elasticities is a difficult task made even more difficult by the fact that elasticity is a concept developed primarily for application to single points on a demand schedule, or at most for small segments of the demand schedule. If the demand schedule turns out to be a simple linear relationship between price and quantity, then the price elasticity will vary from zero to infinity, with a magnitude of one at the half-way point.

To estimate an elasticity properly requires the specification of a point of measurement, the form of the demand schedule, and a set of behavioral observations from which to infer the elasticity. The point at which to measure an elasticity is largely arbitrary, though the measured elasticity is more likely to be accurate when measured at a point which lies within the set of observations. Elasticities can also change with the point of measurement. For example, figure 5-2 shows a demand schedule for which the elasticity varies from negative to zero. The choice of data is limited. There are a relatively small number of observation sets available for many purposes. For example, studies of historical manufacturing energy demand are largely restricted to either time-series data derived chiefly from U.S. census and surveys of manufacturers, time-series Department of Energy data, or cross-sectional data based on U.S. and international (OECD member state) sources.

As this last example points up, data are not collected in either equal qualities or quantities across the world. The highest quality sources belong to the developed

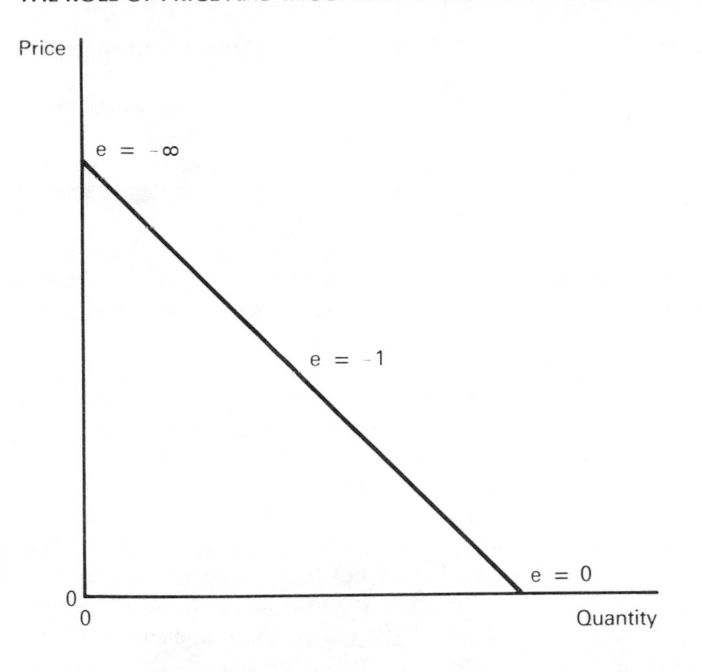

FIGURE 5-2. Price elasticities along a simple linear demand schedule.

OECD member countries (regions 1, 2, 3). There are less data available for the developing world, and almost nothing is available for the centrally planned economies. Freely circulating research is similarly skewed toward OECD members.

One criticism that has been leveled against elasticities which are derived from historical experiences is that they do not incorporate any information about factors which will influence the future, and are therefore not necessarily anything at all like the value of elasticities which will be observed in the future. Attempts to remedy this deficiency have been made. The most fruitful approach developed to date has been to construct economic-engineering models of the relationships affecting energy demand which incorporate process details and mechanisms by which to adopt new processes over time. While this approach makes maximum use of current understanding of events which affect future energy consumption, it makes no allowance for unforeseen events such as new discoveries, innovations, or technological change. As a consequence, these elasticities tend to be conservative, estimating less responsiveness to price and income/production changes than studies based on historical data.

The final element one needs to measure energy elasticities (or energy demand in general) is a behavior equation structure. As one might imagine, the choice of equation structure can be critical in the determination of elasticities, especially if one wishes to measure elasticities at prices or incomes lying beyond the range of historical observation. For example, both the demand equations,

$$E = a/P$$

and

$$E = b - cP$$

where a, b, and c are parameters, E is total energy demand, and P is the price of energy, have price elasticities of demand equal to -1 where $P = b/(2/c)$. But at $P = b/c$, the second equation has an infinite elasticity, while the first equation's elasticity remains equal to -1.

This simple example points up the not well understood fact that, despite all of the work which has been undertaken in recent years to establish general functional forms capable of describing any arbitrary matrix of elasticities at a point, functions impose themselves on the data away from the point of estimation, and in forecasting—and especially very long-term forecasting, it is the area beyond historical observations that is the most interesting.[1]

Finally, it is important to note that the concept of elasticity is a partial adjustment concept. That is, it measures the impact of a change in one variable alone (e.g., price) when all other variables remain constant. To develop elasticities from a statistical base requires either good fortune in selecting a data base in which only price changes, or more realistically, postulating the relationship between all of the important variables and the demand for energy. That is, specify an appropriate form for the function f in

$$E = f(X_1, X_2, \ldots, X_n)$$

where E is the demand for energy, X's are key variables such as prices, income and weather, and f establishes the relationship between these variables. Elasticities are then generated by changing one variable by a single percent and noting the percentage change in the quantity demanded, or, if the relationship is a smooth, continuous one, by computing $d \ln E / d \ln X_i$, which yields an equivalent estimate. Statistically determined parameters for the function f therefore determine elasticity values, along with the point at which values are computed.

PRICE ELASTICITIES

Introduction

The following material briefly surveys the literature on price elasticities of demand for energy. A more thorough discussion can be found in Edmonds' *A Guide to Price Elasticities of Demand for Energy: Studies and Methodologies.*[2] Each section briefly describes the state of understanding of price elasticities in one of three end-use categories: Residential/commercial, industrial, and transport. Five categories of secondary energy are considered in each sector: oil, gas, coal, electricity, and aggregate energy.[3]

Despite the relatively broad spectrum of elasticity categories considered, a few general conclusions are possible. First, the entire range of price elasticities falls within theoretically expected bounds (0 to $-\infty$). Second, price elasticities bridge an enormous range, with the disparity between least and most elastic price elasticities spanning more than two orders of magnitude in some cases. Third, residential/commercial and industrial price elasticities for individual fuels tend to be elastic, with the transport sector's demand inelastic. Fourth, aggregate price elasticities of demand tend to be less elastic than fuel-specific elasticities. It is impossibile to tell whether or not this is due to the theoretically appealing rationale that interfuel substitution is easier than conservation, or rather due to the lesser number of studies of aggregate energy demand. Finally, it should be noted that the preponderance of studies rely on U.S. data and were conducted by U.S.-based researchers.

The Residential/Commercial Sector

There are really only three major fuels considered in the residential/commercial sector: oil, gas, and electricity. Coal is not consumed in great quantities in the United States, from which most of the studies considered here draw data. There are in fact no studies which consider coal as a residential commercial fuel.

There seems to be a consensus that gas and electricity are price elastic in the long term. While studies obviously show both inelastic and elastic results (see table 5-1), the majority of studies find these fuels to be price elastic. The same can be said of oil, though its support is more varied, with some studies showing it to be extremely elastic while several show extreme inelasticity. In contrast to the relatively elastic fuel-specific findings, all sources agree that the aggregate price elasticity is inelastic, though some studies indicate that the in-

TABLE 5-1. The Residential/Commercial Sector: Range of Long-Term Price Elasticities of Demand for Energy by Fuel

	Elasticity Range		
Fuel	*Least Elastic*	*Most Elastic*	*Consensus Elasticity*[a]
Oil	−0.3	−9.5	−1.4
Gas	0.0	−3.9[b]	−1.2
Coal	c	c	c
Electricity	0.0	−2.6	−1.1
Aggregate	−0.3	−0.9	−0.5

Sources:

J. A. Edmonds, *A Guide to Price Elasticities of Demand for Energy: Studies and Methodologies,* Institute for Energy Analysis, Oak Ridge, Tenn., Research Memorandum (ORAU/IEA-78-15(R), 1978).

D. R. Bohi, *Analyzing Demand Behavior: A Study of Energy Elasticities* (Baltimore, Md.: Johns Hopkins University Press for Resources for the Future, 1981).

L. D. Taylor, *The Commercial Demand for Energy: A Review of Existing Knowledge* (Department of Economics, University of Arizona, 1977).

J. A. Edmonds, B. Cohen, and S. Wagner, "Factor Substitution in the Industrial Sector with a Case Study of the Oak Ridge Industrial Model," Draft report, 1981.

T. H. Morlan, D. H. Skelly, and A. P. Reznek, *Price Elasticities of Demand for Motor Gasoline and Other Petroleum Products,* DOE/EIA-0291, U.S. Department of Energy (May 1981).

W. D. Nordhaus, *International Studies of the Demand for Energy,* selected papers presented at a conference held by the International Institute for Applied Systems Analysis (Amsterdam: North-Holland; 1978).

M. S. Commons, "Implied Elasticities in Some United Kingdom Energy Projections," *Energy Economics,* vol. 3 (July 1981).

Energy Modeling Forum, *Aggregate Elasticity of Energy Demand,* vol. 1. Energy Modeling Forum, Stanford University, Stanford, Calif., 1980.

Report of the Working Group on Energy Elasticities, Energy Paper 17 (London: Her Majesty's Stationery Office, 1977).

[a]Consensus elasticities were obtained by averaging values from source studies.

[b]The elasticity value of "large" specified by the Federal Energy Agency for the Project Independence model is not further defined.

[c]No longer a significant residential/commercial fuel.

elasticity is only slight, but with a consensus elasticity of −0.5.

The Industrial Sector

The industrial sector uses all four secondary fuels, and is characterized by elastic demand for all four. Both the elasticity ranges and consensus values are very similar to those found in the residential/commercial sector. Coal and electricity tend to be somewhat less elastic than oil and gas. This is perhaps owing to the relatively lesser ability to substitute for these fuels in the industrial sector. Coal, for example, cannot be substituted for oil or gas without extensive and expensive retrofitting, while the former fuels can be substituted for one another and for coal with relative ease. Similarly, electricity is found in uses for which it is difficult to substitute other fuels, for example, direct electrolytic services, lighting, and motor drive. As with the residential/commericial sector, the aggregate energy elasticity is found to be inelastic, with a consensus value of −0.4. (See table 5-2.)

The Transport Sector

The transport sector is dominated by oil, and while there is some use of electricity and coal, the quantities are so small that elasticity studies completely ignore these fuels. As a result, there should be little difference between the price elasticity for oil and for aggregate energy in this sector. We note that the one aggregate energy elasticity estimate is in fact very close to the average elasticity of oil elasticity studies. In the long term, the price elasticity of de-

TABLE 5-2. The Industrial Sector: Range of Long-Term Price Elasticities of Demand for Energy by Fuel

	Elasticity Range		
Fuel	*Least Elastic*	*Most Elastic*	*Consensus Elasticity*[a]
Oil	−0.0[b]	−4.7	−2.0
Gas	−0.0[b]	−3.9	−1.3
Coal	−0.0[b]	−2.2	−1.1
Electricity	−0.1	−2.0	−1.1
Aggregate	−0.1	−1.1	−0.4

Sources: Sources consist of references listed in table 5-1 plus individual studies.

[a]Consensus elasticities were obtained by averaging values from source studies.

[b]Very small negative value—between 0 and −0.05.

TABLE 5-3. The Transport Sector: Range of Long-Term Price Elasticities of Demand for Energy by Fuel

	Elasticity Range		
Fuel	*Least Elastic*	*Most Elastic*	*Consensus Elasticity*[a]
Oil	−0.2	−1.5	−0.6
Gas	b	b	b
Coal	b	b	b
Electricity	b	b	b
Aggregate	−0.4[c]	−0.4[c]	−0.4[c]

Sources: Sources consist of references listed in table 5-1 plus individual studies.

[a]Consensus elasticities were obtained by averaging values from source studies.

[b]Not a significant transport fuel.

[c]Only one study calculates this magnitude. See J. A. Edmonds, *A Guide to Price Elasticities of Demand for Energy: Studies and Methodologies,* Institute for Energy Analysis, Oak Ridge, Tenn., Research Memorandum (ORAU/IEA-78-15(R), 1978), p. 14.

TABLE 5-4. The Residential/Commercial Sector: Range of Long-Term Income Elasticities of Demand for Energy by Fuel

	Elasticity Range		
Fuel	*Least Elastic*	*Most Elastic*	*Consensus Elasticities*[a]
Oil	−0.3	1.9	0.9
Gas	−0.3	3.1	1.1
Coal	b	b	b
Electricity	−0.2	2.0	0.9
Aggregate	0.3	1.6	0.9

Sources: Sources consist of references listed in table 5-1 plus individual studies.

[a]Consensus elasticities were obtained by averaging values from source studies.

[b]Not a significant residential/commercial fuel.

mand for oil is found to be inelastic, though there are some exceptions to this consensus. (See table 5-3.)

INCOME AND OUTPUT ELASTICITIES

Studies of Developed Economies

When we consider income and output elasticities, the record is somewhat more diverse than for price elasticities. There is not even unanimous agreement on whether or not energy is a normal or inferior good in consumption. Nonetheless, the consensus would consider energy a normal good. Furthermore, the average energy elasticity from statistical studies is very close to unity in *all* cases—that is, for *all* fuels in *all* sectors. This interesting result is an average, however, and it is associated with a reasonably wide variation in statistical findings and a great many industrial studies which have an output elasticity of one by assumption.

It is interesting to note that there is no difference in magnitude of income elasticities between fuels and the energy aggregate. This consistency is remarkable in light of the tendency for lower aggregate price elasticities than fuel price elasticities. Furthermore, it is remarkable that the average of numerous diverse studies agrees with the pre-1973 observation of a constant energy-GNP ratio.[4] The results from elasticity studies are displayed in tables 5-4, 5-5, and 5-6.

Income and Energy in the Developing World

Most studies that discuss income and energy prices consider only the developed na-

TABLE 5-5. The Industrial Sector: Range of Long-Term Output Elasticities of Demand for Energy by Fuel

	Elasticity Range		
Fuel	*Least Elastic*	*Most Elastic*	*Consensus Elasticities*[a]
Oil	−2.8	1.0	1.2
Gas	−0.3	1.0	0.8
Coal	0.6	1.0	0.9
Electricity	0.3	2.6	1.1
Aggregate	0.7	1.0	1.0

Sources: Sources consist of references listed in table 5-1 plus individual studies.

[a]Consensus elasticities were obtained by averaging values from source studies.

tions. Lack of data for the developing countries has precluded the use of the more detailed studies done in the United States and other OECD countries. Nevertheless, there is a clear trend toward income elasticities that exceed one in the developing world. (See table 5-7.)

The tendency toward high energy–GNP elasticities, where energy is defined in terms of commercial fuels, can be partially explained by the nature of the development process itself. This entails a shift from the traditional to modern sector of the economy, which is usually accompanied by the development of a manufacturing sector. The higher commercial energy intensity of the modern sector (even when the modern agricultural sector is compared with its traditional counterpart) implies a higher energy-income elasticity as a simple result of sectoral shifts.

Needless to say, all developing nations are not alike. While the average energy-GNP elasticity rose slightly between the pre- and post-1973 periods, there are nu-

TABLE 5-6. The Transport Sector: Range of Long-Term Income Elasticities of Demand for Energy by Fuel

	Elasticity Range		
Fuel	*Least Elastic*	*Most Elastic*	*Consensus Elasticities*[a]
Oil	−0.1	1.7	1.0
Gas	b	b	b
Coal	b	b	b
Electricity	b	b	b
Aggregate	1.0[c]	1.0[c]	1.0[c]

Sources: Sources consist of references listed in table 5-1 plus individual studies.

[a]Consensus elasticities were obtained by averaging values from source studies.

[b]Not a significant transport fuel.

[c]Only one study calculates this magnitude. See J. A. Edmonds, *A Guide to Price Elasticities of Demand for Energy: Studies and Methodologies,* Institute for Energy Analysis, Oak Ridge, Tenn., Research Memorandum (ORAU/IEA-78-15(R), 1978), p. 14.

TABLE 5-7. Economic Growth, Commercial Energy Consumption, and the Energy-GNP Elasticity in Selected Countries, 1970–78 (percent per year)

	Economic Growth Rates		*Growth in Commercial Energy Consumption*		*Energy-GNP Elasticity*	
	1970–73	*1973–78*	*1970–73*	*1973–78*	*1970–73*	*1973–78*
Algeria	5.3	5.5	14.4	11.3	2.7	2.1
Brazil	12.8	7.2	12.8	7.8	1.0	1.1
Columbia	7.2	5.6	5.4	3.9	0.8	0.7
Egypt	4.4	11.1	5.6	11.9	1.3	1.1
India	1.8	4.4	4.3	4.8	2.4	1.1
Indonesia	8.7	6.7	13.4	18.4	1.5	2.7
Jamaica	4.6	−2.5	4.5	−0.6	1.0	0.2
Kenya	5.5	4.8	8.8	3.7	1.6	0.8
Korea	9.7	10.3	5.8	9.1	0.6	0.9
Mexico	6.0	3.5	7.3	6.9	1.2	2.0
Nigeria	7.0	7.0	20.1	8.5	2.9	1.2
Philippines	6.8	6.1	3.9	4.5	0.6	0.7
Portugal	8.9	1.6	15.5	2.3	1.7	1.4
Thailand	6.9	7.3	11.8	5.1	1.7	0.7
Turkey	7.7	6.1	11.5	7.7	1.5	1.3
Venezuela	4.4	7.7	7.1	4.0	1.6	0.5
All LDCs	6.7	5.3	8.0	7.3	1.2	1.4

Source: Data from J. Dunkerley, W. Ramsay, L. Gordon, and E. Cecelski, *Energy Strategies for Developing Nations* (Baltimore, Md.: Johns Hopkins University Press for Resources for the Future, 1981), pp. 89–90.

merous counter examples in the sixteen-country sample shown in table 5-7. Similarly, there is a wide variation about the general trend. Elasticities range from 0.6 to 2.9 in the pre-1973 period and from 0.2 to 2.7 in the post-1973 period.

The GNP Feedback Elasticity

The relationship between energy and the economy is a two-way one. The economy affects the demand for energy through the influences of income and output. Energy in turn affects the general economy, since it too is a factor of production and its higher price and lesser availability reduce the overall productive capacity and real income of the society. But, in the long term, how much impact does energy have on the economy?

While the issue of energy-economy interactions has received attention in the literature, it is surprising how little reference there is to the impacts of energy price on the economy. Energy feedback elasticities are rare finds in the literature.[4] Where these elasticities are reported, they are subject to the same difficulties of interpretation as any price elasticity measure.[5] For example, energy is frequently disaggregated by fuel type and hence a change in the price of energy may have different impacts, depending on which fuel price changed the price of energy.[6] In addition, there are short-term and long-term elasticities, with the former being more elastic than the latter.[7]

Despite these difficulties, we have attempted to construct a survey table which displays magnitudes of energy feedback elasticities. In most cases these elasticities have been constructed out of information available from the various sources and were not computations made by the authors. The methodology for constructing elasticities for studies in which this magnitude is not reported independently varies in detail but is similar in approach. Two cases for a given model are chosen in which the only variation consists of an energy supply or price assumption. The two resulting GNPs are then compared with the corresponding energy prices. Results are reported in table 5-8.

The magnitudes of these elasticities tend to be small. All of them are under 0.15 in absolute value, while some range down as far as 0.00 when rounded to two decimal places. All of these elasticities, it must be noted, assume that a real price increase (none assume a price decrease) is phased in smoothly over the period to 2000, and that there are no unexpected or discontinuous changes in the price level. This has not been the history of energy (and particularly oil) price increases in the past. The 1973–74 price increase was abrupt and unexpected. The 1980–81 increase was also abrupt, though its shock was somewhat cushioned by the expectation of higher energy prices and by the associated inventory stockpiles which had been accumulated during the period. These numbers tend to represent lower bounds for energy feedbacks. The Rowen-Weyant (1980) study gives an energy feedback estimate predicated on the assumption that energy prices rise abruptly over a short period (one year), but are not unanticipated and are again cushioned by the existence of a strategic petroleum reserve.[8] These values are not outside the range of uncertainty associated with the long-term, smooth price trajectory estimates.

All of this lends support to the notion that price expectations play a key role in the determination of the energy feedback elasticity.[9] In this study, in which the focus is on the long term, it makes most sense to ignore the short-term impacts on GNP of a set of unexpected price increases (or decreases)[10] and focus instead on the long-term aspects of the problem. As Pindyck has argued, even if there are relatively severe short-term impacts, the long-term consequences are similar since the economy gradually accommodates itself to the new reality of higher prices, leaving a result similar to the case of smooth and anticipated energy price increases.[11]

TABLE 5-8. Long-Term Energy Feedback Elasticities

Study	*Elasticity*	*Notes*
Manne	−0.03	Computed from information supplied within the paper for the year 2000.
Ridker–Watson–Shapanka	−0.004 to −0.008	Computed from information supplied within the paper for the year 2000.
Reister–Edmonds	−0.10	Computed from information supplied within the paper for the year 2000.
Hudson–Jorgenson	−0.03	As reported in Bopp and Lady.
Pindyck	−0.03	
Rowen–Weyant	−0.03 to −0.08	Computed from information on oil price impacts on GNP, and assumes that the energy price elasticity with respect to a change in the oil price is 0.35. The elasticities are for a 1-year change in price and availability with petroleum resources and so are implicitly short-term elasticities.
Jorgenson	−0.11 to −0.14	The study reports oil price feedback elasticities in the range of −0.042 to −0.048. The assumed relationship between oil price and energy price is the same as used for Rowen–Weyant. Results from the Dale W. Jorgenson Associates Dynamic General Equilibrium Model (DGEM) are reported in U.S. DOE.
Mork-Hall	−0.05	As reported in Bopp and Lady.
Bopp-Lady	−0.04	Elasticity computed from data as reported in Bopp and Lady.

Sources:

H. S. Rowen, and J. P. Weyant. "An Integrated Program for Surviving an Oil Crisis," Department of Operations Research, Stanford University, Stanford, Calif. (November 1980).

U. S. Department of Energy. *Interrelationships of Energy and the Economy,* DOE/PE-0030 (Washington, D.C.: U.S. Government Printing Office, July 1981), p. 45. (For Jorgenson's results.)

R. S. Pindyck, *The Structure of World Energy Demand* (Cambridge, Mass.: MIT Press, 1979), pp. 272–73, fn. 3.

A. S. Manne, "ETA-MACRO: A Model of Energy-Economy Interactions," in *Modeling Energy–Economy Interactions: Five Approaches,* C. Hitch, ed. (Baltimore, Md.: Johns Hopkins Press for Resources for the Future, 1977), pp. 1–45.

R. G. Ridker, W. P. Watson, Jr., and A. Shapanka, "Economic, Energy, and Environmental Consequences of Alternative Energy Regimes: An Application of the RFF/SEAS Modeling System," in *Modeling Energy–Economy Interactions: Five Approaches,* pp. 135–199.

D. B. Reister and J. A. Edmonds, "A General Equilibrium Two-Sector Energy Demand Model," in *Modeling Energy–Economy Interactions: Five Approaches,* pp. 199–246.

A. Bopp and G. M. Lady, "On Measuring the Effects of Higher Energy Prices," *Energy Economics,* vol. 4, no. 4 (October 1982), pp. 218–224.

NOTES

1. For a further discussion of this issue, see D. B. Reister and J. A. Edmonds, "Energy Demand Models Based on the Translog and CES Functions," *Energy,* vol. 6, no. 9, 1981, pp. 917–926. In addition, one might consult two volumes by Lester Taylor, *The Commercial Demand for Energy: A Review of Existing Knowledge;* and *A Review of Load Forecasting Methodologies in the Electric Utility Industry,* both University of Arizona, Department of Economics.
2. See J. A. Edmonds, *A Guide to Price Elasticities of Demand for Energy: Studies and Methodologies,* Institute for Energy Analysis, Oak Ridge, Tenn., Research Memorandum (ORAU/IEA-78-15(R), 1978). Also of interest would be J. A. Edmonds, B. Cohen, and S. Wagner, "Factor Substitution in the Industrial Sector with a Case Study of the Oak Ridge Industrial Model," EIA/DOE draft working

paper. T. H. Morlan, D. H. Skelly, and A. P. Reznek, *Price Elasticities of Demand for Motor Gasoline and Other Petroleum Products,* DOE/EIA-0291 (AR/IA/EUA/81-03), U.S. Department of Energy Information Administration (Washington, D.C.: U.S. Government Printing Office, 1981). D. R. Bohi, *Analyzing Demand Behavior: A Study of Energy Elasticities* (Baltimore, Md.: Johns Hopkins University Press for Resources for the Future, 1981). *Report of the Working Group on Energy Elasticities,* Energy Paper (London: Her Majesty's Stationery Office, 1977).

3. Secondary energy refers to refined fuels. In contrast, primary energy refers to energy at the stage of extraction, and tertiary energy refers to the energy which is actually applied to a process (e.g., steam used to make paper).
4. See, for example, *Modeling Energy–Economy Interactions: Five Approaches,* C. J. Hitch, ed. (Baltimore, Md.: Johns Hopkins University Press for Resources for the Future, 1977); U.S. Department of Energy, *Interactions fo Energy and the Economy,* DOE/PE-0030, July 1981; A. Bopp and G. M. Lady, "On Measuring the Effects of Higher Energy Prices," *Energy Economics,* vol. 4, no. 4 (October 1982), pp. 218–224; K. Mork and R. Hall, "Energy Prices and the U.S. Economy in 1978–1981," *The Energy Journal,* vol. 1, no. 4, p. 44; and E. Hudson and D. Jorgensen, "Energy Policy and the U.S. Economy, 1972–1976," *Natural Resources Journal,* vol. 18, no. 877, 1978; H. S. Rowen and J. P. Weyant, "An Integrated Program for Surviving an Oil Crisis," Department of Operations Research (Stanford University, Stanford, Calif.: November 1980); R. S. Pindyck, *The Structure of World Energy Demand* (Cambridge, Mass.: MIT Press, 1979), pp. 272–273, fn. 3.
5. By energy feedback elasticity, we mean the percentage change in GNP resulting from a 1 percent change in the price of energy.
6. See J. A. Edmonds, *A Guide to Price Elasticities of Demand for Energy: Studies and Methodolgies,* or D. R. Bohi, *Analyzing Demand Behavior* for a full discussion of these problems.
7. Most typically it is the price of oil which is varied. See, for example, U.S. DOE, *Interrelationships of Energy and the Economy,* DOE/PE-0030 (Washington, D.C.: U.S. Government Printing Office, July 1981), pp. 44–46.
8. In the short term, an increase in the price of energy leaves users with fewer alternatives. Producers of goods have a limited capacity for changing their techniques of production since capital stocks are fixed and contractual arrangements bind decision makers. The impact of a change in the price of energy is maximized in this case, since at the limit the economy can operate as an input–output structure. Thus heavy impacts in one sector may have multiplier effects throughout the economy. This is reminiscent of constant energy–GNP ratio theory. In the long term, capital stocks must be replaced and contracts expire. Production processes become more flexible and substitution possibilities expand. The impact of an increase in the price of energy becomes smaller.
9. Undocumented experiments with the Oak Ridge Industrial Model (ORIM) showed that great differences in aggregate energy price elasticities for industry can be traced to differences in expectations about future prices. Since the industrial energy price elasticity is closely related to the aggregate energy feedback elasticity, these experiments lend support to the hypothesis that price expectation assumptions matter in the determination of the feedback elasticity.
10. Note that for unexpected price decreases, feedback elasticities would tend to be smaller in absolute magnitude than those shown in table 5-8, assuming that the economy has acquired capital and technology which economizes under a high energy price scenario. This capital stock is obviously not as efficient as one acquired in expectation of low energy prices should low energy prices materialize. The short-run increase in GNP for a fixed decline in energy price is not as large as the long-run increase in GNP when the capital stock can be readjusted. Thus, the short-term energy feedback elasticity $e_s(-)$, is smaller than the long-term energy elasticity, e, if prices fall. By the same logic, the decline in GNP resulting from an unexpected energy price increase is larger than for an increase in price, when the capital stock can be varied. One would expect that

$$e_s(-) < e < e_s(+)$$

where $e_s(+)$ is the short-run energy feedback elasticity corresponding to an energy price increase.

11. See R. S. Pindyck, *The Structure of World Energy Demand* (Cambridge, Mass.: MIT Press, 1979), pp. 271–274.

Energy and Resource Classifications 6

Future supplies of energy are a vital concern to all nations. The interaction of supply availability—that is, costs, technologies, and resource constraints with demand (consumer preferences, population, income, and technology)—will shape the world's future in terms of total energy use and fuel shares for each nation. Any long-term consideration of energy supply must be broad. Any study also faces the question of boundaries; it must decide on fuel sources, geographical area, and specificity in describing energy uses, to name a few. The choice of boundaries will in turn limit the conclusions that can be made. For example, the economic viability of a shale oil industry in the United States cannot be judged without considering global energy resources. The continued availability of imported oil at prices less than the cost of producing shale oil will defer or preclude the development of oil shale in the United States (and elsewhere) unless governments subsidize an oil shale industry or limit oil imports. Similarly, oil shale development may be precluded or deferred by the availability of other liquid fuels at lower costs—enhanced oil recovery, offshore oil, or coal or biomass liquids—or substitution away from expensive liquid fuels, for example, substitution of electric for gasoline-powered vehicles.

The next nine chapters deal with global resource availability. The regional breakdowns are those described in chapter 1. The energy sources considered include nine major fuel categories and their subcategories, as shown in table 6-1.

Each chapter in this section develops estimates of the total potential amount of energy available from the particular source being considered, and discusses the technology and costs of exploiting the resource. Other important supply considerations, which might be included in a broad interpretation of costs, such as pollution and transportation, are reviewed only where they affect the viability of the resource as a major contributor in satisfying future needs. Our consideration of supply is conditional on the demand for the resource at a given price.

TYPES OF ENERGY

Forms and Carriers

There are several ways in which energy resources can be conceptually categorized. On one level it is possible to distinguish between primary, secondary, and final or useful energy, primary energy being the energy in its raw form—the energy in a ton of coal or barrel of oil as it comes out of the ground, final energy being a measure of the actual useful work done or useful heat supplied, and secondary energy being a stage between primary and useful energy

TABLE 6-1. Energy Forms

1.	Conventional oil
2.	Unconventional oil:
	Enhanced oil recovery and deep offshore and polar region oil
	Shale oil
	Tar sands/heavy oil
	Coal liquids
	Biomass liquids
3.	Conventional gas
4.	Unconventional gas:
	Tight gas formations
	Gas from coal seams
	Gas from shale deposits
	Geopressured brine
	Hydrate zones
	Coal gas
	Biogas
5.	Coal
6.	Nuclear:
	Fission
	Breeder
7.	Solar and other:
	Solar thermal electric
	Photovoltaic
	Wind
	Geothermal
	Fusion
8.	Hydro
9.	Biomass:
	Biomass waste
	Energy crops

that represents conversion of a primary energy form to one that can be used in capital equipment doing the work or supplying the heat.

Understanding the differences among primary, secondary, and useful energy is the starting point for understanding why the supply of energy is a concern even though there are enough coal resources to supply all the world's energy needs for hundreds of years or even though the energy value of global insolation (i.e., heat from the sun) is tens of thousands of times more than global energy consumption. The current cost of converting primary energy to useful secondary energy products varies from almost zero as in natural gas to infinite in the case of fusion. A short-run view of the world is concerned with finding primary energy forms that can be converted with available technologies into the secondary forms that are required for existing plant and equipment. A long-run view, on the other hand, looks at both the primary-to-secondary transformation technologies and capital stock, and the work or heat-producing technologies and capital stock as malleable. In the very short run there is a demand for specific fuels, with very little interfuel substitution possible; in the very long run, the demand for energy offers almost infinite substitution potential between secondary fuel types. Both long- and short-run demand for energy are derived from some demand for a specific product such as transportation, heat or light, or the operation of radios and televisions.

Primary-secondary-final energy forms can also be classified according to their source, or carrier. At the primary fuel level, there is crude oil, natural gas, falling water, solar heat, and so on. At the secondary energy level, four basic carrier types can be identified—liquids, gases, solids, and electricity. Finally, useful energy is delivered as mechanical drive or heat.

Traded and Nontraded Energy

Most energy analysts agree that all energy sources, such as fuelwood and dung, should be properly included in a country's energy balance. However, it is not always clear what types of energy should be included in an energy accounting system. Classifying energy into commercial/noncommercial categories makes a useful distinction that can be applied to the developed as well as the developing countries. Commercial energy is taken to mean oil, coal, gas, and electricity and noncommercial is taken to mean such fuels as firewood, solar heat, industrial waste, dung, and crop residues that are not purchased.

Traded versus untraded energy is a useful distinction and probably the only clear distinction possible. Yet it is not particularly satisfying because the availability or lack of untraded energy clearly affects the demand for traded energy, and vice versa.

Rising prices for traded energy encourage conservation of those types of energy, which may mean overall less energy consumption or more reliance on nontraded energy—human power, fuelwood, or solar heat. Advances in utilizing nontraded energy will reduce the demand for traded energy, and nontraded energy fuels can become traded fuels. For example, the production of solar electricity and fuelwood, or the conversion of biomass to methane or alcohol can be centralized, requiring markets or other distributional systems, or decentralized, user-oriented systems.

RESOURCES

Exhaustible and Renewable Energy Resources

Much of the debate concerning long-run sources of energy centers around the distinction between exhaustible energy resources, e.g., oil and gas—and renewable forms of energy—e.g., solar and the breeder reactor. As generally posed, the issue at hand is one of reducing the world's dependence on exhaustible sources of energy—which *must* "eventually" run out—and substituting renewable sources of energy, which by definition can provide a flow of energy over all time. Before accepting this dichotomy as the critical determinant of future fuel use, it is instructive to examine the conceptual underpinnings of the exhaustible and renewable energy resource classes.

The prototypical exhaustible energy resource is one which is a nonaugmentable stock; when used up, it is gone. The prototypical renewable energy resource is a nonaugmentable resource flow; one can tap into the resource flow at any time without reducing the amount available in the future. By specifically designating the stock and flow resources as *nonaugmentable,* the question arises as to whether there are *augmentable* energy resources of either the stock or flow variety. Before attempting to assign energy resources to one of the four categories, it is necessary to fix on the meaning of augmentable. One might include as processes which augment the current stock or flow discovery, technological improvement, installation of capital equipment capable of extracting or converting the energy in new resource occurrences, or increased intensity of capital, labor, and materials in known or presently utilized resource occurrences.

The necessity to distinguish and define augmentable resources as distinct from nonaugmentable resources forces a reconceptualization of the problem. Actual exhaustion of resources is unlikely to occur under any circumstances. All energy resources are augmentable by any one of the processes listed above, including discovery, technological improvement, development, and higher extraction and production expenditures. For oil and gas (in conventional deposits), the extent to which augmentation is possible is limited enough to present binding constraints. On the other hand, coal resources are less likely to pose a binding constraint on coal use.

In light of these characteristics, the nonaugmentable stock concept is at best applicable only as an approximation. As noted, to obtain even a reasonable approximation, one needs to define oil and gas very carefully. For example, if oil is defined broadly as crude liquids from fossil fuels, then resource limits are quickly pushed out so that constraints are meaningless over a period of 100 years or more into the future. Similarly, if one expands the category of conventional gas of biological origin to include abiogenic gas, geopressured gas, coal gasification, and the other sources, the potentially available resource could be huge. Thus, for fossil fuels as a group and for liquids or gases individually (given the availability of conversion technologies), total exhaustion of the resource will not occur over at least the next 100 years. This conclusion does not make the concept of exhaustion necessarily useless. High-grade (low-cost) resources such as cheap conventional oil and gas will be used up and once they are used up, they are gone.

In contrast, low-cost hydro installations will continue to supply low-cost hydro even though it has become economically attractive to harness a higher cost flow. Hydro, solar, and wind energy are nonaugmentable in the broadest sense; the amount of flowing water (excepting pumped storage) is given, as is wind speed and the level of insolation, but knowledge of total annual flows from these sources offers little guidance to future energy supply unless the resource concept is given an economic dimension. Once this is done, augmentation is possible through improved technology or higher prices.

If we wished, now, to classify energy resources by their conceptual categories at the broadest level, the fossil fuels and nuclear fission energy are or are based on nonaugmentable stock. The breeder and fusion energy might be labeled an augmentable stock. Biomass would be labeled on augmentable flow but is limited by insolation and geographic constraints. Solar, hydro, wind, and a variety of other renewable energy sources would be labeled nonaugmentable.

Terminology

It is of utmost importance to set down a standard system for determining when a deposit or energy resource is, in fact, countable as part of the resource base, resources, or reserves.

Despite many efforts to develop uniform sets of terms and classifications, confusion and misuse of terms continue to appear in the energy literature. This abuse is most apparent in the use of the word "resources" in comparisons of mineral supplies.

Figure 6-1 shows a mineral classification system developed by V. E. McKelvey. It has been widely accepted as a useful way to classify energy resources and is used by the

FIGURE 6-1. Resource base. (From V. E. McKelvey, "Concepts of Reserves and Resources," appearing in *Methods of Estimating the Volume of Undiscovered Oil and Gas Resources*. Edited by John P. Haun. American Association of Petroleum Geologists, Tulsa, Oklahoma, 1975.)

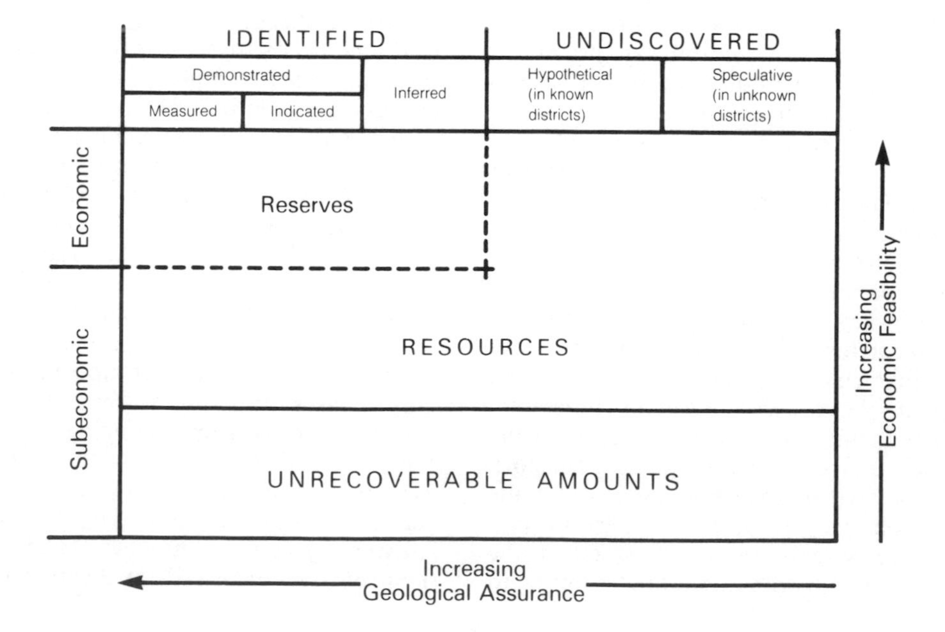

TABLE 6-2. Energy Terminology

Exhaustible Energy Terminology		*Renewable Energy Terminology*	
Resource base	The broadest definition of the amount existing of a mineral. Includes produced, unproduced, discovered, and undiscovered amounts. The term carries no connotation of being producible under any reasonable set of costs and technologies.	Resource base	The maximum flow of energy per time period (normally per year). No presumption of being able to practically utilize the full amount is necessarily implied.
Recoverable resources	That part of the *resource base,* including discovered, undiscovered, produced, and unproduced amounts, that is producible under a specified set of costs and technologies. As a shorthand, *recoverable resources* will be generally referred to as simply *resources.*	Recoverable resources	Economic and technologic feasibility criteria are specified, implying conditions under which some share of the *resource base* could be used; the term will be referred to as simply *resource* in most discussions.
Reserves	*Resources* that are known to exist and are remaining in the ground at a specified date. Reserve figures include only those amounts of resources that are producible with technologies existing at the specified date at total costs not exceeding the market price at the specified date.	Production capacity	The maximum amount of energy that could be used during a given period with the energy-utilizing capital stock taken as fixed.
Produced	*Resources* that have been extracted by a specified date and have been consumed or stockpiled.	Production	The amount of energy actually produced and used or stored in a useful form during a given time period.
Remaining resources	*Resources* not *produced* by a specified date.	Remaining resources	*Resources* not under *production* by a specified date.
Undiscovered resources	*Resources* that have not been *produced* by a specified date and are not classified as *reserves* on that date.	Unexploited resources	An existing *resource* flow for which there is no production capacity (capital stock).

U.S. Geological Survey and the U.S. Bureau of Mines. It illustrates the two fundamental dimensions of fossil resources: geological certainty—from well known to speculative—and economic feasibility—from recoverable at present prices with existing technologies to subeconomic.

Table 6-2 contains a list of terms used for exhaustible and renewable resources. While the specific aspects of the definitions differ somewhat between the two resource types, they have corresponding concepts derived from three basic dimensions: quantity, time, and producibility. The resource base is measured only along the quantity dimension; recoverable resources are measured along the quantity and producibility dimensions; reserves (production capacity) are measured along all three dimensions; and production is measured along the time and quantity dimension (and only implicitly along the producibility dimension). While other useful terms can be derived from those discussed here, they are not needed for our purposes.

Before moving on, it is useful to emphasize the economic and technological dimensions in our definition of resources. There has been an unfortunate tendency in the literature to only vaguely describe the price and technology assumptions for resource estimates. For some analysts, "present technological and economic conditions" suffices as a description. Others suggest that resource estimates include amounts that will be recoverable in the future with the "likely" evolution of price and technology. Possibly one of the more important contributors to the misperception of the resource definition is the term "ultimately recoverable resource," which suggests that there is an amount of the resource that will be recovered regardless of price and technology assumptions.

As will become evident in the next nine chapters, the ultimate amount of the resource, which we will refer to as the resource base, has little relevance in answering the question of how much will be ultimately produced. Typically, technological constraints (which are in fact cost constraints) will block a large share of the resource base from ever being exploited. In addition, it is by definition *impossible* to estimate ultimately recoverable resources without completely specifying costs and availabilities of other energy sources and the ability to substitute away from the resource in question.

CONCLUSION

In reporting and comparing energy resource amounts, it is our hope to reduce, or at least not add to, the confusion surrounding such estimates. The definitions set out here are used consistently throughout the text. Chapter 7 reviews conventional oil resources; it is followed by chapters on unconventional oil, conventional gas, unconventional gas, coal, nuclear, solar and wind, hydroelectric and ocean thermal energy conversion systems, and biomass.

Conventional Oil 7

Eventually the world must face relatively scarce supplies of conventional oil products. The long-term questions about oil then are not whether crude oil will become scarce eventually, but rather center on when and how the world will make the transition to alternative forms of energy. It is therefore critical to know how long the world's crude oil resources will last.

The term "conventional oil" will be taken here to mean oil available using recovery techniques known as of 1980 that would make unsubsidized production of the oil profitable at 1980 prices (roughly 30–40 dollars per barrel). Oil recovered with experimental technologies (enhanced oil recovery), oil shale deposits, tar sands, deep oil, and deep offshore oil are excluded from this category. Oil recovered from these deposits or with these techniques will be termed unconventional oil. (See chapter 8 for a discussion of unconventional oil.)

It is doubtful that oil prices will exhibit a long-term falling trend over the period of our analysis. As a result, conventional oil resources, as defined, will be oil resources that are profitable to produce.[1]

The distinction between conventional and unconventional oil can be viewed as a horizontal line on the resource diagram of figure 6-1 separating economic and subeconomic resources. Petroleum resource deposits above the horizontal line fall into the conventional oil resource category, while those below are unconventional. In addition, we include produced amounts with conventional oil resources as well as discovered and undiscovered amounts.

Total resource availability of conventional oil acts as an ultimate constraint on cumulative production. However, two additional factors place more immediate constraints on production. These are the discovery rate and the production rate. Discovered oil, as defined here, will be taken to mean oil that has been classified as proved reserves. In terms of figure 6-1, this definition includes only measured reserves, which are "essentially synonymous" with proved reserves.[2] Excluded from the broad category of reserves are those listed as indicated and inferred. The inverse of the reserves-to-production ratio is the production rate. Resources, discoveries, reserves, and the production rate are all related and their interaction in the case of oil is discussed in the following sections.

WORLD OIL RESOURCES

Methodologies

The amount of undiscovered conventional oil resources in a region is known only with considerable uncertainty. However, geologic evidence and extrapolation of trends in oil exploration success combine to yield estimates that are probably an order of magnitude less uncertain than many behavioral parameters that are required (either explicitly or implicitly) in any attempt to characterize long-run energy use. Among recent estimates of the amount of conventional oil resources, the highest is about three and one-half times the lowest. However, the great majority of estimates are clustered within a fairly narrow range.

Table 7-1 summarizes and contrasts five general techniques for estimating undiscovered oil resources. The first three depend more on actual geologic exploration and mapping that may take place even before exploratory drilling begins. They follow from oil company techniques of assessing promising sites for exploratory wells. They are site specific in that they assess particular petroleum basins on the basis of the site's geologic characteristics. The last two methods approach the problem from a much different perspective. Here, the region is defined and one infers a functional form that fits historical discovery data for the region. These data are then used to estimate the parameters of the functions.

Each method has shortcomings, but the techniques work in a complementary fashion. The geological analogy techniques can be used even if the region has only recently been explored and has not been drilled, while the extrapolation techniques are more applicable in regions that have reached a mature state of exploration.[3] The more exhaustive global estimates of world oil resources must rely on a combination of estimating procedures and cross-checking of techniques.[4]

O'Carroll and Smith have examined probabalistic methods for estimating undiscovered petroleum resources.[5] In these types of models, the discovery of reservoirs is viewed as a random sampling from the underlying population of reservoirs and the discovery of any existing reservoir is random, with the probability of discovery dependent on reservoir size.[6] These models require a fairly lengthy history of exploration in the resource "play" (the associated geologic structures that are thought to contain oil) and data on each well drilled in the area. Thus, the method requires more information than is generally available for many areas, especially only partially explored areas. This model has been applied to the North Sea oil play, yielding results that compare favorably with judgmental estimates of the area's resources.

Estimates

Conventional Wisdom

Table 7-2 lists past estimates of global oil resources, beginning in 1946. The table combines two reviews of the oil resource estimate literature—a recent estimate by the International Institute for Applied Systems Analysis (IIASA)[7] and a review by Hubbert.[8] Different authors emphasize the use of individual techniques reported in table 7-1, but as noted in the IIASA study, "Apart from a few recent exceptions, none of them (global oil and gas resource estimates) has given the data used with which a crosschecking of the results would be possible. Only a few groups have stated the assumptions or explained the methodology used, usually in very general terms."[9]

Figure 7-1 illustrates the historical trend in estimates of oil resources. Weeks's estimates have tended to be on the high side, Hubbert somewhat lower. Weeks and Hubbert have the longest series of estimates. In addition, the estimate by Grossling is interesting not only because of the potentially large amount of oil but also the large range given. The World Energy Conference (WEC) estimate was based on an assessment of estimates by other researchers.

Despite the variety of methodologies employed, there has been surprising agreement, at least since 1965, about the general range of resource estimates. The mean of estimates in table 7-2 that were made from 1965 through 1973 is approximately 2,270 billions barrels of oil (bbos), including both high and low estimates where a range is given. The standard deviation is about 780

TABLE 7-1. Methods of Petroleum Resource Estimation

Technique	*Data Needed for Estimation*	*Critical Assumptions or Parameter Values*	*Problems/Criticisms of the Technique*
Sediment volumetric method	Must estimate the total volume of sediments in the oil basin.	Proportion of sediment value that is a petroleum reservoir; proportion of petroleum that is likely to be recoverable (Together these are the proportionality factor.)	A "reasonable range" for the proportionality factor varies from 0.5–0.10 to 0.05–0.10.[a] This obviously results in tremendous differences in resource estimates.
Geological parameter method	Compare geological parameters (structural characteristics, geometry of the basin, deposition of source, and reservoir rocks) of producing fields with geological parameters of unexplored regions.	Choice of parameters to rely on. Assumption that similar geological formations hold similar amounts of petroleum.	A large number of parameters, not all independent, need to be considered.
Probabalistic exploration engineering analysis	Same as geological parameter method.	Same as geological parameter method, but probability distributions must be assigned to independent variables.	Same as geological parameter method, but in addition the choice of specified probability distributions is somewhat arbitrary.
Production and reserve extrapolation	Past production and reserve data for given region.	The particular functional form fitted to past data.	Different functional forms can lead to widely different resource estimates. Not appropriate for largely unexplored regions.
Analysis of the discovery index and exploration success	Volume of resources discovery per foot of wildcat drilling (discovery index) for a given region.	How the discovery index changes as exploration of the region proceeds.	The functional form used and parameters of the functional form describing the change in the discovery index as exploration proceeds are somewhat arbitrary. As with production/reserve extrapolation, the region must be fairly well along the exploration/production path to provide base data.

Source: Based on discussion of these five techniques in David Levine, "Oil and Natural Gas Resources of the Soviet Union and Methods of Estimation," *Project Interdependence: U.S. and World Energy Outlook Through 1990,* Congressional Research Service (Washington, D.C.: U.S. Government Printing Office, November 1977), pp. 821–848; and Wolf Hafele, Project Leader, *Energy in a Finite World: A Global Systems Analysis,* Report by the Energy Systems Program Group of the International Institute for Applied Systems Analysis (Cambridge, Mass.: Ballinger, 1981), table 7.2, p. 208.

[a]The U.S. Geological Survey considered the 0.5–1.0 range appropriate. Oil company geologists and M. King Hubbert believed this to be wildly optimistic; suggesting the 0.05–0.10 range; National Academy of Sciences, *Mineral Resources and the Environment* (Washington, D.C.: National Academy of Sciences, 1975).

TABLE 7-2. Historical World Oil Resource Estimates

Year	*Name*	*Company*[a]	*Estimate* ($\times 10^9$ *bbl)*
1946	Duce	Aramco	500
1946	Pogue		615
1948	Weeks	Jersey	617
1949	Levorsen	Stanford	1,635
1949	Weeks	Jersey	1,015
1958	Weeks	Jersey	1,500/3,000
1959	Weeks		2,000/3,500
1962	Hubbert	NAS	1,250
1965	Hendricks	USGS	1,984/2,480
1968	Weeks	Weeks	2,200/3,350
1969	Ryman	Jersey	2,090
1969	Hubbert	USGS	(1,350–2,000)
1970	Moody	Mobile	1,800
1971	Warman	B.P.	(1,200–2,000)
1971	Weeks	Weeks	2,290/3,490
1972	Jodry	Sun	1,952
1973	Vermer	Shell	1,930
1973	Odell	UNIV.	4,000
1973	Warman	B.P.	1,915
1974	Kirby, Adams	B.P.	(1,600–2,000)
1974	Hubbert	USGS	2,000
1975	Moody	Moody	(1,705-2,030-2,505)
1976	Grossling	USGS	(1,960-2,200-3,000-5,600)
1977	Parent, Linden	IGT	2,000
1977	Delphi	IFP (WEC)	2,200/2,500
1978	Nehring	Rand (CIA)	(1,700-2,000-2,300)

Sources: Wolf Häfele, Project Leader, *Energy in a Finite World: A Global Systems Analysis,* Report by the Energy Systems Program Group of the International Institute for Applied Systems Analysis (Cambridge, Mass.: Ballinger, 1981), p. 210; and M. King Hubbert, "World Oil and Natural Gas Reserves and Resources," in *Project Interdependence: U.S. and World Energy Outlook Through 1990,* Congressional Research Service (Washington, D.C.: U.S. Government Printing Office, November 1977), p. 638.

[a]Company or organization sponsoring the researcher, as applicable. The abbreviations used refer to the following organizations: Jersey, Standard Oil of New Jersey; NAS, National Academy of Sciences; USGS, United States Geological Survey; B.P., British Petroleum; UNIV, University Sponsorship; IGT, Institute for Gas Technology; IFP (WEC), Institut Francais du Petrole for the World Energy Conference; Rand (CIA), Rand Corporation for the U.S. Central Intelligence Agency.

bbos. For estimates made after 1973, the mean is about 2,380 bbos and the standard deviation is 990 bbos. However, if the very high estimates by Grossling (3,000 and 5,600 bbos), Odell (4,000 bbos), and Weeks (3,490 and 3,350 bbos) are excluded, the mean is about 2,000 bbos, with a variance of only 310. Figure 7-2 indicates the degree of uncertainty connected with oil resource estimates and is based on the Delphi study of data of the World Energy Conference.[10] The community of research organizations participating in the WEC survey is very similar to that listed in table 7-2. The figure indicates the clustering of estimates around the 2,000 bbo figure but also the splinter group holding the view that conventional oil resources may be much higher. The analysis of the mean and variance of recent estimates of global resources may be suggestive of the amount of conventional resources and the uncertainty

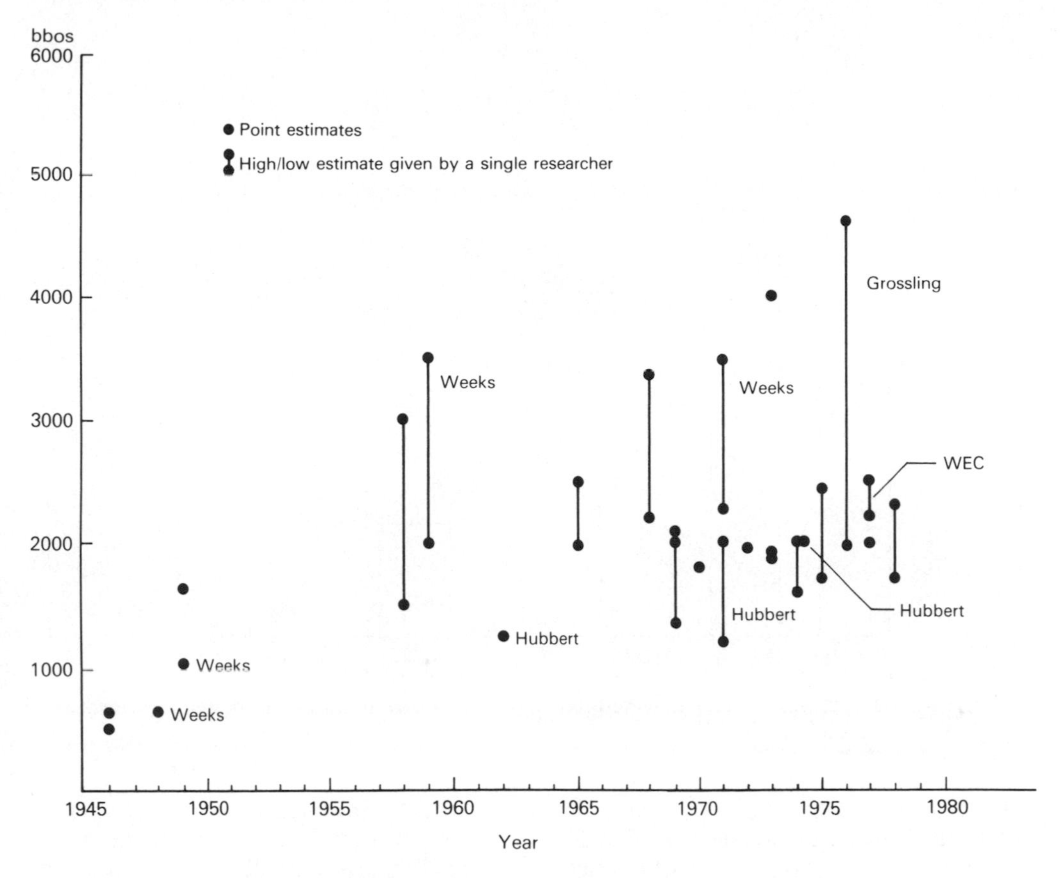

FIGURE 7-1. Historical estimates of conventional oil resources.

connected with the estimates. However, such an interpretation cannot be rigorously defended.

It is highly unlikely that global estimates are truly independent, since a considerable amount of discretion and judgment remain in determining parameters for the estimation procedures. Rather, "reasonable" parameter values are likely to be, in part, determined by whether they give reasonable results, and "reasonable" results depend largely on previous estimates. Such interdependence is likely to result in a smaller variance in estimates than is justifiable and possibly a centering of estimates around a value that is far from a true value. In addition, most estimates may tend to be somewhat on the conservative side because they rely on analogy techniques, either geological or in terms of exploration success, that will not account for truly unknown developments such as the discovery of oil in geological structures not previously thought to contain oil. There appears to be no good way of accounting for such possibilities or absolutely constraining the potential additions to conventional oil resources if such discoveries should occur. Nehring contends that monitoring geologic structure characteristics as exploration continues is the best, and perhaps the only, way to assess the possibility of such occurrences.[11]

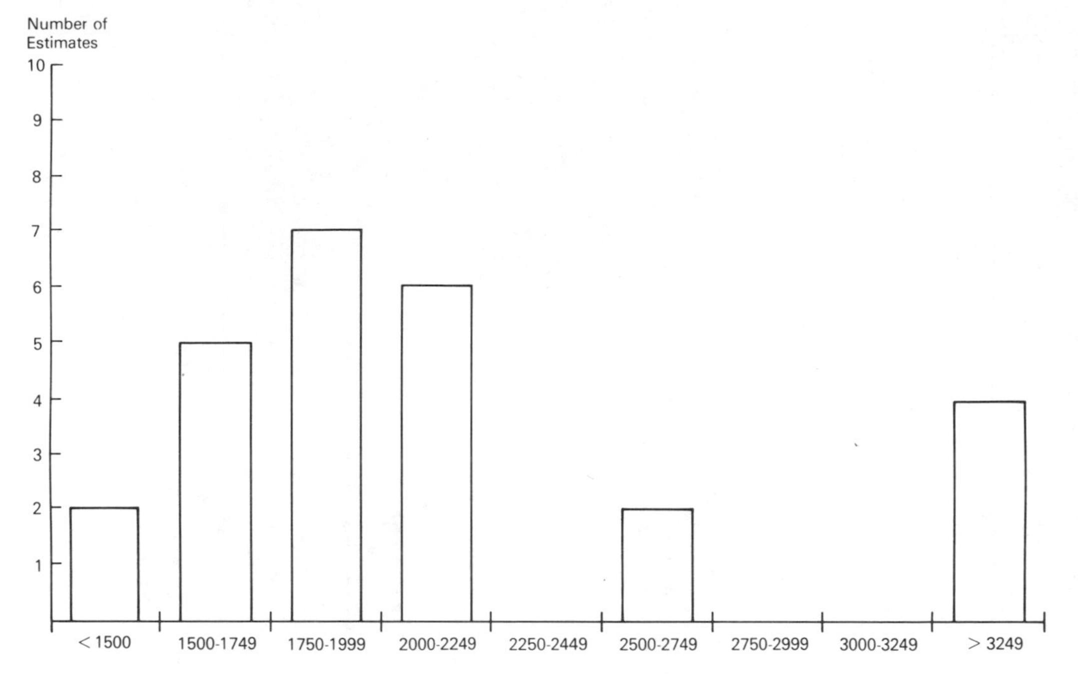

FIGURE 7-2. Distribution of estimates of total resources of conventional oil, based on World Energy Conference Delphi study.

Unconventional Estimates

Grossling is among the most extreme in arguing for at least the possibility of a much larger estimate of conventional resources. His contention is that "the petroleum potential of about half the world's prospective area has been grossly underestimated by the current conventional wisdom. The area referred to is in non-Communist, non-OPEC developing nations of Latin America, Africa, and South and Southeast Asia Extended."[12] Table 7-3 contrasts Grossling's estimates for the developing country regions with "conventional wisdom" estimates for these regions. The 1981 IIASA study was persuaded by this argument, concluding that "the resources of three geological areas—South America, Africa, and East and South Asia have been underestimated . . . considering these prospective areas."[13]

Analysts who remain skeptical of higher estimates of world oil resources generally cite the importance of giant oil fields (greater than 500 million barrels). Illustrative of their importance is the fact that 2 percent of the world's petroleum fields contain 79 percent of all known oil and gas reserves.[14] In this view, the crude exploration techniques of the past were more likely to have uncovered giant oil fields. A slump in the discovery of new oil provinces containing giant fields is expected to continue and, as a result, most remaining

TABLE 7-3. Alternative Oil Resource Estimates for Developing Country Regions (billions of barrels)

Region	*Conventional Wisdom Estimate*	*Grossling's Estimate*
Latin America	166	336–960
Africa	160	350–1,000
Southeast Asia and Far East	131	70–200
Total	457	756–2,160

Source: Adapted from tables presented in Bernardo F. Grossling, *Window on Oil: A Survey of World Petroleum Sources* (London: *Financial Times,* 1976), pp. 646 and 654.

oil must be found in smaller fields. Past experience suggests that small fields, while numerous, can contribute only a small amount to total reserves. The recent large finds in Mexico, with estimates of 120–200 bbos,[15] suggest that giant fields will continue to play an important role in future supplies. Advocates of lower estimates argue that giant fields will continue to account for the major share of additional oil discoveries (either extensions of existing fields or new discoveries), but unless large amounts of oil are found to exist in unlikely geologic structures or techniques are developed for extracting a much larger share of the oil in place, there is no reason to revise conventional global resource estimates.

Economic and Technical Sensitivity of Estimates

Table 7-4 presents base case conventional oil resource estimates and summarizes cumulative production and finds through 1979. The resource estimates for the centrally planned and developed country regions are representative of "conventional wisdom" estimates.[16]

The estimates for the three developing country regions are Grossling's low estimates and, therefore, reflect a somewhat more optimistic view of conventional oil resources in the developing countries. Estimates for region 3 (JANZ) are based on a review of the WEC regional resource estimates.

The figures in table 7-4 can generally be described as recoverable within present economic and technical limitations. A more explicit statement of present economic conditions is impossible, given state-of-art estimation techniques. If interpreted literally, the resource estimates should be revised upward since oil prices have risen considerably since the date the estimates were made. However, no further

TABLE 7-4. Conventional Oil Resources, Reserves, and Production (bbo)

Region	*Resource*	*Proved Reserves January 1, 1980*	*Cumulative Production Through 1979*	*Cumulative Share Produced*	*Cumulative Share Discovered*
1. US	227	24.9	115.9	0.510	0.619
2. OECD West	162	28.9	13.9	0.086	0.265
3. JANZ	6	2.2	1.3	0.224	0.607
4. EUSSR	343	66.5	62.6	0.183	0.377
5. ACENP	111	19.0	6.1[a]	0.055	0.226
6. MIDEST	630	341.9	110.7	0.176	0.723
7. AFR	329	53.6	27.8	0.084	0.247
8. LA	326	54.8	51.5	0.158	0.326
9. SEASIA	60	16.1	11.6	0.194	0.462
Total	2,194	610.0	401.9	0.183	0.461

Note: The standardization assumes 5.800×10^6 Btus/barrels at standard temperatures and pressures. Country-specific energy contents of gas and oil are based on production for a given year. A changing mix of production from various deposits within a country as new fields are opened and old fields are closed can change the average figure. Thus, these figures are not strictly applicable to resource or reserve estimates. However, application of the factors to reserve and resource estimates is preferable to allowing raw figures to remain, which, as one can observe from the table, would result in a fairly consistent overestimation of the energy content of oil of about 5 percent.

The adjustment factors are computed as a reserve weighted average of the Btu content of individual country crudes. Btu content from Federal Energy Administration, *Energy Interrelationships: A Handbook of Tables and Conversion Factors for Combining and Comparing International Energy Data* (Springfield, Va.: National Technical Information Service, June 1977), p. 22. Reserves (for weights) are from the *Oil and Gas Journal,* December 28, 1979.

The following adjustment factors were used to convert barrels of oil to barrels of oil equivalent: US-crude oil 0.94, natural gas 1.03; OECD West-crude oil 0.95, natural gas 1.03; JANZ-crude oil 0.94, natural gas 1.00; EUSSR-crude oil 0.95, natural gas 1.03; ACENP-crude oil 0.95, natural gas 1.03; MIDEST-crude oil 0.95, natural gas 1.08; AFR-crude oil 0.94, natural gas 0.89; LA-crude oil 0.97, natural gas 1.09; and SEASIA-crude oil 0.94, natural gas 0.99.

[a]Estimated.

upward revisions have been made for several reasons.

First, analysts did not noticeably revise resource estimates upward after the 1973–74 oil price hikes. The IIASA study suggests that an equally large rise in the cost of production may have negated the price impact. Similar forces may have been at work in the 1979 price hike.[17] Second, the phrasing of statements accompanying resource estimates suggests a fairly loose interpretation of present economic and technological conditions. In particular, some researchers look forward to unspecified higher prices or general cost reductions in experimental technologies that will allow access to lower grades of oil resources at existing prices. A definition of "present technological conditions" in operational terms usually begins by assuming average recovery factors existing today and then may assume an increase in the share of in-place oil recovered as a result of the widespread adoption of recovery techniques proven to be cost effective in limited applications, or an even greater increase if an assumption is made that some experimental techniques will eventually prove out.

Third, the world total of 2,194 bbos is already on the high side of generally accepted estimates. The only major attempt to explicitly control for cost and technology in deriving resource estimates was the WEC Delphi study in which participants were asked to report estimates of recoverable oil available at less than 5, 12, and 20 1976 dollars per barrel.[18] (This is equivalent to prices of 6.16, 14.78, and 24.65 dollars per barrel in 1979 constant dollars.)[19]

The results of this effort indicate that there is a wide divergence of opinion as to the recoverability of oil under different cost assumptions—for example, anywhere from 10 to 70 percent of the total resource was estimated to be available at less than five 1976 dollars per barrel.[20] Thus, it is difficult to adjust world conventional oil resources specifically for rises in price. Accepting a total estimate on the high side of recent estimates can be interpreted, at least in part, as a reflection of recent price rises.

Regional versus Global Resource Estimates

A final note on regional versus global resource estimates is necessary. Regional estimates vary tremendously and maturation of the production region has not tended to increase certainty of the resource estimate. As evidence, one need only point to the World Energy Conference's Delphi Study.[21] The high and low estimates for the United States and Canada, the most extensively explored and developed countries, varied by a factor of 8; whereas the high and low estimates for the USSR and China, where very little information is available, varied by a factor of only 3.5. Similarly, estimates for the developing country regions appear, in general, no less uncertain than for maturely developed regions, if one uses the standard deviation of various estimates for the region as a measure of uncertainty. However, most analysts will argue that their estimates for undeveloped regions are considerably more uncertain. The higher degrees of uncertainty attached to reported estimates for the developing countries and the small variance among estimates from different authors for these same regions, appear inconsistent. This inconsistency supports the view that estimates are still largely a matter of judgment and, especially in cases where data are fragmentary, apparently independent estimates may be very dependent on the first or "generally accepted" estimate for the region.

THE RATE OF OIL EXPLORATION AND DISCOVERY

Figure 7-3 illustrates the history of conventional oil production from its beginning in the United States in 1862 to the present. Important dates associated with the discovery and development of the world's oil fields are noted. The listing is not comprehensive but is an attempt to note the start

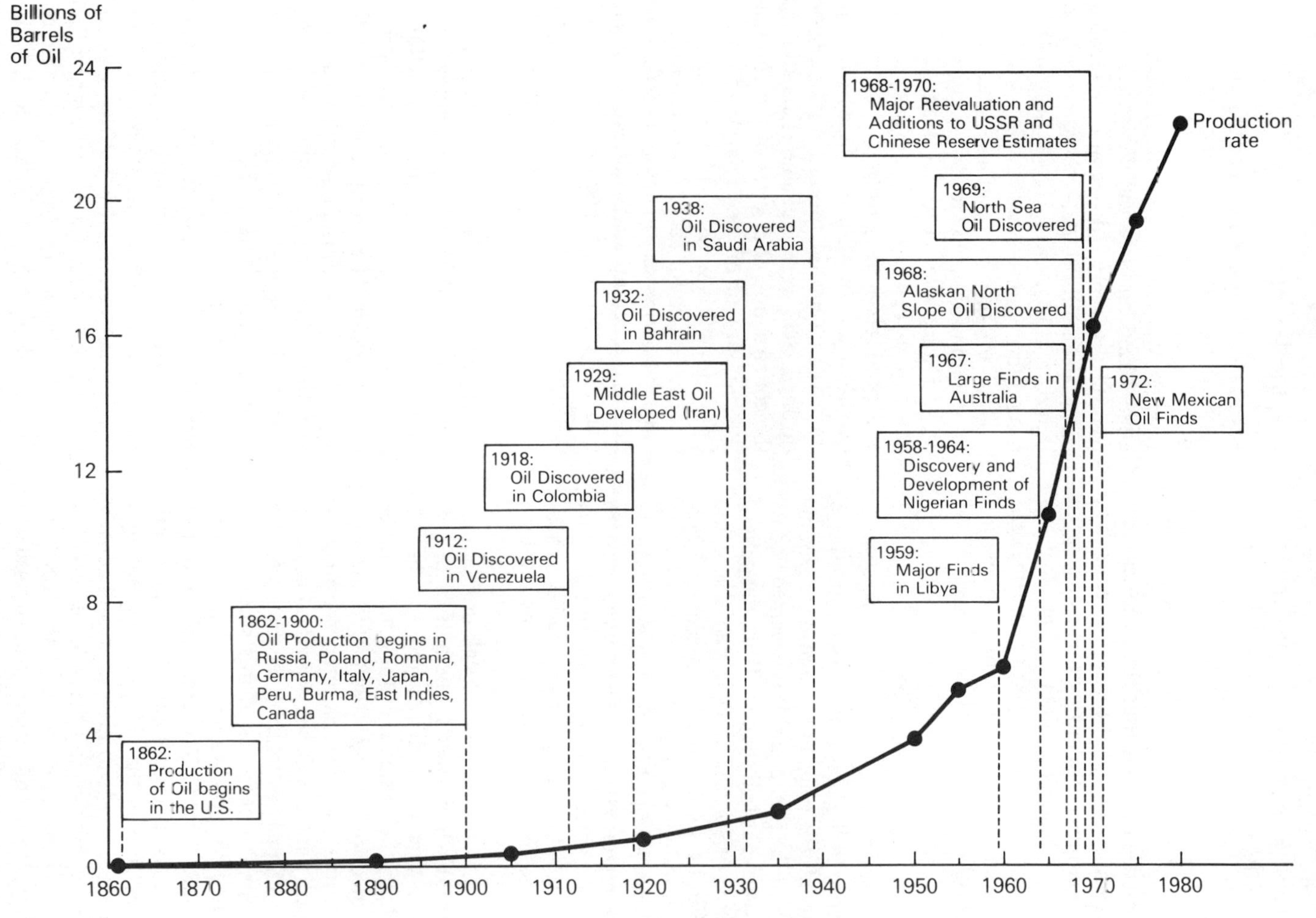

FIGURE 7-3. History of global oil production.

of major oil production in most important oil-producing countries or regions.

The Logistic Curve and Oil Discoveries

It has been argued that the historical rate of resource exploitation can be explained more concisely by fitting a logistics curve through historical data.[22] The logistics curve has several properties that make it desirable for describing oil exploitation over time. The curve is formally given by

$$\frac{f}{(1 - f)} = e^{(a+bt)}$$

where f is the cumulative share of oil discovered, t is time, and a and b are parameters of the function.[23] It is S shaped and its derivative is bell shaped. Thus, the path of discovery (production) is one where the additions to cumulative discovery (production) increase initially and then fall off. Hubbert's work has shown a very close fit for the United States. The curve is easy to estimate since it is log-linear and the data necessary to estimate it are readily available. The major constraint imposed by the curve is that the exploitation path is symmetric; thus peak finds (peak production) occur(s) when exactly one-half of the resource has been discovered (produced). Further, the estimate of the total resource accepted for the region will affect the fit of the curve and the parameter estimates. Hubbert has developed a methodology for endogenously estimating the total resource but, for the purposes of this study, it was deemed of more value to accept resource estimates based on a review of the results of various methodologies.[24] In this regard, Levine argues that Hubbert's estimation procedure tends to give relatively conservative estimates and "should be regarded as the least amount we can expect to find."[25] Root and Attanasi have used the Hubbert methodology to forecast peak global oil production and conclude that it is likely to occur around 2000.[26]

Historical Analysis

Parameters of the logistics curve can be estimated for each of eight aggregate Oak Ridge Associated Universities/Institute for Energy Analysis (ORAU/IEA) world regions based on historical data of cumulative oil discovery. All regions correspond to the usual ORAU/IEA categories except that data limitations forced a combined estimate for regions 4 and 5.

Data on reserve estimates and yearly production from 1949 to 1979 were assembled. Data were obtained from annual worldwide reports of oil and gas reserves and production appearing in the *Oil and Gas Journal.* In addition, figures for cumulative global oil production by region through 1953 were available. This allowed the construction of a series of the cumulative amount of oil discovered by region. Combining this series with the estimates of total resources for each region reported in table 7-4 allows one to compute a time series of the cumulative share of conventional oil discovered. The specific equation estimated was

$$\ln(f/1 - f) = a + bt$$

where f is the cumulative share, t is time in years, and a and b are the parameters to be estimated. An ordinary least squares (OLS) technique was used to estimate the regression equation. The results are reported in table 7-5. These estimates serve as a guide to future discovery rates and can be applied to the oil production path.[27]

The estimates in table 7-5 show varying rates of discovery. The rates for OECD West and JANZ are most rapid, reflecting the preponderant share of total resources accounted for by a single oil province. Discovery can proceed fairly rapidly once the province has been located and a major find has been made. In contrast, the oil resources of regions 1 (US) and 8 (LA) are spread among a fairly large number of fields and oil provinces, and exploitation has been much slower. The rate of region 7

TABLE 7-5. Estimated Logistics Curves for Cumulative Oil Discovery[a]

Region	*a*	*b*	*Peak Production Year*	R^2
1. US	−103.13	0.052393	1968	0.991
2. OECD West	−236.36	0.118995	1986	0.983
3. JANZ	−361.41	0.182874	1976	0.868
4. EUSSR 5. ACENP	−208.99	0.105425	1982	0.964
6. MIDEST	−223.73	0.113616	1969	0.960
7. AFR	−275.60	0.138971	1983	0.861
8. LA	−119.62	0.060011	1993	0.947
9. SEASIA	−185.30	0.094348	1964	0.924

[a]Equation estimated was $\ln(f_t/1 - f_t) = a + b$ (year) where f_t is the cumulative share of total resources discovered by year t.

(AFR) is probably considerably more rapid than can be expected to hold over the long run because of the relatively short history of oil production and the predominance of the few large North African finds in the discovery history to date. The combined estimate for regions 4 and 5 (EUSSR and ACENP) may also be somewhat high, but this is largely because of data problems. Available data for recent years include categories other than proved reserves, while the reserve definition for earlier years is unspecified. Thus, the data series may include only proved reserves in early periods and a broader category of reserves in later periods.

The peak find period is given by $t = -a/b$. The estimated peak find year for each region is reported in table 7-5. As can be seen, most regions have peaked or are very near peaking in terms of finds. If production lags discovery, as is the case with Hubbert's estimates for the United States, the estimates in table 7-5 suggest a peak in global oil production within the next 25 years. One way of orienting the production curve in relation to the discovery curve is to force it to pass through the most recent cumulative production figures. Such an analysis suggests a global peak in production between 1995 and 2005.

RATE OF PRODUCTION

In the normal course of events, production rates should present a lagged reflection of discovery rates, though as OPEC has demonstrated, the relationship is not irrevocable. In the short term, the problem is compounded not only by a variety of political and economic considerations, but also by technical considerations.

To begin with, the reserves-to-production ratio and reserve estimates are not, in theory, independent because the rate at which oil is extracted from a field influences the amount of oil ultimately recoverable. Rapid extraction leads to a rapid fall in the field's natural pressure, thereby reducing the length of time the field will produce without artificial repressurization.

A choice of profit-maximizing combinations of extraction rates and recovery techniques depends on the present and expected future price of oil, the discount rate, and the entire range of relevant physical characteristics of the field. Thus, the reserves-to-production ratio is partly determined by economic factors and partly by geological-physical characteristics. Furthermore, the natural field pressure and recovery techniques employed vary from field to field. Aggregate reserves-to-production ratios (production rates) are likely to vary considerably because of these factors and also because regions such as the Middle East have shut-in fields or fields only in the process of being developed, tending to give a high reserves-to-production ratio, while in the conterminous United States, few fields are shut in and most fields are at a mature stage of development. Hubbert posits a relationship between rate of discovery and rate of production that implies a gradual rise in the production-to-reserves ratio.[28]

Figure 7-4 is a plot of production-to-reserve ratios for seven world regions. The regions are those used throughout this study, except that ACENP and EUSSR are combined as centrally planned (CP); JANZ is excluded. If anything, the plot indicates a lack of regularity. Only the US and MIDEAST regions show the expected gradual rise over the entire period. A major problem in attempting to discern trends in the movement of production rates appears to be the erratic nature of oil discovery. For those regions with major oil fields relative to the size of proved reserves during the observation period (1950–79), the production rate tends to drop sharply and then rise slowly. This is witnessed in the case of

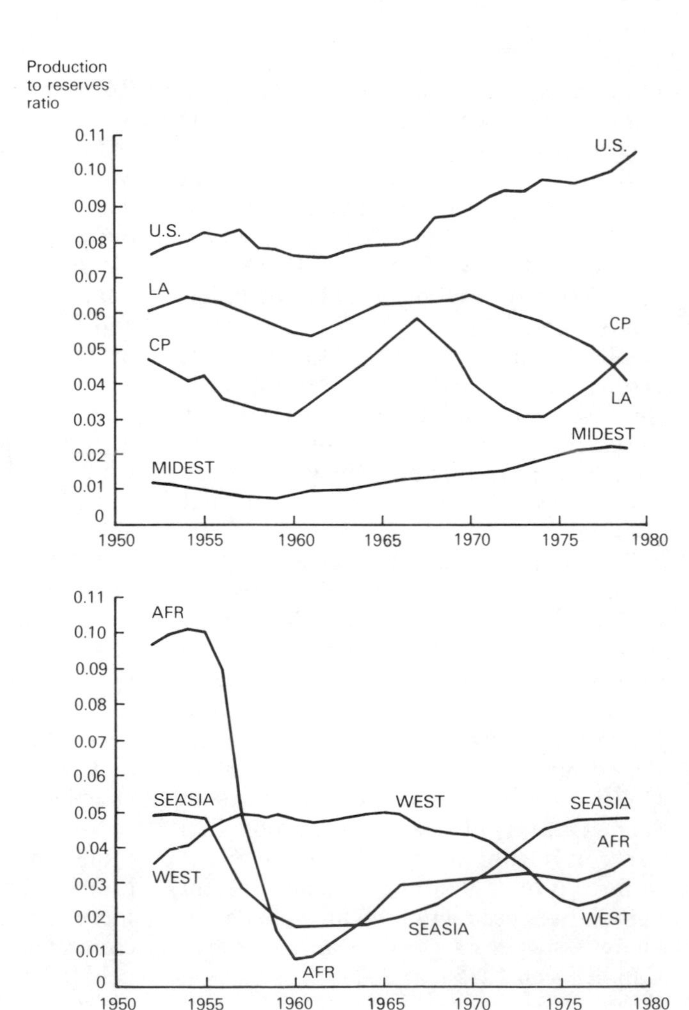

FIGURE 7-4. Plot of five-year moving average of production-to-reserves ratios. *Note:* The region labels are those used throughout the study except that data constraints forced a combined reporting for ACENP and EUSSR, labeled CP (Centrally Planned). (Base data from the *Oil and Gas Journal.*)

Africa in the late 1950s with large finds in North Africa; in OECD West with the North Sea finds in the earlier 1970s; in the centrally planned economies with large revisions and additions to reserves in the USSR in the late 1960s; in SEASIA with additions to reserves in Indonesia in the mid-1950s; and most recently with Latin America and the additions to Mexican reserves. In contrast, the United States and the Middle East have had large finds, but they have been steady additions rather than single finds large enough to double reserves of the region from one year to the next. Thus, the lack of regularity in figure 7-4 appears to be largely the result of irregularity in the finding rates.

MODELING OIL PRODUCTION

Generally, one would like to model oil production as sensitive to price, whether one views the supply responsiveness as very sensitive or very insensitive. In terms of modeling production, several theoretical considerations are important and, depending on the particular assumptions made, lead to considerably different model specifications and results. The basic theoretical economic model yielding implications for supply of a constrained exhaustible resource can be traced to Hotelling.[29] The deterministic result of the theoretical model relates to the rent path (price received above production costs) associated with exhaustion of the resources. Hotelling showed that the optimal resource exhaustion would be at a rate such that the rent element of the oil price increases at the interest rate. The intuition of such a result is transparent; rents that increase more rapidly than the interest rate would cause suppliers to hold more oil for the future since appreciation in the value of the oil reserves would give a greater return than could be obtained in financial investments. Conversely, a tendency for the oil price to increase slower than the interest rate would lead to greater production in earlier periods since holding financial assets would be more remunerative than holding the oil reserves. Positing such responses gives rise to the deterministic price path noted earlier. Complications are introduced in making the model operational, e.g., the appropriate present and future interest rate must be determined and expectations concerning demand must be accounted for. The simple theoretical model poses a known future demand curve although a rational expectations argument concerning future demands could be used to give the model more appeal. While such considerations are of interest, the critical assumption embedded in the model is the view of the resource as a fixed, known resource with no physical constraints on the rate of production. In the extreme, the full resource could be available tomorrow if suppliers found such an exhaustion path in their economic interest.

Nordhaus develops this model, adding somewhat more realism.[30] The basic addition is different resource grades, recognizing the different costs of accessing and producing oil from deposits with different geologic, geographic, and physical characteristics. Such a model allows one to posit a low-grade resource as offering a practical price backstop. Nordhaus shows that higher grade resources will be fully produced before lower grade ones as a direct extension of the Hotelling theorem.

The theoretical model provides a basic insight into the likely exhaustion path of oil; however, it is lacking as a description of the world since, in practice, a wide range of resource grades are produced simultaneously. One can begin to realistically describe oil production by applying the model to drilling rather than oil production and imposing constraints on the rate of production from drilled reserves.[31] However, a more important consideration is finding rates for various grades. Modeling of the oil finding process is confounded by the fact that one cannot plausibly extend the Nordhaus result to discoveries. Discovery clearly has a random element associated with it. It is difficult to predict a field size before discovery (although geologic evidence may give a clue as to the likely potential); thus if one views the main element of cost (grade differentiation) as the field size, it is likely that some lower grade resources will be discovered and produced before higher grade resources are found. There are other important grade factors, e.g., viscosity, rock porosity, geographic location as it affects cost of access to the field, offshore water depth, and climate. The argument of inseparability of the finding process among oil grades holds in varying degrees for each of these grade dimensions.

Recognizing the discovery element of oil production takes one some distance from the basic assumptions underlying the Hotelling model approach. It has been noted elsewhere that discovery can be viewed as sampling without replacement, giving rise to fewer oil finds as discovery proceeds—the key relationship being barrels per foot drilled. Price considerations enter in terms of intensity of the search, i.e., feet drilled. Presumably an expected price response would be to increase exploratory drilling as price increased. Since discovery is generally related to drilling footage, one could, under the simple assumptions of the sampling without replacement view, compact the bulk of the discovery process into a very short time period. A basic problem in implementing such a model is obtaining a reading on reasonable parameter values for the drilling response to price increases. Worldwide geophysical activity dropped from 1972 levels in the developing country regions of the world despite the oil price rises in 1973–74, suggesting a perverse price response.[32]

Approaches

Various approaches have been used in modeling oil production. Cherniavsky suggests three categories of models: extrapolation, econometric, and discovery process models.[33] Of these, process models are the most satisfying in terms of a description of how the world works; oil is found, proved, and production occurs as constrained by a reserves-to-production ratio. The total resource is implied by the integral of the function describing finds per foot drilled. Several problems are encountered in implementing such models. Cherniavsky notes that while constraints on drilling levels exist, it is not clear how to set them up in the model.[34] Additional critical concerns include data availability (even in the United States), the functional form of the finding rate curve, and the reserves-to-production ratio. Cherniavsky adds that "the discovery process model relies on curve fitting just as heavily as any of the curve fitting extrapolation methods."[35]

The economic/econometric models reviewed by Cherniavsky build on the Hotelling theoretical work already described. Some fail to incorporate explicit resource constraints and in general the models have had a poor prediction history.

The extrapolation models are the simplest of the three types. Hubbert's work is the best example of these models. They have been relatively successful as prediction tools[36] but have been criticized on several grounds—primarily because the function used cannot be derived from any physical considerations nor are relationships causal. Economic forces are not considered and discovery depends on time rather than exploratory effort.[37]

The Energy Modelling Forum (EMF) reviewed ten medium-term international oil models.[38] Key to all the models was a differential treatment of OPEC and the rest of the world. Beider[39] was able to distinguish two broad categories—simulation and optimization models. The optimization models relied on either maximizing discounted utility of consumption or discounted profit streams from production over time. Such models require a specification of expectations. In all the models, OPEC entered as a separate producer, in some cases subdivided further to consider Iran and Saudi Arabia separately or Mexico as a producer separate from the rest of the world and acting similar to OPEC. OPEC behavior was modeled as a price reaction function in the simulation models where OPEC adjusted prices to achieve a desired utilization level of exogenously specified capacity. Thus the long-run or equilibrium behavior has production completely determined by the exogenously determined capacity. No dynamic optimization occurs in these models. The three optimization models also specified OPEC capacity exogenously. Only one modeled OPEC as an optimizer (though constrained by exogenously determined capacity). The others specified OPEC behavior exogenously or as a price reaction function; dynamic optimization entered in non-OPEC production or consumption.

Non-OPEC production was handled in a variety of ways in the various models. Several modeled oil supply as a supply elasticity with or without explicit or endogenous resource depletion. The optimization models used a resource cost curve that, combined with backstop technology specifications, implies the level of ultimately recoverable conventional oil. Some of the simulation models provided greater detail for non-OPEC production—often only for the United States—including some specifications of new discoveries and constraints in reserve-to-production ratios. Table 7-6 is taken from Beider, and compares the ten oil models reviewed by the Energy Modelling Forum.[40]

The "correct" or best approach to modeling oil production is far from obvious. The dynamic optimization techniques have a strong theoretical basis to the degree one is willing to accept a description of the world that minimizes the impact on production of the dynamics of the finding process and constraints on production given reserves. A basic result of the modeling approach is that higher resource grades are exhausted first and completely before the next resource grade is exploited. Such an assumption is refuted by existing real world evidence. Further, such a modeling approach implies problems of discontinuity, particularly in transition from conventional oil to a backstop technology (shale oil). The transition occurs in one time period (a year). Such a rapid transition is not likely or physically possible. To provide a more realistic transition, various constraints must be imposed on conventional oil production and possibly on the rate of expansion of the backstop technology.

The more detailed discovery process models can provide more realistic descriptions of the world and offer insights into the discovery/production aspects of oil supply, but price response parameters are difficult to estimate and have not been examined extensively in the literature. Simple price response mechanisms are unlikely to be consistent with dynamic profit maximization. The construction of a dynamic profit maximization model with realistic discovery and production detail must face the problem of specifying expectations. Rational expectation, and perfect in the sense that producers' expectations specified in the model are realized in the model, is the most reasonable specification or at least the most reasonable point of orientation. However, solving for the rational expectations solution results in a complex model, requires iteration over time, and would be very costly. Given the large number of uncertainties existing in parameter values for such a model, it is unlikely that such an exercise would necessarily provide reasonable exhaustion paths or narrow the range of plausible paths.

Hubbert's simple extrapolation technique gives no insight into the economic aspects of oil production, yet it has given good results in the past. The technique can offer nothing in terms of analyzing the impact of regulation, taxes, or pricing on oil production.

CONCLUSIONS

Remaining conventional oil resources represent a fairly large energy resource. While its size is subject to considerable uncer-

TABLE 7-6. Summary Description of 10 World Oil Market Models as Implemented in EMF 6 Study

	Models									
Features	*IEES/OMS*	*OILMAR*	*WOIL*	*Cately/Kyle/ Fisher*	*IPE*	*OILTANK*	*Opeconomics*	*ETA-MACRO*	*Kennedy/Nehring*	*Salant/ICF*
Model type	Recursive simulation[a]	Recursive simulation	Recursive simulation	Recursive simulation	Recursive simulation	Recursive simulation	Recursive simulation	Intertemporal optimization	Intertemporal optimization	Intertemporal optimization
Foresight	None	None	None	None	None	None	None	Consumers[e]	Non-OPEC conventional oil producers	OPEC and non-OPEC producers
Regional aggregation	All EMF OECD regions, OPEC, and LDCs	U.S., OPEC, and rest of WOCA	U.S., other OECD, OPEC, Mexico, and other LDCs	OPEC and WOCA	8 production and consumption regions	16 production and consumption regions	Iran, Saudi Arabia, other OPEC, OECD, LDCs, and CPEs	OECD and OPEC	9 production regions; U.S., other OECD, and ROW consumption regions	All non-CPE EMF regions
OPEC										
Production capacity	Exogenous	Exogenous	Exogenous	Exogenous	Endogenous, based on expected future demand	Exogenous (except in high price scenario)	Exogenous	Exogenous	Exogenous	Exogenous
Price or quantity determination	Price reaction[b] function	Price reaction function	Price reaction function including expected future capacity utilization; costs	Price reaction function including capacity utilization at time $t - 1$	Price reaction function including R/P ratio; costs, exogenous royalties	Price reaction function including capacity utilization at time $t - 1$	Hybrid price reaction function/ supply function minimum revenue targets	Price determination function of OECD imports	Exogenous production	Intertemporal optimization to maximize discounted profits
Consumers										
Direct[c] price influence on demand	Via crude-price elasticities	Via crude-price elasticities	Product-price response via capital stock	Via crude-price elasticities	Via crude-price elasticities	Via product-price elasticities	Via crude-price elasticities	Via intertemporal optimization to maximize discounted utility of consumption	Via product-price elasticities for old and new capital	Via crude-price demand curve
Energy-GNP feedback	One-parameter feedback in OMS; other adjustments in econometric models	Not included	One-parameter feedback with short lag	Not included	Not included	Not included	One-parameter feedback with long lag	Aggregate production function for OECD economy	Not included	Not included

⅔ adjustment time (years)	5	5	10; depends on actual GNP	14	5	10	Approx. 2	14 pre-2000, 16.4 post-2000; depends on actual GNP	10–12.3 pre-2000, 12.5–14 post-2000; varies by region	0
Oil vs. total energy[d]	OMS is oil-only, but explicit fuel switching occurs in IEES	Oil only	Several energy forms are considered	All energy demands and non-OPEC supplies are aggregated	Oil only	Oil only	Demands for several energy forms are aggregated	Several energy forms are considered	Oil only	Oil only
Non-OPEC										
Producers' price influence on conventional supply	Via elasticities	None-oil production is exogenous	Via investment (U.S.), elasticities (ROW)	Via supply curve	Via elasticities	Via new discoveries and the recovery factor; oil production occurs at constant R/P ratios	Oil production is exogenous; oil price affects only coal and natural gas production	Via intertemporal optimization to maximize discounted utility of consumption	Via intertemporal optimization to maximize discounted profits	Via temporal optimization to maximize discounted profits
Depletion	Resource cost curves	Implicit in exogenous production path	Minimum R/P ratios; resource cost curve for U.S.	None—technological growth may be considered net of depletion	Depletion multipliers (except in Egypt and Mexico)	Constant R/P ratios	Implicit in exogenous production path	Resource cost curves	Resource cost curves	Resource cost curves
Use of EMF backstop quantities	No backstop; time horizon is 1995	Upper bounds	Upper bounds	No backstop is explicitly identified	Base projection	Not used; production depends on price via investment	Included exogenously	Base projection	Upper bounds	Upper bounds

Source: Perry Beider, *Comparison of EMF 6 Models,* Working Paper EMF 6.10. Energy Modelling Forum, Stanford, Calif.

[a]OMS is of the recursive simulation type; IEES is a "recursive optimization" model.

[b]All price reaction functions use capacity utilization at time t as one of the determinants of percent price change from t to $t + 1$; the other explanatory variables are as shown.

[c]Excluding energy-GNP feedback.

[d]This item applies to both the consumer and non-OPEC producer sectors

[e]Consumers in ETA-MACRO also make the production and investment decisions.

tainty, a reasonable base estimate of the total conventional oil resource is 2,309 billion barrels of oil. Adjusting this figure for regional variations in energy content per barrel of oil yields a world total of 2,194 $\times$ 10^6 barrels of oil equivalent (5.800 $\times$ 10^6 Btus/barrel) of 13,431 exajoules (EJ). Of this, only 18.2 percent had been produced through 1979.

Despite the fact that less than 20 percent of estimated conventional oil resources have been produced, oil production is likely to peak around 2000 or shortly thereafter, based on a Hubbert curve analysis.

The regional distribution of conventional oil heavily favors the MIDEST region, with nearly 29 percent of total resources and a like share of remaining resources. US, EUSSR, AFR, and LA are each estimated to have a similar share of the world's total resource—between 10 and 16 percent. However, of these regions, US has produced 51 percent of estimated conventional resources while the others have produced less than 20 percent. Thus, US's remaining resources represent less than 7 percent of the world's remaining conventional oil; the figure is 16 percent for EUSSR, 15 percent for LA, and 17 percent for AFR. Among the remaining four regions, ACENP, OECD West, and SEASIA, estimated resources fall between 2.7 and 8.2 percent of total resources; JANZ's total resources are negligible in terms of estimated global resources.

Both the global total and regional estimates of oil resources are subject to considerable uncertainty. The estimates accepted for the developed and centrally planned regions are representative of "conventional wisdom" while the estimates for the three developing country regions are somewhat higher than "conventional wisdom" estimates.

NOTES

1. Most likely there will be a rent element associated with conventional oil production (price will exceed cost of production). Such a rent exists for most oil produced today, the exception being the most difficult to produce areas. Unconventional oil may or may not be produced or have rents associated with production, depending on the price path for energy, advances in extraction technology, environmental concerns, and government programs.
2. V. E. McKelvey, "Concepts of Reserves and Resources," *Methods of Estimating the Volume of Undiscovered Oil and Gas Resources,* American Association of Petroleum Geologists, Tulsa, Okla., 1979, p. 11.
3. M. King Hubbert, "Degree of Advancement of Petroleum Exploration in the United States," *American Association of Petroleum Geologists Bulletin,* vol. 51 (November 1967), p. 2211.
4. John D. Moody and Robert E. Geiger, "Petroleum Resources: How Much Oil and Where?" *Technology Review* (March/April 1975), pp. 37–40.
5. See specifically James L. Smith, *A Probabilistic Model of Oil Discovery,* MIT Energy Laboratory Working Paper, MIT-EL80-008WP, Cambridge, Mass., 1980, p. 3; but also Frank M. O'Carroll and James L. Smith, *Probabilistic Methods for Estimating Undiscovered Petroleum Resources,* MIT Energy Laboratory Working Paper, MIT-EL80-008WP, Cambridge, Mass., 1980.
6. O'Carroll and Smith, *Estimating Undiscovered Petroleum Resources.*
7. Wolf Häfele, Project Leader, *Energy in a Finite World: A Global Systems Analysis,* Report by the Energy Systems Program Group of the International Institute for Applied Systems Analysis (Cambridge, Mass.: Ballinger, 1981), p. 210.
8. M. King Hubbert, "World Oil and Natural Gas Reserves and Resources," in *Project Interdependence: U.S. and World Energy Outlook Through 1990,* Congressional Research Service (Washington, D.C.: U.S. Government Printing Office, 1977), p. 638.
9. Häfele, *A Global Systems Analysis,* p. 190.
10. World Energy Conference, *Oil and Gas Resources* (New York: IPC Science and Technology Press, 1978), p. 24.
11. Richard Nehring, *Giant Oil Fields and World Oil Resources,* prepared for the Central Intelligence Agency (Santa Monica, Calif.: Rand Corp., 1978), p. xi.
12. Bernardo F. Grossling, "A Critical Survey of World Petroleum Opportunities," in *Project Interdependence: U.S. and World Energy Outlook Through 1990,* Congressional Research Service (Washington, D.C.: U.S. Government Printing Office, 1977), p. 657.
13. Häfele, *A Global Systems Analysis,* p. 198.
14. Arthur A. Meyerhoff, "Economic Impact and Geopolitical Implications of Giant Petroleum Fields," *American Scientist,* vol. 64 (1976), p. 536.

15. Congressional Research Service, *Mexico's Oil and Gas Policy: An Analysis* (Washington, D.C.: U.S. Government Printing Office, 1978), p. 2.
16. Grossling, "A Critical Survey of World Petroleum Opportunities," p. 646.
17. Häfele, *A Global Systems Analysis,* p. 188.
18. WEC, *Oil and Gas Resources,* p. 140.
19. To convert to 1979 dollars, the U.S. GNP deflator was used, as reported in U.S. Department of Commerce, *Survey of Current Business,* vol. 6 (December 1980), p. 18.
20. WEC, *Oil and Gas Resources,* p. 140.
21. World Energy Conference, *World Energy: Looking Ahead to 2020* (New York: IPC Science and Technology Press, 1978), p. 24.
22. See, for example, Hubbert, "Degree of Advancement of Petroleum Exploration in the United States."
23. Note that a can be considered an orientation parameter, since when $t = 0$

$$a = \ln[f/(1 - f)]$$

Similarly, b can be thought of as a rate of exploitation parameter, since

$$b = \frac{d \ln[f/(1 - f)]}{dt}$$

24. Hubbert, "Degree of Advancement of Petroleum Exploration in the United States."
25. David Levine, "Oil and Natural Gas Resources of the Soviet Union and Methods of Estimation," *Project Interdependence: U.S. and World Energy Outlook Through 1990,* Congressional Research Service (Washington, D.C.: U.S. Government Printing Office, 1977), pp. 826–827.
26. David H. Root and E. D. Attanasi, "An Analysis of Petroleum Discovery and a Forecast of the Date of Peak Production," unpublished, n.d. Authors' work for U.S. Geological Survey.
27. Hubbert, "Degree of Advancement of Petroleum Exploration in the United States."
28. Ibid.
29. The original statement, by Hotelling, appears in Harold Hotelling, "The Economies of Exhaustible Resources," *Journal of Political Economy,* vol. 39, pp. 137–175. For a recent and more developed statement of this theory, see William D. Nordhaus, *The Efficient Use of Energy Resources* (New Haven, Conn.: Yale University Press, 1979), pp. 1–21.
30. Nordhaus, *Efficient Use of Energy Resources,* p. 1–21.
31. Ibid., p. 9, points this out in a footnote.
32. For data on oil exploration, see *1980 Statistical Review,* Energy Economics Research Ltd. (Berkshire, United Kingdom: Guild Gate House, 1980).
33. E. A. Cherniavsky, *Long-Range Oil and Gas Forecasting Methodologies: Literature Survey* (Upton, N.Y.: Brookhaven National Laboratory, 1980).
34. Ibid., p. 14.
35. Ibid., p. 11.
36. Ibid., p. 6.
37. Ibid.
38. Energy Modelling Forum, *World Oil: EMF6 Summary Report* (Draft 4), Stanford, Calif., sponsored by the Electric Power Research Institute, U.S. Department of Energy, and the Gas Research Institute. Particular model structures are discussed in more detail in Perry Beider, *Comparison of EMF 6 Models,* Working Paper EMF6.10. Energy Modelling Forum, Stanford, Calif.
39. Beider, *Comparison of EMF 6 Models,* p. 5.
40. Ibid.

Unconventional Oil 8

Unconventional oils are, for the most part, not unconventional at all but are established technologies that were replaced by cheap crude oil from deep wells. Various forms of unconventional oil represent a wide range of technologies for directly producing crude oil, crude oil-like substances, and liquid secondary fuels. These technologies include enhanced recovery of oil in conventional deposits, shale oil, tar sands and heavy oils, liquids from coal, and liquids from biomass. The technologies are new only in the sense that they are being or will be applied at an unprecedented scale and that variations on the basic technology are being tried in an effort to reduce cost.

Oil shales are shale deposits that contain a high concentration of organic materials (kerogens) which when heated (retorted) can be converted to oil and gas. Tar or bituminous sands are deposits of sand and sandstone containing hydrocarbons that are high in specific gravity and in sulfur and metal. Bitumens cannot flow at normal ambient temperatures whereas heavy oils can. Oil can be recovered from both substances. Coal liquids are the product of liquefaction processes which convert coal to liquid fuels. Similarly, liquids can be produced from biomass. Finally, recovery methods which are capable of recovering oil from conventional type deposits, but at costs that are greater than $30 to $40 constant 1979 dollars, are also included as unconventional. Two types of deposits are included in this last category: oil in place in conventional oil deposits but requiring costly technologies to extract, and oil in conventional-type deposits except that the deposits are in areas costly to reach—primarily deep offshore areas and arctic and antarctic regions.

Among this group of technologies, oil shale and coal represent potentially unlimited supplies of feedstock for refined liquids up to the year 2050. Tar sands and heavy oils represent a possible source of oil at least as large as conventional oil resources. Costly recovery of oil from conventional oil deposits would represent a resource that is likely to be, at most, equal to conventional oil resources. Biomass and mineral-based resources are not easily compared because biomass is a renewable source of liquids. Resource considerations for biomass are examined in chapter 15.

Technologies for utilizing these resources have been proved experimentally and in the field. All unconventional oil resources have in the past been exploited on a commercial level. Methanol from wood has been produced for thousands of years; alcohol was mixed with gasoline to provide a transport fuel in the midwestern United States in the 1930s[1] and was developed further in Germany during World War II.[2] Coal conversion was developed in Germany during World War II and is presently operating at a commercial level in South Africa. The use of energy in oil shale predates liquid oil recovery.

Widely available and cheap conventional crude oil largely displaced "unconventional" sources of energy liquids. The major reason for this was cost, and the obstacle to development of alternative sources of liquids in the near future continues to be cost. Generally, shale, tar sands, biomass, and coal conversion technologies are very capital intensive. For significant cost reductions, capital requirements must be reduced. *In situ* technologies for extracting shale and tar sands provide the most promising prospect for cost reduction but are only now being tested in demonstration-size plants. The history of cost estimates for shale and coal liquids has been notorious for escalating so as always to be somewhat more costly than conventional oil. Based on this history, the usual learning curve analysis—implying cost reductions as the technology is adopted and matures—appears inoperative.

THE RESOURCE BASE

Shale Oil

The global shale oil resource base is huge. The most well-known and most widely quoted estimates of its size are Duncan and Swansen's 1965 estimates, given in table 8-1.[3] Duncan and Swansen referred to the estimates as "orders of magnitude." The higher grade resource base estimates of 4,000,000 $\times$ 10^{15} Btus are, for perspective, well over 10,000 times 1979 global energy production. The regions in table 8-1 do not completely conform with those used throughout the study. Neither refinement in regional breakdowns nor in the estimates themselves can be considered an important goal given the ubiquity of large amounts of organic shale. There is little reason to dispute Duncan and Swansen's estimates in terms of the resource base. Research interest in global resources of shale oil has since centered on the exploitability of the energy locked in organic shale. Discussion of these aspects of shale oil is left for the sections on resources and extraction technologies.

TABLE 8-1. Global Oil Shale Resource Base—Orders of Magnitude Estimates (10^{15} Btus)

	Energy Content of Organic Shale[a]	
Region	*10–65 Percent Organic Matter*	*5–10 Percent Organic Matter*
North America	570,000	2,900,000
United States	310,000	1,600,000
Europe	310,000	1,600,000
Australia and	230,000	1,200,000
New Zealand	230,000	1,200,000
Asia	1,300,000	6,500,000
Africa	960,000	4,900,000
South America	470,000	2,300,000
Total	4,000,000±	20,000,000±

Source: Adapted from O. C. Duncan and V. E. Swansen, "Organic Rich Shale of United States and World Land Areas" (U.S. Geological Survey Circular 523, 1965).

Note: The regions do not conform to those used throughout the study.

[a]Estimates rounded.

Tar Sands and Heavy Oil

There are no estimates of a tar sand resource base comparable to those for oil shale. Estimates of tar sand resources collected by the World Energy Conference are incomplete in geographical coverage—there are few countries reported, and when they are reported, estimates are more appropriate for resources or reserves rather than the resource base.[4] Such estimates indicate that the resource base for global tar sands and heavy oil is at least as large as that for conventional oil. These estimates are more fully discussed under recoverable unconventional oil resources.

Enhanced Recovery, Deep Offshore, and Polar Regions

The resource base for enhanced recovery consists of all oil in place that is unrecoverable with conventional methods. Primary recovery techniques (relying on natural field pressure alone) normally result in recovery of between 5 and 20 percent of oil in place.[5] Secondary recovery, using water or gas flooding, enables recovery of 20 to 50 percent of oil in place.[6] The world average recovery rate is largely speculative, but is thought to be on the order of 25 to 30 percent,[7] with the more frequently cited figure being 25 percent. In the Soviet Union, 85 percent of the oil presently produced is from fields using water flooding methods; the present recovery coefficient is estimated at 45 percent.[8]

The focus of oil resource estimates has typically been on recoverable resources, even though, as discussed in chapter 7, estimation procedures may require estimation of in-place oil. The Delphi poll carried out by the committee on oil resources for the World Energy Conference is one of the few published sources of estimates of oil in place. Their sampling of experts indicated that opinions on the amount of oil in place ranged from roughly 3,600 bbos to 15,000 bbos. Of the twenty-six respondents, fifteen believed oil in place was less than 5,000 bbos, three believed it to be between 5,000 bbos and 7,300, and eight felt the total was more than 7,300 bbos.[9]

Among the ten Delphi study respondents who gave estimates for oil in place and recoverable resources, the estimated share of in-place oil expected to be ultimately recovered ranged from 25 to 50 percent. A total of seventeen observations were available from the ten respondents as a result of seven giving high and low estimates. Three of the estimates implied an overall recovery factor of less than 30 percent, nine implied a recovery factor of 30 to 40 percent, and five implied a recovery factor of 40 percent or more.[10] The simple mean value was 36 percent. This information suggests that most researchers who estimate recoverable oil resources assume some improvement from the presently existing average recovery factor. Such an improvement need not necessarily mean further increases in oil prices or improvements in extraction technology. As already noted, the Soviet Union, by its own estimates, recovers 45 percent of oil in place. In the United States, ultimate recovery of oil in place using only primary and secondary recovery methods is estimated as likely to be about one-third.[11]

Table 8-2 gives low, middle, and high estimates of the resource base of different sources of expensive, conventional type oil. The middle estimate is derived from the estimated conventional oil resources accepted as a base case in chapter 7. The estimates reflect the fact that only about a third of the oil in place is recoverable using

TABLE 8-2. Resource Base for Enhanced Oil Recovery and Other Expensive Conventional-Type Oil (bbos)

	Total Oil in Place[a]	*Conventional Oil Recovery*[b]	*Recoverable Oil—Deep Offshore*[c]	*Recoverable Oil—Polar Regions*[c]	*Column 1 Minus Columns 2, 3, and 4*
Low	3,600	900	0	0	2,700
Middle	6,400	2,309	15	5	4,071
High	15,000	6,950	180	50	7,820

[a]Low and high are from WEC, 1978. *Oil and Gas Resources,* Delphi study results, p. 122. Middle estimate assumes estimate of conventional oil of 2,309 bbos from chapter 7 represents 0.36 percent recovery of oil in place (Delphi study average—see text).

[b]Low and high are estimates made by the same researchers as high and low estimates of oil in place. Middle estimate is from chapter 7.

[c]Median estimates for the eleven respondents giving separate estimates for off-shore and polar regions in WEC, 1978. *Oil and Gas Resources,* p. 129. The median estimates for twenty respondents including those giving offshore and polar region estimates are 23 and 24.9 bbos.

conventional primary and secondary methods. Thus, the resource base for enhanced recovery is considerably larger than the estimated conventional oil resource. It must be noted, however, that geological constraints and physical characteristics of oil fields prevent the average oil recovery factor from reasonably approaching anything near 100 percent. A 70 to 80 percent recovery factor, suggested by two Soviet scientists as possible in the Soviet Union, is among the highest estimated recovery factors cited as approachable on a broad scale.[12]

Liquids from Coal and Biomass

The resource base for coal liquids is, potentially, all coal reserves, though quite clearly direct use of coal as a solid fuel and conversion to gas will vie for available coal. Regional estimates of the coal resource base are given in chapter 11.

Estimates of potentially available biomass are given in chapter 15. As in the case of liquids from coal, the production of biomass liquids must compete for available feedstocks with biogas and direct use of biomass as a solid fuel.

RECOVERABLE AMOUNTS OF UNCONVENTIONAL OIL

Shale Oil

Several characteristics of shale oil deposits are important in determining recoverability. Marland[13] has recently reviewed these factors, which include yield per ton, shale thickness, lateral extent, depth, and water content. Duncan and Swansen in their 1965 estimates[14] relied solely on yield per ton to establish recoverability of oil from organic shale; any shale yielding 10 gallons per short ton was considered an oil shale.[15] Various net energy analyses of oil shale processing led Marland to conclude "10 gallons per [short] ton to be the practical cutoff for oil shale but to preserve the category of 5–10 gallons per ton as a hedge for realization of the most optimistic of possibilities."[16] Another important characteristic is shale thickness—for inclusion as a recoverable resource Donnell[17] required the shale deposit to be at least 10 feet thick. The large share of the shale oil resource base is undiscovered. Among the identified deposits, few have been carefully appraised and, as a result, little is known about their recoverability characteristics.[18]

Table 8-3 classifies shale oil resources for six world regions along the lines of the resource diagram in chapter 6. The resource is classified along dimensions of geological certainty, from known to speculative and along economic dimensions, given that the most important economic dimension is oil yield per ton. The resource estimates are based on an assumed 50 percent recoverability factor.

Marland notes that the estimates quoted are "only the crudest first approximations. Better estimates for the United States, and certainly for areas outside the United States, will follow only from more thorough field appraisals of the known and suspected deposits—appraisals that will probably not be made without stronger economic incentive to do so. It is immediately evident that the state of knowledge advanced very little between 1965 and 1978."[19]

Even the most restricted shale oil grade definition listed in table 8-3—resources recoverable under present conditions (known and economically and technologically recoverable)—represents an oil resource of more than 60 times the world's 1979 oil production.[20] Oil contained in the highest yielding shales (greater than 25 gallons per ton) represents a resource of 16,720 bbo, or over seven times the estimated total global conventional oil resource.

The World Energy Conference included oil shale in their 1980 survey of world energy resources. However, they note that the figures obtained from the questionnaire are very incomplete.[21] Only thirteen countries responded, indicating a total of approxi-

TABLE 8-3. Shale Oil Resources of the United States and World Land Areas, by Resource Grade (billions of barrels by grade, grade in gallons of oil yielded per ton of shale)

		Marginal and Submarginal (oil equivalent in deposits)								
		Known Resources			*Order of Magnitude of Possible Extensions of Known Sources*			*Order of Magnitude of Undiscovered and Unappraised Resources*		
	Recoverable under Present Conditions	*Grade*								
Area		*25–100*	*10–25*	*5–10*	*25–100*	*10–25*	*5–10*	*25–100*	*10–25*	*5–10*
Africa	10	90	Small	Small	N.E.	N.E.	N.E.	4,000	80,000	450,000
Asia	20	70	14	N.E.	2	3,700	N.E.	5,400	106,000	586,000
Australia and New Zealand	Small	Small	1	N.E.	N.E.	N.E.	N.E.	1,000	20,000	100,000
Europe	30	40	6	N.E.	100	200	N.E.	1,200	26,000	150,000
North America	80	520	1,600	2,200	900	2,500	4,000	1,500	45,000	254,000
South America	50	Small	750	N.E.	N.E.	3,200	4,000	2,000	36,000	206,000
Total, world	190+	720	2,400	2,200	1,000	9,600	8,000	15,000	313,000	1,740,000

Source: Gregg Marland, *Shale Oil: U.S. and World Resources and Prospects for Near Term Commercialization in the United States,* Institute for Energy Analysis, Oak Ridge, Tenn., Technical Report (ORAU/IEA-79(R), 1979).

N.E. = No estimate.

Estimates and totals rounded.

mately 290 bbos of proved recoverable oil (over 70 percent of it in the United States) and additional resources of 1,860 bbos (almost 95 percent of it in the United States).[22] Such an estimate cannot be considered a "resource" estimate as defined in chapter 6, but rather is a category more appropriately labeled inferred and speculative reserves. Perhaps of more interest in the survey than the total amounts reported are the countries that reported, indicating that they have shown enough interest in shale oil to begin assessing their resources. The thirteen countries responding were the United States, Australia, New Zealand, Thailand, the Soviet Union, Argentina, Brazil, Jordan, Morocco, Zaire, West Germany, Sweden, and Spain.

Tar Sands and Heavy Oils

Deposits of tar sands and heavy oils represent a potential source of petroleum that is likely to be as large or larger than conventional oil resources. Available estimates of global resources are limited to summing across estimates of oil contained in a relatively few known fields. Such estimates must be viewed with considerable care when comparing them with estimates of conventional oil and oil shale where global resource estimates include undiscovered resources.

Table 8-4 is based on World Energy Conference estimates of tar sand resources. The 1978 figures reported by WEC are for oil in place in known tar sand deposits. In order to compare 1978 and 1980 figures, a range of estimates of recoverable oil were made for the 1978 oil-in-place estimates using different recoverability assumptions. The lower estimate is based on 5 percent recovery of oil in place, assuming only surface mining technologies are used. The high estimate is based on 50 percent recovery (an optimistic estimate) and is applicable only if *in situ* technologies are used in conjunction with any open-pit mining operation.

The 1980 WEC estimates are for recoverable oil. Estimates of recoverable oil for Venezuela and the USSR that are higher than those computed from the 1978 WEC study, plus the inclusion of recoverable oil from Jordanian tar sands, boost the world total above the high estimate computed from the 1978 data.

The 1980 WEC estimate indicates an amount of oil of somewhat more than half the estimated total conventional oil resources. However, several factors suggest that the estimates in table 8-4 are a very incomplete global inventory. First, only six countries responded to the 1980 WEC survey queries concerning tar sands and several of the nonrespondents were identified as having tar sand deposits, though relatively small ones, in the 1978 conference survey. Second, the 1978 estimates were based on only twenty-one fields. Third, in conjunction with the second point, these oils are known to "exist in abundance on the earth with wide geographic distribution."[23] And fourth, the economics of recovery of oil from tar sands has not justified intensive exploration for tar sand deposits.

These considerations suggest that if economic considerations justified extensive exploration and development, oil from tar sands could represent a source of oil as large or larger than conventional oil resources.

Recoverability of Conventional Oil Using Extraordinary Techniques

Primary and secondary recovery methods—using only natural field pressure and water flooding—are likely to recover approximately one-fourth of the oil in place. Thus three-fourths of conventional oil is left as a target for enhanced oil recovery (EOR) techniques.[24] Among the (EOR) techniques presently used or being experimented with are thermal methods (steam injection or *in situ* combustion of part of the oil), injection of hydrocarbons or gases miscible with petroleum (mainly CO_2), and injection of water improved by chemical

TABLE 8-4. Identified Deposits of Tar Sands (billions of barrels of oil)

	WEC (1978)[a]		
Region/Country	*Oil in Place*	*Recoverable*[b]	*WEC (1980)*[c]
1. United States	27 (9 fields)	1–14	<1
2. OECD Europe			
Canada	920 (4 fields)	46–460	264
Federal Republic of Germany	N.A.	N.A.	<1
3. OECD Asia	N.A.	N.A.	
4. Communist Europe			
Albania	<1 (1 field)	<1	N.A.
Rumania	< (1 field)	<1	N.A.
Soviet Union	600 (2 fields)	30–300	365
5. Communist Asia	N.A.	N.A.	N.A.
6. Mideast			
Jordan	N.A.	N.A.	79
7. Africa			
Madagascar	2 (1 field)	<1	N.A.
8. Latin America			
Trinidad	<1 (1 field)	<1	N.A.
Venezuela	700 (2 fields)	35–350	519
9. South and East Asia	N.A.	N.A.	N.A.
Total	2,250	112–1,125	1,228

Note: Columns may not add to totals due to rounding; totals computed from unrounded data.

N.A. = Not available.

[a]From Executive Summaries of Reports on Resources, Conservation and Demand to the Conservation Commission of the World Energy Conference, *World Energy Resources* 1985–2020 (New York: IPC Science and Technology Press, 1978), p. 46.

[b]Reported in billions of barrels. The source did not specify whether barrels were standardized for energy content. WEC, 1978, states that 5 to 10 percent of in-place oil in tar sand deposits is recoverable with open-pit mining while *in situ* technologies could recover 30 to 50 percent. The high and low estimates are based on the extreme values of recoverable fractions under the open pit and *in situ* technologies.

[c]From World Energy Conference, *Survey of Energy Resources 1980,* by the Federal Institute for Geosciences and Natural Resources, Hanover, Fed. Rep. of Germany (London: Alan Armstrong & Associates, 1980), part B, tables 2.4 and 2.6. Figures reported in barrels of oil equivalent at 6.8 $\times 10^6$ Btus per barrel, converted from joules, as reported in the source.

additives (surfactants, polymers, alkaline solutions).[25]

This definition of enhanced recovery includes some oil which is considered conventional within the context of this study. The WEC Delphi study results, previously discussed, suggested an overall eventual recovery rate of roughly 36 percent for conventional oil. Such an estimate implicitly includes some oil recovery beyond that possible using natural pressure and water flooding techniques. The U.S. Department of Energy (DOE)[26] considers thermal recovery (mainly steam flooding) an already established and economically attractive technology in the United States. The DOE estimates that oil recovered using thermal methods will be a major component of enhanced oil recovery prior to 2000; beyond 2000 more expensive, advanced technologies will be applied.[27] Our definition of conventional oil would include most oil recovered using thermal techniques. It is impossible to establish a direct correspondence between technology types and the conventional/unconventional categories as defined; an extreme example is recovery of oil from polar regions. Partial recovery of this oil is possible by relying on natural field pressure (a conventional recovery method) but because of the harsh climates, the cost of recovery leads to classifying the oil as unconventional oil.[28]

Estimates of the resource base for extraordinary recovery (using extraordinary to distinguish oil recovered using enhanced recovery methods that are economic at present prices) developed in table 8-2, ranged from 2,700 to 7,820 bbos. Various assumptions of the amounts of oil in place ultimately recoverable through the full application of advanced technology have been used.

The WEC suggested that enhanced recovery could increase the global average recovery factor to 45–50 percent.[29] The National Petroleum Council (NPC), in a study for the U.S. Department of Interior, estimated that enhanced recovery could increase U.S. oil reserves by as much as 24 billion barrels or 85 percent by the year 2000 in the lower 48 states.[30] If one assumes that primary and secondary recovery methods achieve recovery of 25 percent of oil in place, the additional recovery from enhanced technologies would raise the recovery factor to 46 percent. Another study of enhanced recovery of oil in the United States was more optimistic, suggesting an increase of 35 billion barrels in reserves, yielding a possible recovery factor of 56 percent.[31] The NPC study was broader in terms of number of reservoirs assessed and applied stricter criteria for accepting reservoirs as potentially suitable for enhanced recovery. The IIASA study considered an ultimate recovery factor of 40 percent for the world.[32] The Soviet Union currently estimates their recovery factor at 45 percent and has plans to increase it to 58–63 percent by 2000. Eventually it could reach 70 to 80 percent.[33]

Several factors must be assessed concerning a possible world average recovery rate. The field size is likely to affect the economics of enhanced recovery. The United States has a large number of oil fields in the small giant size category (500 to 2,500 million barrels). In the Soviet Union and Middle East, a large amount of the known oil is in very large giant (2,500–10,000 million barrels) and super giant fields (10,000 million barrels or more).[34] Other characteristics, such as the type of rocks containing the oil, viscosity, and the depth of the field are all important in determining the economic viability of recovery through the advanced technologies. In addition, the location of the field may affect recoverability; in harsher climates it may be uneconomic to operate oil wells at the very low production rates normally achievable using advanced technology. Location may also be important in determining the economic availability of gases in chemicals needed for enhanced recovery. Presently, the possibility of using CO_2 produced as a byproduct of coal gasification for injection into oil wells is being explored by a consortium of companies planning a coal gasification plant in North Dakota.[35] Carbon dioxide could also be recovered from oil shale deposits in carbonate rock located in the Four Corners region of the western United States.[36] Carbon dioxide from either of these regions could be piped to oil fields in the southwestern United States, for example. In such cases, the cost of pipeline construction would be the determining factor. Regions without ready access to CO_2 may find that the cost of enhanced oil recovery based on carbon dioxide injection is unfavorable.

Finally, most experience with enhanced recovery techniques has been in applications to producing fields as the natural pressure has dropped off. Less experience has been gained in reopening fields that have produced in the past but were shut down before enhanced oil recovery technologies were known or economical. The cost of simply reopening the field will add to the cost of recovery in such cases. The somewhat greater uncertainty and cost involved in reopening old fields will result in greater selectivity in applying enhanced re-

TABLE 8-5. Oil Potentially Recoverable Using Enhanced Recovery Technologies (bbos)

	Unrecoverable with Conventional Technologies[a]	*Enhanced Recovery Potential*			*Offshore and Polar*[a]	*Conventional Oil Estimates*[a]
		40[b]	*50*[b]	*70*[b]		
Low	2,700	540	900	1,620	0	900
Middle	4,071	251	891	2,171	20	2,309
High	7,820	—[c]	550	3,550	230	6,950

[a]From table 8-2.

[b]Assumed rates of ultimately recoverable oil in place, in percent.

[c]Conventional oil estimates assumes recovery of 46 percent of oil in place.

covery technologies, thereby tending to reduce the overall recovery rate achievable under a given oil price regime.

Table 8-5 contains straightforward calculations of the potential oil available from enhanced recovery under different estimates of oil in place and ultimately achievable rates of recovery. Under the low resource base estimate, a considerable amount of oil could be recoverable using enhanced recovery methods because the base recovery rate for conventional oil is only 25 percent (no improvement from the present rate). The high resource base, in contrast, is associated with a conventional recovery factor of 46 percent.

Ultimate recovery rates in the range of 40 to 50 percent appear likely given existing technologies. These would be at costs that would justify recovery attempts under likely future price scenarios given estimates of the costs of backstop liquids (oil shale or coal liquids). To achieve recovery rates much beyond 50 percent of in-place oil, further technological advances, lowering the cost of enhanced recovery, and relatively expensive substitutes would be necessary. A global average recovery of 70 percent must be considered very optimistic and achievable only in a limiting case involving full use of available technologies in all oil fields, regardless of cost.

In the middle case, some improvement in average recovery factors (from 25 to 36 percent) is included in conventional oil as a result of broader application of existing technologies that are cost effective at present prices. The most likely contribution of enhanced oil recovery to unconventional oil supplies appears to be in the range of 250 to 900 bbos. At the extreme, enhanced recovery could nearly double the amount of oil available from conventional deposits.

TECHNOLOGIES AND COSTS

Cost estimates of conventional oil substitutes vary greatly. The estimates of synfuels, particularly shale oil, always seem to remain just above conventional oil prices. Stobaugh and Yergin reported in 1979 that "three decades ago, chemical engineering students were taught that if oil prices went up 20 percent, oil from shale would become profitable. Since then, the price of oil has increased by a good deal more than 20 percent in real terms, but shale oil apparently still remains unprofitable. As the price of oil has risen, so has the estimated cost of producing oil from shale. Many engineers and promoters say that the price of oil must go up just another 20 percent for oil from shale to be economic."[37]

Costs and technology descriptions for each fuel will be examined separately. At this point it is useful to report a few estimates of costs of unconventional liquids. Table 8-6 reports the estimates of several different studies or reports which attempted to compare the cost of alternative unconventional liquids.

TABLE 8-6. Comparative Costs of Unconventional Oil (1979 dollars per barrel of oil equivalent)

Source/Date	*Shale Oil*	*Oil Sands*	*Coal Liquids*	*Biomass*
Cameron Engineers/1979[a]	14.5–25	25–32	29–37.7	N.E.
Pace/1979[b]	26	31	36–37	N.E.
ICF Incorporated/1979[c]	32	N.E.	38	38
Bechtel/1981[d]	39	52	N.E.	N.E.
Mellon Institute/1980[e]	17–25	N.E.	52–81	81–102
Project Independence/1974[f]	10	16	16	N.E.
CONAES/1979[g]	23–30	N.E.	33–39	37

N.E. = No estimate.

[a]From Cameron Engineers, 1979, "Overview of Synthetic Fuels Potential to 1990" in *Synthetic Fuels,* U.S. Senate, Committee on the Budget, Subcommittee on Synthetic Fuels (Washington, D.C.: U.S. Government Printing Office, September 27, 1979), p. 177. Prices per barrel are converted from prices per barrel at 6×10^6 Btu per barrel to the equivalent price per barrel at 5.8×10^6 Btu per barrel. Cost ranges encompass different estimates of investment costs per barrel for a single technique, but also differential costs for different techniques. Reported as the selling price.

[b]Senate testimony of S. Frank Culbertson, President, Rocky Mountain Division, Pace Co., Consultants and Engineers, Inc., in U.S. Congress, Senate, Committee on the Budget, *Costs and Economic Consequences of Synthetic Fuels Proposals,* 96th Cong., 1st sess. (Washington, D.C.: U.S. Government Printing Office, 1979), p. 13.

[c]Report in ICF Incorporated, 1979, "Oil Import Reduction: An Analysis of Production and Conservation Alternatives," in *Synthetic Fuels,* U.S. Senate, Committee on the Budget, Subcommittee on Synthetic Fuels (Washington, D.C.: U.S. Government Printing office), p. 278. Reported as expected costs under the administration's proposals for synfuels.

[d]From H. F. Brush, Bechtel Corp., "Readiness: Where Do We Really Stand?" paper presented at the Government Research Corporation Conference, "Synthetic Fuels: Worldwide Outlook for the 80s," San Francisco, Calif., February 19–20, 1981, p. 24.

[e]From Richard H. Shackson and H. James Leach, *Using Fuel Economy and Synthetic Fuels to Compete With OPEC Oil* (Arlington, Va.: Mellon Institute, Energy Productivity Center, 1980). Shale costs are based on Bechtel estimates. Costs are for secondary liquids (transportation fuels). The coal liquids technology is direct conversion to methanol. Biomass technology is ethanol from corn.

[f]From Federal Energy Administration, *Project Independence Report* (Washington, D.C.: U.S. Government Printing Office, 1975), p. 49. Converted from 1974 dollars. Values reported are oil prices estimated as necessary to elicit supplies from the various synthetic fuels sources.

[g]From National Academy of Sciences, *Energy in Transition 1985–2010,* the Final Report of the Committee on Nuclear and Alternative Energy Systems (San Francisco: W. H. Freeman, 1979), pp. 138–139, 374. Biomass cost is for methane (gas) from municipal waste. It is implicitly assumed in reporting this figure under liquids that the cost of producing methanol (liquids) from methane is very small. All cost estimates were converted from 1978 dollars.

While there is considerable difference in the estimated costs for individual fuels, the ranking in terms of costs is consistent over the various studies. Oil from shale is estimated to be the cheapest of the four alternatives. Oil from tar sands tends to be the next cheapest, though the cost range overlaps with coal liquids. Biomass tends to be most expensive although it overlaps with the upper range estimates of coal liquids. The estimated costs of coal liquids and biomass are especially sensitive to the assumptions concerning the cost of coal or biomass input. Generally, coal liquefaction or methanol production from coal assumes the plant is located near the coal source, thereby minimizing transportation costs. Little detail is given with the ICF figures for biomass, but in the case of the Mellon Institute figures, it is important to note that the biomass liquid assessed is ethanol from corn. Corn grain is among the most expensive biomass feedstocks that can be used to produce ethanol; methanol production from biomass is likely to prove cheaper than ethanol production since it utilizes cheaper biomass feedstocks (wastes and wood). The high cost reported by the Mellon Institute must be considered unrepresentative of biomass liquids in general. In contrast, the CONAES study reported methane (gas) costs from municipal wastes, thus the feedstock cost is zero. Further processing to obtain a liquid fuel

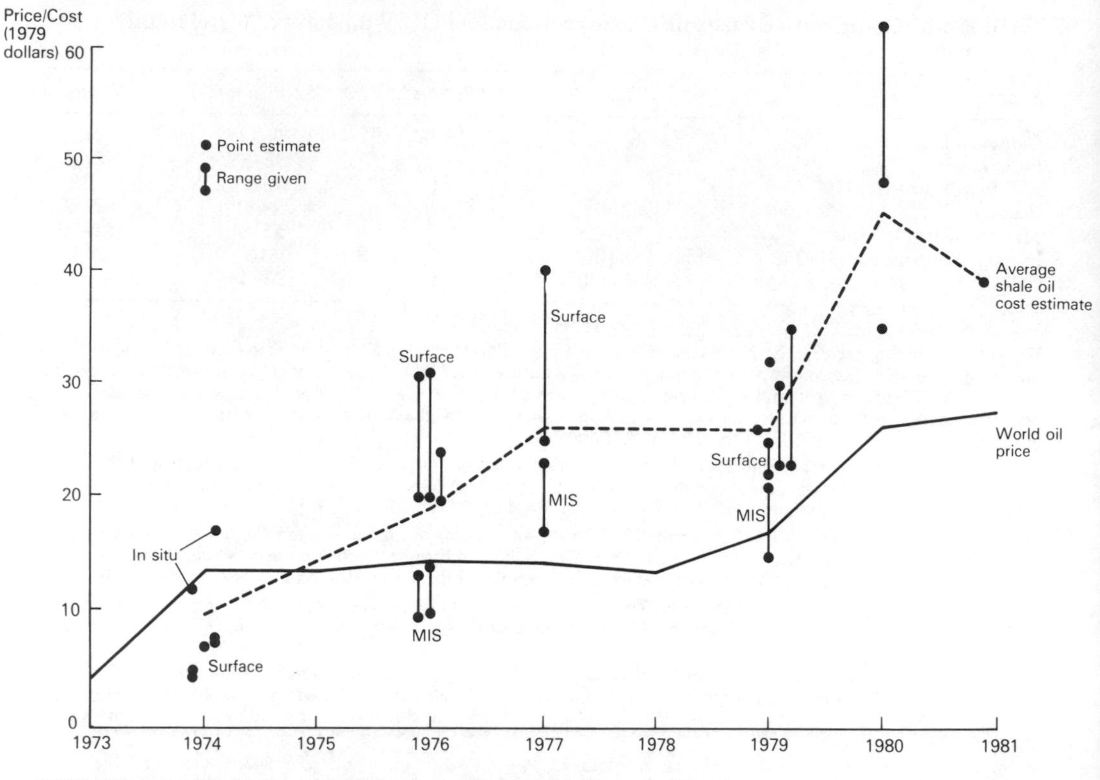

FIGURE 8-1. Oil shale cost estimate and the world price of oil.

Notes: 1974: 4.38–4.88 dollars for surface retort, 12 dollars for *in situ* (cost estimate as opposed to selling price); 7.30–7.80 dollars, surface techniques; 17 dollars, *in situ* techniques (estimated selling price). Interagency Oil Shale Task Force, 1974, cited in University of Oklahoma Science and Public Policy Program, *Energy Alternatives: A Comparative Analysis* (Washington, D.C.: U.S. Government Printing Office, 1975), pp. 2–42. 1974: 7 dollars per barrel, Federal Energy Administration, *Project Independence* (Washington, D.C.: U.S. Government Printing Office, 1974), p. 49. Estimated breakthrough price of shale. 1976: 10–14 dollars for modified *in situ,* 20–30 dollars for surface retort. Philip White in U.S. Congress, Senate, Committee on Interior and Insular Affairs, "Future Prospects for Oil Shale Development—Oversight: Prototype Oil Shale Leasing Hearings Before the Subcommittee on Minerals, Materials and Fuels," 94th Cong., 2d sess. (Washington, D.C.: Government Printing Office, 1976). Converted to 1979 dollars. 1976: 10–13 dollars, modified *in situ,* 20–30 dollars, surface retort. Energy Research and Development Administration (ERDA), cited in the Full Reports to the Conservation Commission of the World Energy Conference, *World Energy Resources 1985–2020, Oil and Gas Resources,* by Pierre Despraires, W. T. McCormick, Jr., L. W. Fish, R. B. Kalisch, and T. J. Wander (New York: IPC Science and Technology Press, 1978), p. 68. 1976: 20–24 dollars, unspecified extraction technique. Energy Research and Development Administration, *A Preliminary Social and Environmental Assessment of the ERDA Solar Energy Program 1975–2020,* Volume I—Main Report by SRI International, formerly Stanford Research Institute, Menlo Park, California (Washington, D.C.: Energy Research and Development Administration, July 1976), p. iv–12. 1977: 17–23 dollars, *in situ,* 25–40 dollars, surface techniques, U.S. Department of Energy, reported in Edward L. Allen, "Energy Supply and Demand Scenarios in 2000 and the Estimated Impact of Newer Technologies," Contract Report to Office of Technology Assessment in support of OTA's *Direct Use of Coal* assessment. Converted to 1979 dollars. 1979: 15–21 dollars, modified *in situ;* 22–25 dollars surface extraction. From Cameron Engineers, "Overview of Synthetic Fuels Potential to 1990," in U.S. Congress, Senate, Committee on the Budget, *Synthetic Fuels,* Report by the Subcommittee on Synthetic Fuels, 96th Cong., 1st sess. (Washington, D.C.: U.S. Government Printing Office, 1979), p. 177. Converted from dollars per barrel of oil equivalent at 6×10^6 Btus per barrel to dollars per barrel at 5.8×10^6 Btus per barrel. 1979: 32 dollars, unspecified extraction technique. From ICF Incorporated "Oil Import Reduction: An Analysis of Production and Conservation Alternatives" in *Synthetic Fuels,* ibid., p. 278. 1979: 23–30 dollars, unspecified extraction technique. From National Academy of Sciences, *Energy in Transition,* the Final Report of the Committee on Nuclear and Alternative Energy Systems (San Francisco: W. H. Freeman, 1979), p. 138. Costs are converted from 1978 dollars. 1979: 26 dollars, unspecified technique, Pace Consulting, appearing in U.S. Congress, Senate, Committee on the (methanol) would add to the cost of production. For these reasons, this estimate must be considered relatively low.

Oil from Shale

Technologies

The extraction of oil from oil shales is not a new technology. Indeed, the process has been in use for hundreds of years. England has a patent to produce "oyle from a kind of stone" which dates back to 1694.[38] Oil was being distilled from shale in the United States prior to its discovery in deep wells in 1859.[39] The promise of shale as a vast source of oil was recognized by the U.S. Geological Survey as far back as 1918.[40] The two general classes of extraction technologies are *in situ* and surface processes. Surface techniques involve mining the shale and then retorting it above ground, whereas *in situ* techniques involve retorting the shale underground. The oil recovered from these processes must also be prerefined to produce a substance similar to crude oil. In surface retorting, the shale is mined either above ground or underground and then heated to extract oil. Since the oil accounts for no more than 5–10 percent of the shale weight,[41] large quantities of spent shale must be disposed of. This process also uses large amounts of water to capture the dust created by crushing the shale. Problems of waste disposal and water usage can be considerably reduced by using *in situ* retorting, which shatters the deposit by explosives and then retorts the shale underground. Research on true *in situ* (TIS) processes has been slow, as companies have focused on modified *in situ* (MIS). MIS is in an advanced stage of development; it involves a combination of the TIS and surface processes. A portion of the shale is deep mined and is (or can be) retorted above ground. The removal of a portion of the shale through mining creates an underground retort, allowing oil shale remaining in the ground to be retorted *in situ*. This process recovers less oil per ton of shale than surface retorting, but more than TIS.[42]

Figure 8-1 illustrates the recent trends in oil shale cost estimates. Several studies have attempted to explain the parallel movement of shale oil cost estimates and the world price of oil.[43] The basic microanalytic reason is that oil or energy is a direct input into shale oil production and is also embodied in the equipment and material required for the shale processing plant. However, even an extreme estimate of the possible increase in the cost of shale oil associated with an increase in the cost of oil cannot explain the parallel historical movement of shale oil costs and oil prices.[44]

Another reason cited for this phenomenon is that increases in oil prices lead to general inflation, carrying along with them the cost of shale oil.[45] Such an argument is

---

Budget, *Costs and Economic Consequences of Synthetic Fuels Proposals,* 96th Cong., 1st sess. (Washington, D.C.: U.S. Government Printing Office), p. 13. 1979: 23–35 dollars, range over all techniques. Gas Research Institute, reproduced in U.S. Congress, House, Committee on Interstate and Foreign Commerce, *The Energy Factbook,* by the Congressional Research Service, Library of Congress, 96th Cong., 2d sess. (Washington, D.C.: U.S. Government Printing Office, November 1980), p. 739. Converted from dollars per 10^6 Btus to dollars per barrel at 5.8×10^6 Btus per barrel. 1980: 35 dollars, unspecified technique. From World Energy Conference, *Survey of Energy Resources 1980,* by the Federal Institute for Geosciences and Natural Resources, Hanover, Fed. Rep. of Germany (London: World Energy Conference, 1980), p. 162. Converted to 1979 dollars. 1980: 48–62, range depending on rate of return/technique used. From U.S. Congress, Office of Technology Assessment, *An Assessment of Oil Shale Technologies,* Report No. OTA-M-18 (Washington, D.C.: U.S. Government Printing Office, 1980), p. 16. Converted to 1979 dollars. World Oil Price: World price of oil is represented as the average Saudi contract price for the year and converted to constant 1979 dollars using the U.S. GNP deflator. Oil prices are from International Monetary Fund (IMF), *International Financial Statistics Yearbook* and July 1981 monthly edition, IMF, Washington, D.C. U.S. GNP deflators are from U.S. Department of Commerce, *Survey of Current Business,* December 1980 and June 1981 (Washington, D.C.: U.S. Government Printing Office).

flawed, however, since general inflation will also have a tendency to decrease the real price of oil.

An apparently legitimate reason for shale oil cost increases during the 1970s was differential inflation in the plant and industrial materials sectors. To the extent that these sectors are more energy intensive than the economy in general, the differential inflation would have been caused by rising oil prices. However, there is no evidence to support this argument. Rather, the increases appear to be due to other causes which were coincident with oil price increases.[46]

The two approaches for explaining the increase in shale oil cost estimates already discussed implicitly assume that past estimates have been correct. A third explanation is that early estimates were overly optimistic. These were calculated with little actual data on costs of running a full-scale plant. Rather, they were based on extrapolation from similar processes, crude scaling up of data from small experimental plants, and optimistic assumptions about cost reductions due to learning as the technology matured. The engineering plant costing procedure involves four stages: *initial estimates* which are back-of-the-envelope calculations; *preliminary design estimates* in which the plant's subsystem flows, but not the component subprocesses, are costed out; *detailed design estimates* of costs for specific materials and components, and *final estimates* of precise costs for a specifically designed and located plant, including labor, materials, and all components.[47] One study found that improvements in estimates, from initial estimates to detailed design, accounted for 40 to 50 percent of the cost estimate increase between 1971 and 1977.[48]

Finally, it appears that those making estimates recently may have had a tendency to be pessimistic about costs or include add-ons above those explicitly accounted for. Such pessimism is often reflected as an assumption of a relatively high real rate of return (15 or 20 percent) above taxes, which would be far out of line with returns of other industries. For example, utilities are guaranteed a normal rate of return, which in practice has been a 12 to 15 percent nominal return on capital in recent years. With inflation rates at or above 10 percent, the real return has been less than 5 percent. Since capital costs are the major costs of shale refining, the price of shale oil is very sensitive to the assumed rate of return.

The motive for pessimistic assessment of shale oil costs appears to be twofold. First, estimators are well aware of the recent history of apparent underestimation of shale oil costs. Second, the absence of commercial shale oil production is taken as prima facie evidence that the full cost of shale oil production is greater than whatever the existing oil price happens to be. Such an assumption is not necessarily correct. A considerable body of theoretical literature (within the industrial organization literature) exists which argues that limit pricing by existing producers of a particular good can preclude entry by competitors.[49] Under certain conditions, the extant producers may charge a price above the potential competitors' breakeven price for the product while still effectively preventing entry. In order to maintain such a position, the extant producer must be prepared to lower the price below the potential competitor's costs, thus forcing losses on the competitor. The general applicability of the theory to oligopoly situations has been questioned because in order to force the price below the competitor's cost, the extant producer must carry considerable reserve capacity. In addition, if both the extant producer and the potential competitor have equal access to technologies and resources, the extant producer will undergo losses during periods of price cutting, making it unclear which firm will stay in the market the longest. Even if the extant producer is able to charge a higher price for his product than his potential competitor's price, he may still be no better off than if he had simply allowed entry since he must incur the cost of holding excess capacity to make the threat of price cutting credible.

The limit pricing argument goes through much easier, however, if there is asymmetry in access to factors of production. Viewing the extant producer as OPEC (with considerable excess capacity and little cost of holding excess capacity because the capital cost of producing conventional oil from OPEC's fields is minimal compared with the capital cost of shale oil production) and the potential competitors as companies producing shale oil, gives the needed asymmetry to allow a limit price above the cost of shale oil. The threat of price cutting by OPEC—to maintain revenues or meet some revenue requirement perhaps—is very credible given the size of OPEC reserves. To the extent that such an analysis is correct, it is unlikely that shale oil, or other costly synthetics, will enter the market at appreciable commercial levels until basic conditions in the oil market have changed. The changes required represent changes that reduce OPEC's power over oil prices—basically increasing world demand for OPEC oil relative to OPEC reserves. Specific changes include a general increase in worldwide oil demand (due to expanding income and populations), a decrease in non-OPEC oil production (thereby focusing world demand on OPEC), and exhaustion of OPEC reserves. Under such circumstances OPEC would be able to sell as much oil as it could physically produce without depressing the world oil price below the cost of producing synthetic fuels or shale oil.

The eventual long-run cost of shale oil may be very different from current estimates. However, considerable caution is justified in extrapolating from recent cost estimate escalations into the future. The fact that shale oil is not presently being produced on a commercial level cannot be taken as prima facie evidence that the cost is greater than present oil prices. The long-run best guess concerning shale oil appears to be that it is marginally competitive with conventional oil, with oil prices in the $30 to $40 (1979 constant dollar) range. Short-run cost escalation above this range is possible and probable if very rapid expansion of the industry occurs, putting pressure on the costs of hiring trained personnel and the price of specialized equipment. A degree of uncertainty remains in cost estimates, but "it is probable, but not certain, that current cost estimates are fairly realistic and that there will be no further substantial increases, other than normal inflation."[50] The present range of estimates, however, does not preclude the possibility that shale oil may eventually be available at less than the current price of conventional oil.

Other Considerations

The potential environmental impacts of oil shale development have been used as an argument suggesting that shale oil development will be limited in the United States. Water availability has also been cited as a potentially major constraint on shale development in the western United States. Table 8-7 contains estimates of the cost of controlling environmental pollution and obtaining water for shale oil production in the western United States as reported in 1980.[51]

While the control of air and water pollution represent nonnegligible costs, additional cost per barrel would not appear to prohibit development. The problem of water appears to present no major constraints since diversion from basins such as the Columbia or Missouri could provide essentially unlimited water at a relatively small additional cost per barrel. The environmental control costs are those that must be incurred to meet existing standards. The specific control technologies have been proved in other industries. The existence of class I pollution areas near the prime oil shale deposits in Colorado could constrain development since, as the laws are presently formulated, no degradation of air quality is allowed in such areas. However, in the long run such policies are

TABLE 8-7. Oil Shale: Cost of Environment and Water Inputs in the Western United States (1979 dollars per barrel)

	Cost	*Limits/Concerns*
Air pollution	0.90–1.15	Nearby class I pollution areas (no degradation allowed under clean air act) could limit development.
Water pollution control	0.25–1.25	Present standards do not control nonpoint source pollution. Major groundwater pollution from leaching may result in new standards.
Land reclamation	0.01–0.04	No major constraints with in situ technologies.
Water supplies	0.00–0.79	
Region	(0.00–0.36)	Constraints possible at operations level of 1 $\times$ 10^6 barrels per day.
Interbasin diversion	(0.40–0.79)	No practical constraints.
Total	1.16–3.23	

Source: Congress of the United States, Office of Technology Assessment. *An Assessment of Oil Shale Technologies,* Report No. OTA-M-18 (Washington, D.C.: U.S. Government Printing Office, 1980), pp. 32, 285–418.

subject to change. Other factors may produce more stringent policies, however. For example, concerns have arisen over potential groundwater pollution caused by the leaching of spent shale. The health and economic hazards of groundwater pollution from shale oil are likely to be observable and directly attributable to the shale oil industry. As a result, there is likely to be strong public demand to limit groundwater pollution from the industry. Such limits may take the shape of direct limits on the size of the industry or imposition of costly control technologies.

In this regard, the cost estimates in table 8-7 assume a relatively small shale oil industry rather than an industry of a size needed to supply a major share of domestic or possibly global energy requirements. Pollution problems that seem manageable under production scenarios of a half to a million barrels per day may become difficult and expensive to manage under production scenarios of 20 to 30 million barrels per day or more. Thus, the figures presented in table 8-7 may substantially underestimate pollution control costs under large production scenarios. A much more pessimistic but probably more realistic scenario might posit an absolute limit on shale oil production in the range of perhaps 1 to 5 million barrels per day in the western United States.

Pollution costs and consequences in other areas of the United States (where poorer quality shales exist) and worldwide are largely unknown. Present U.S. pollution policies are generally more stringent than anywhere else in the world and, therefore, pollution problems are unlikely to act as a constraint on private developers in the near future. Rising incomes abroad, which will alter the tradeoff between environment and growth, or degenerating environments, or both, may lead to pollution control requirements that are as costly or more so than those estimated for the United States. But, once again, it is impossible to say little more than that such constraints may or may not exist in the future.

Status of Oil Shale Development Projects Around the World

The oil shale deposits in the United States are the largest known and contain some of the richest shales, but oil shale development has not been limited to this country. Among western European countries, shale oil was produced as far back as 1850 in the United Kingdom and 1838 in France. As estimated 250 million barrels of oil was produced in Europe (including eastern European countries) between 1961 and 1975. Production of oil from shale was most important before 1900; after World War II production continued in France only because of government subsidies.[52] Interest in shale oil has developed again in West Germany; oil shale was being burnt directly in fluidized-bed-fired power stations on a small scale in 1979.[53]

In China, exploration and development of oil shale goes back to the 1920s.[54] The Japanese began production in the 1930s and present production estimates range from 12 to 70 million tons of shale per year, with some being used directly. Six retorting plants were reported to be operating in 1977, with shale oil production reaching 30,000 b/d.[55] South Africa supported a small shale oil industry from 1935 through 1960; cumulative production was about 3.5 million barrels.[56] Shale deposits exist over much of Africa, though many of them are unappraised. More recently, Morocco has begun building a pilot shale oil plant, scheduled for completion in mid-1982. The plant will combine shale mining with a surface retort. Plans call for a first-stage commercial facility capable of producing 8,000 barrels per day (b/d).[57]

The Soviet Union uses oil shale directly as a feedstock for power plants. In 1979, an experimental processing plant began producing shale liquids as a feedstock for a power plant in Estonia, replacing direct use of oil shale. Total oil shale production in 1979 is estimated to have been 40 million tons.[58] Assuming an effective average yield of 20 gallons of oil per ton of shale puts the energy value of the shale at just under 20 million barrels of oil (approximately 110×10^{12} Btus) in 1979.[59] However, the Soviet oil shale industry has not achieved planning targets of the 1975–80 five-year plan.[60] Yugoslavia is planning to begin producing shale oil in 1982 and is expected to reach 60,000 barrels per day. The operation is expected to be profitable, with oil prices above $20 (1979 dollars) per barrel.[61]

In the Middle East, Jordan has begun seriously exploring its oil shale reserves as a potential source of liquids for power plants. Present activities are limited to feasibility studies with the aid of West German and Soviet technicians. Shale is not expected to contribute to Jordan's energy supply until after 1990.[62]

In Australia, shale oil was produced during the period 1865 to 1952.[63] Plans to develop a demonstration plant producing 15,000 to 18,000 b/d by 1985 were recently cancelled but the sponsors may go directly to a commercial-sized plant. Original plans called for commercial operations of 180,000 to 240,000 b/d by 1990.[64]

In Latin America, Brazil and Argentina have significant oil shale resources. Brazil has had a demonstration plant (1000 b/d) operating since 1973. Plans are underway to build a plant capable of producing 51,000 b/d.[65]

Tar Sands and Heavy Oils

Technologies

Technologies for recovering tar sands and heavy oils are similar to those used to recover oil from shale even though their geological histories are quite different. Geologists believe tar sands and heavy oils were once conventional oil that, as a result of contact with the atmosphere, lost its light fractions, whereas oil shale is oil in the process of formation.[66] Mining tar sand with surface retorting is a known technology. Canada has been producing oil from oil sand deposits using surface retort technology since 1967. *In situ* techniques are in the pilot stage in Canada, Venezuela, and the Soviet Union.[67]

While the technologies for producing oil from tar sands and shale are similar at the simplest level, the actual production processes for extracting oil from tar sands tend to be more complex. Shale oil technologies are relatively simple; pyrolysis at a given temperature yields a liquid which can be upgraded to a crude oil-like product. The oil sand extraction technologies tend to be much more site specific and require more effort to control the retort process. Considerable effort must be invested to develop an appropriate process for a given site and it is not fully transferable to other deposits. Thus (as seen in table 8-6), oil from tar sands tends to be more costly to produce than shale oil.[68]

Costs

There are fewer estimates of the cost of extracting oil from tar sands than for shale oil costs. As noted above, however, the technology is such that costs will tend to be somewhat greater than the cost of extracting oil from shale. Table 8-8 lists several estimates.

Not surprisingly, the estimates tend to show the same general tendency to rise over time as do shale oil cost estimates. Production that has occurred in Canada has been under government guarantees. Present research and development of tar sand production technologies are centered in Canada, Venezuela, and the USSR, where large easily accessible deposits exist. The United States has relatively small known deposits of tar sands; efforts have focused on developing shale oil resources instead. Heavy oil is being produced in the United States primarily in California, using steam or other sources of heat to bring the heavy oil to an ambient temperature that allows it to flow to the surface.

Coal Liquids

Technologies

Two basically different processes exist for obtaining liquids from coal. Direct liquefaction yields a substance like crude oil from which the usual range of petroleum derivatives can be produced. The process requires adding hydrogen directly to the coal to produce a liquid. Indirect processes work by heating the coal in the absence of oxygen to produce methane. The methane (a gas) is then liquefied to produce methanol. Methanol can be used without further refining, but automobiles require modification to accept methanol. Processes also exist to produce gasoline from methanol.

The production of methanol is a commercially proved process developed in Germany during World War II and used today to produce liquids in South Africa at a commercial level. Direct liquefaction technologies have not been proved commercially. Other than South Africa, the United States has been particularly interested in developing coal liquefaction technologies as a way of using its vast coal resources. The only coal synfuel project near a commercial level is the Great Plains gas-

TABLE 8-8. Cost Estimates of Tar Sand Oil Recovery (1979 dollars)

Source/Year	*Cost Estimate*
Hottel & Howard/1971	3
WEC/1978	31–33
Cameron Engineers/1979	25–30
WEC/1980	35
Bechtel/1981	52

Sources:

H. C. Hottel and J. B. Howard, *New Energy Technology: Some Facts and Assessments* (Cambridge, Mass.: MIT Press, 1971), cited by the Science and Public Policy Program, University of Oklahoma, *Energy Alternatives, A Comparative Analysis* (Norman, Okla.: University of Oklahoma, May 1975, pp. 5–15.

The Full Reports to the Conservation Commission of the World Energy Conference, *World Energy Resources 1985–2020, Oil and Gas Resources,* by Pierre Despraires, W. T. McCormick, Jr., L. W. Fish, R. B. Kalisch, and T. J. Wander (New York: IPC Science and Technology Press, 1978), p. 74.

Cameron Engineers, "Overview of Synthetic Fuels Potential to 1990," in U.S. Congress, Senate, Committee on the Budget, *Synthetic Fuels,* Report by the Subcommittee on Synthetic Fuels, 96th Cong., 1st sess. (Washington, D.C.: U.S. Government Printing Office, 1979), p. 177.

World Energy Conference, *Survey of Energy Resources 1980,* by the Federal Institute for Geosciences and National Resources, Hanover, Red. Rep. of Germany (London: Alan Armstrong & Associates, 1980), p. 162.

H. F. Brush, Bechtel Corp., "Readiness: Where Do We Really Stand?" paper presented at the Government Research Corporation Conference, "Synthetic Fuels: Worldwide Outlook for the 80s," San Francisco, Calif., February 19–20, 1981, p. 24.

ification project. At one time, it was hoped that it would be operational by 1984 but delays in obtaining licenses and problems in finding a guaranteed market for the gas have resulted in considerable setbacks.[69]

With the exception of the South African plant, which processes 33,000 tons per day (TPD) of coal, all other planned plants are at the demonstration or pilot plant size (1 to 6,000 TPD capability).[70] The Polish, West German, and Belgian governments have or plan to support development of gasification or liquefaction plants.[71] The United States, Japan, and West Germany had been cooperating on developing a solvent refining process—SRC-II (a direct liquefaction process) but two years ago canceled the project, citing cost escalation.[72] A smaller scale plant (250 TPD) using the Exxon Donor Solvent process of direct liquefaction completed a successful run in 1981.[73]

The basic problem that must be overcome in producing liquids from coal is increasing the hydrogen content of the final product. Crude oil is normally 12–14 percent hydrogen by weight whereas coal is 2–6 percent. The two generic coal liquid processes differ in their approaches; indirect liquefaction effectively removes solid material, thereby increasing the hydrogen content of the remaining gas. The direct liquefaction processes add hydrogen, which is expensive and complex.[74]

Costs

Estimates of the cost of producing liquids from coal vary considerably and, like shale oil, have shown a tendency to escalate over time. Table 8–9 lists several estimates. The cost of producing liquids from coal will tend to be more than producing shale oil or oil from tar sands because coal normally has a low hydrogen content whereas oil shale (Green River) and tar sands are approximately 10 percent hydrogen by weight.[75] Thus hydrogen is needed as a material input (in the case of direct liquefaction processes) and the capital costs are greatly increased. Indirect liquefaction,

TABLE 8-9. Estimates of the Cost of Producing Liquids from Coal (1979 dollars per barrel of oil equivalent)

Source/Year	*Cost*
NAS/1977	28–48
IGT/1978	34
AGA/1979	44
Booz-Allen/1979	29–35
Cameron Engineers/1979	29–38
U.S. DOE/1979	25–32
EPRI/1979	33
ESCOE/1979	28
AERG/1979	22–32
CRS/1980	29–41
IIASA/1981	32

Sources:

National Research Council, *Assessment of Technology for the Liquefaction of Coal,* Commission on Sociotechnical Systems, Committee on Processing and Utilization of Fossil Fuels, Ad Hoc Panel on Liquefaction of Coal (Washington, D.C.: National Academy of Sciences, 1977), pp. 131–152, cited by National Academy of Sciences, *Energy in Transition 1985–2010* (Washington, D.C.: National Academy of Sciences, 1979), pp. 176–180.

Institute for Gas Technology, IGT, *Energy Topics,* June 19, 1978, p. 2.

American Gas Association, *Fact Book* (Washington, D.C.: American Gas Association, September 1979), p. 2.

Booz-Allen & Hamilton, "Analysis of Economic Incentives to Stimulate a Synthetic Fuels Industry," in U.S. Congress, Senate, Committee on the Budget, *Synthetic Fuels,* Report by the Subcommittee on Synthetic Fuels, 96th Cong., 1st sess. (Washington, D.C.: U.S. Government Printing Office, 1979), p. 102.

Cameron Engineers, "Overview of Synthetic Fuels Potential to 1990," in U.S. Congress, Senate, Committee on the Budget, *Synthetic Fuels,* Report by the Subcommittee on Synthetic Fuels, 96th Cong., 1st sess. (Washington, D.C.: U.S. Government Printing Office, 1979), p. 155.

Based on interim costs used in analysis of National Energy Plan II, Report to Congress, prepared by U.S. Department of Energy, May 1979, cited by the Department of Energy, *Technical and Economic Feasibility of Alternative Fuel Use in Process Heaters and Small Boilers* (Washington, D.C.: U.S. Government Printing Office, 1980), pp. 4–10.

D. F. Spencer, M. J. Gluckman, and B. N. Looks, *A Comparative Analysis of Implications of Various Incentives for Mature Commercial Synthetic Fuel Plants* (Palo Alto, Calif.: Electric Power Research Institute, July 1979).

K. A. Rogers and R. F. Hill, *Coal Conversion Comparisons,* ESCOE, Inc., Document # FE 2468-51, July 1979.

American Energy Research Co., *Opportunities for Coal to Methanol Conversion,* prepared for DOE (McLean, Va.: American Energy Research Co., April 1980), pp. 47–48.

U.S. Congress, Senate, Committee on the Budget, *Synthetic Fuels,* Report by the Subcommittee on Synthetic Fuels, 96th Cong., 1st sess. (Washington, D.C.: U.S. Government Printing Office, 1979), p. 19.

Wolf Häfele, Project Leader, *Energy in A Finite World: A Global Energy Systems Analysis,* Report by the Energy Systems Program Group of the International Institute for Applied Systems Analysis (Cambridge, Mass.: Ballinger, 1981).

while not requiring hydrogen as an input, requires a large capital investment since it is basically two processes: initial gasification and then liquefaction of the gas. One study, in comparing costs of the two technologies, found indirect liquefaction somewhat more expensive than direct liquefaction (37 versus 36 1979 dollars).[76]

Feedstock cost is of some importance, giving coal somewhat of an advantage over aboveground retorting of oil shale and tar sands. While the mining technologies for shale and coal are not that different, suggesting like costs per ton mined, shale has a higher mineral content, thus requiring mining and handling a much larger volume of material for the same amount of energy. This advantage will largely disappear since *in situ* (and MIS) technologies for producing oil from shale and tar sands are being developed, eliminating the need to mine or handle the shale or dispose of the spent shale.

Cost considerations dictate that coal liquefaction plants be located near a source of coal to reduce transportation costs. Since these costs represent a major part of the price of coal, synfuel plants are able to obtain coal at lower costs than average industrial consumers or utilities located some distance from coal mines. Mine mouth costs of coal in the United States are as little as $3–10 (1979 dollars) per ton.[77] Countries such as the Soviet Union, China, Australia, and South Africa have similar grade resources. Thus, at a yield of 2½ barrels of oil per ton, feedstock costs would account for 1 to 4 dollars of the estimated cost per barrel of synthetic crude.[78]

Enhanced Oil Recovery

Technologies

Enhanced recovery of oil from conventional fields, or in line with definitions set out in this study, extraordinary recovery of oil from conventional deposits, must face one of two physical features of oil deposits. Either the drilling location is in a hostile environment—deep ocean or polar areas—or the oil is trapped in the rock. Technologies for getting oil from conventional deposits in harsh environments differ from conventional drilling technologies only in the surface support structures and efforts. Offshore drilling has become commonplace; in 1979 offshore oil production was 13 million barrels per day or roughly 20 percent of total world production.[79] Several technologies are available. The jack-up rig is limited to water depths of 300 feet.[80] Semisubmersible platforms are used in deeper water; 100 are in operation worldwide and have drilled in water depths of about 2,500 feet, with one rated at 6,000 feet.[81] Drill ships are also capable of depths up to 6,000 feet.[82] Currently, efforts are aimed at developing a fully submerged platform, operating completely by remote control.[83] Such a system would eventually allow deep ocean drilling.

The technology for offshore drilling has pushed ahead rapidly in recent years. Water record depths have been pushed from less than 1,500 feet in 1973 to near 5,000 feet in 1979.

Methods used to speed the flow of oil from conventional deposits and increase the total recovery of oil include heating the oil, fracturing the rock, repressuring the well, or injecting chemicals to speed the flow.

Costs

Many of the enhanced recovery technologies are cost effective in at least some applications at existing oil prices (30 to 40 1979 dollars per barrel). By definition, these are included as part of the conventional oil resource. No global studies of the potential for enhanced recovery at various oil prices exist. The few that exist for the United States are dated by the fact that the highest considered price was exceeded by the rise in oil prices that occurred in 1979.[84] Should oil prices continue to rise, there would be an impetus to apply existing enhanced recovery techniques in marginal fields as well as to develop recovery techniques which, if available today, would not be cost effective. In addition, cost-reducing technological advances are likely to play a role in the future. Because the range of techniques and conditions under which they might be applied are so varied, it is impossible to say much more about costs of enhanced recovery without surveying individual oil fields. The existence of backstop technologies such as shale oil or coal liquids is likely to place an eventual cap on such applications, depending on the introduction date and cost of the technology.

Liquids from Biomass

Technologies

Ethanol (grain alcohol) is produced through fermentation of sugars, starches, or cellulose material; methanol is produced from wood through the use of thermochemical conversion techniques[85] or anaerobic digestion. Methanol production from biomass follows the same general process as methanol production from coal—conversion to methane (gas) and further conversion to methanol (liquid). Presently, most biomass energy liquids are ethanol. The often-cited Brazilian biomass liquids program produces ethanol from sugar cane. In the United States, ethanol is produced primarily from corn grain. Either ethanol or methanol can be produced from any type of biomass. However, to produce ethanol from wood (or other cellulose), preprocessing (acid hydrolysis) is necessary to break the wood into sugars which can be converted to ethanol. Such plants exist in the Soviet Union, the United States, and Switzerland, and are being designed in Brazil.[86] Methane can be produced more directly from cellulose, but requires subsequent liquefaction to produce methanol. Ethanol production is highly efficient—80–90 percent efficiencies are obtainable.

Costs

Costs of producing liquids from biomass are high (see table 8-10). However, the cost cannot be fairly compared with crude oil since the final product of ethanol and methanol is a transportation quality fuel. For this reason, the costs are reported per gallon at an energy equivalent of gasoline.

The ethanol production process is less costly than the methanol process but requires as inputs sugars, which tend to be expensive. Grains can be easily converted to sugars through the familiar technologies of producing beverages.[87] Converting cellulose to a sugar suitable for ethanol production adds to the low cost of cellulose products such as wood and waste. Methanol production, on the other hand, can use lower grade inputs—waste and wood—but is more expensive. Most observers believe that biomass liquids, if they are to contribute a substantial amount of energy to the global energy balance, must be derived from lower cost wood and waste cellulose. Grains and sugars and land suitable for growing such crops have too high an alternative value in food and fiber agriculture to compete with other unconventional liquids.[88] Waste, by definition, has no alternative uses and silviculture can use land unsuitable for grain production, thus making these biomass sources the prime candidates for energy production. Ethanol production has valuable by-products, including a mash with a high value as agricultural feed. Conversion of wood to sugars produces methanol, furfural, and lignin residues.[89] Anaerobic digestion of waste to produce methane also produces by-products suitable for agricultural feedstocks.

CONCLUSIONS

A large number of technologies exist for producing liquid fuels from unconventional sources, with varying costs and resource abundance. Shale oil and coal liquids are least constrained in terms of abundance of resources. Tar sands, enhanced oil recovery, and biomass resources are more limited. A ranking of abundance from most to least would be as

TABLE 8.10. Cost of Liquids From Biomass (1979 dollars per U.S. gallon produced)

Author/Date	*Fuel Type*	*Cost of Feedstock*[a]	*Other Costs*[b]	*By-Product Credit*	*Unit Cost*
Poole/1979	Ethanol	1.38 (cane)	0.79	0.11	2.06
Yang/1979	Ethanol	1.53 (cane)	2.04	0.11	3.46
		1.72 (casseva)	1.26	0.13	2.85
Lipinsky/1978	Ethanol	1.72 (cane)	1.16	—	2.88
Katzen/1975	Ethanol	0.61[c] (woodwastes)	21.37	0.41	1.57
		1.37[d] (woodwastes)			2.33
Scheller/1977	Ethanol	1.34[e] (corn)	2.129	0.72	1.91
		2.01[f] (corn)		1.09	2.21
IGCE/1977	Ethanol	0.91 (wood)	1.71	0.53	2.09
Poole/1979	Methanol	1.60–2.25/GJ[g] (wood)	N.A.	N.A.	1.13–1.30
ERDA/1976	Methanol	1.30/GJ[g] (wood)	N.A.	N.A.	0.64–1.28
		2.60/GJ[g] (wood)	N.A.	N.A.	0.96–1.44
Katzen/1975	Methanol	$20/ODT[g] (wood)	N.A.	N.A.	0.83
		$45/ODT[g] (wood)			1.34

Source: Joy Dunkerley et al., *Energy Strategies for Developing Nations,* (Baltimore, Md.: Johns Hopkins University Press for Resources for the Future, 1981). Converted from 1975 dollars per gigajoule. Dunkerley et al., adjusted original author's estimates in some cases.

[a]Cost of feedstock per gallon of fuel produced unless otherwise noted.

[b]Capital, labor, operation and maintenance.

[c]Corresponds to wood at $20/ODT (ODT: oven dry ton).

[d]Corresponds to wood at $45/ODT.

[e]Corresponds to corn at $2.60/bu.

[f]Corresponds to corn at $3.90/bu.

[g]Cost of input in original units as opposed to input per unit of output.

follows: oil shale, coal liquids, tar sands, enhanced oil recovery. Biomass liquid is not comparable because it is a flow concept rather than a stock.

In terms of cost, a ranking from least to most costly would be as follows: oil shale, tar sands, coal liquids, and biomass. Enhanced oil recovery techniques potentially span the full range of costs, depending on the specific technology and conditions under which they are applied.

The cost and abundance of shale oil suggest that it may form a liquid fuel backstop which will relegate other unconventional liquids to a relatively small role, dependent on local conditions (which may give the resource a cost advantage at low levels of output) or government support.

NOTES

1. U.S. National Alcohol Fuels Commission, *Fuel Alcohol: An Energy Alternative for the 1980s* (Washington, D.C.: U.S. Government Printing Office, 1981), p. 50.
2. Denis Hayes, *Rays of Hope: The Transition to the Post-Petroleum World* (New York: W. W. Norton, 1977), p. 194.
3. D. C. Duncan and V. E. Swansen (1965), "Organic Rich Shale of United States and World Land Areas," U.S. Geological Survey Circular 523. Gregg Marland in *Shale Oil: U.S. and World Resources and Prospects for Near-Term Commercialization in the United States* (Oak Ridge, Tenn.: Oak Ridge Associated Universities, 1979) provides a recent review of global shale oil estimates. Other researchers have made estimates of global oil shale *resources* but these have used the Duncan and Swansen *resource base* estimates as a starting point, refining estimates for certain regions (the United States) or changing assumptions of the share recoverable or the minimum share of organic matter from which oil will be recovered.
4. Executive Summaries of Reports on Resources, Conservation and Demand to the Conservation Commission of the World Energy Conference, *World Energy Resources 1985–2020* (New York: IPC Science and Technology Press, 1978), p. 46; World Energy Conference, *Survey of Energy Resources 1980,* by the Federal Institute for Geosciences and Natural Resources, Hanover, Fed. Rep. of Germany (London: World Energy Conference, 1980), part B, tables 2.4 and 2.6.
5. World Energy Conference (WEC), *Oil and Gas Resources* (New York: IPC Science and Technology Press, 1978), p. 77.
6. Ibid.
7. WEC, *World Energy Resources 1985–2020,* p. 40.
8. WEC, *Oil and Gas Resources,* p. 78.
9. Ibid., p. 122. Lowest and highest estimates and break points for ranges have been converted from figures given in gigatons (GT) and rounded to the nearest 100 bbos. The survey asked for a minimum response of whether the world total oil in place was 750 GT or less, 750 to 1,000 GT, or greater than 1,000 GT. Of the 26 respondents, 9 offered actual estimates or ranges of estimates.
10. Ibid., calculated from data on p. 122.
11. Based on a study by the National Petroleum Council, cited in WEC, *World Energy Resources 1985–2020,* p. 78.
12. Reported in WEC, *Oil and Gas Resources,* p. 78.
13. Marland, *Shale Oil.*
14. Duncan and Swansen, "Organic Rich Shale."
15. Marland, *Shale Oil,* p. 2.
16. Ibid. Marland cites G. Marland, A. M. Perry, and D. B. Reister, "Net Energy Analysis of In-Situ Oil Shale Processing" *Energy,* vol. 3 (1978), pp. 31–41; E. R. Van Artsdalen, "Minimum Energy (Thermodynamic Limit) to Obtain Oil from Oil Shale," Institute for Energy Analysis, Oak Ridge, Tenn., 1976; (ORAU/IEA(M)-76-6) and J. R. Donnell (1977), "Oil Shale Resource Investigations of the U.S. Geological Survey," U.S. Geological Survey Open File Report 77-637. The thermodynamic limit estimated by Van Artsdalen was 1.6 gallons per ton with a practical limit several times this value. The net energy limit for modified *in situ* was found to be well above 3 gallons per ton. Donnell cited experimental retorting experience suggesting net energy loss at shales yielding less than 8 gallons per ton.
17. Donnell, "Oil Shale Resource Investigations," cited in Marland, *Shale Oil.*
18. Marland et al., "Net Energy Analysis," cited in Marland, *Shale Oil.*
19. Marland, *Shale Oil,* pp. 9–10.
20. Assuming 5.8×10^3 Btus per barrel of shale oil. In 1979, world oil production was 18.1 quads, U.S. Department of Energy, *1980 Annual Report to Congress,* vol. II (Washington, D.C.: U.S. Government Printing Office, 1981).
21. World Energy Conference 1980. *Survey of Energy Resources,* prepared for the 11th World Energy Conference, Munich, September 8–12, 1980, p. 156.
22. Ibid., table 2–6.
23. W. N. Sande, "Tar Sands and Heavy Oils," Executive Summary in WEC, *World Energy Resources 1985–2020.*
24. WEC, *Oil and Gas Resources,* p. 79.
25. U.S. Department of Energy, *1980 Annual Report to Congress,* vol. III (Washington, D.C.: U.S. Government Printing Office, 1981).
26. Ibid., p. 171.
27. Application of thermal methods, as is presently being done, was estimated to be economically justified at prices as low as $5 (1976 constant dollars or approximately $6.20 in 1979 dollars). See National Petroleum Council, *Enhanced Oil Recovery: An Analysis of the Potential for Enhanced Oil Recovery from Known Fields in the United States, 1976 to 2000* (1976), study conducted for the U.S. Department of the Interior, Washington, D.C.
28. No direct information exists for the costs of drilling for oil in polar regions. Such deposits were considered by the WEC, *Oil and Gas Resources,* p. 88. The study concluded that drilling in Alaska, northern Canada, and Siberia will provide initial experience which can be applied to the far northern arctic regions and the Antarctic but suggests that costs may be in the same range as for oil shales and greater than deep offshore oil.

29. WEC, *World Energy Resources 1985–2020.*
30. National Petroleum Council, *Enhanced Oil Recovery,* cited in WEC, *Oil and Gas Resources,* p. 80.
31. Lewin and Associates, Inc., 1976, "The Potential and Economics of Enhanced Oil Recovery," prepared for the Federal Energy Administration, Washington, D.C., cited in WEC, *Oil and Gas Resources,* p. 80.
32. Wolf Häfele, Project Leader, *Energy in a Finite World: A Global Systems Analysis,* Report by the Energy Systems Program Group of the International Institute for Applied Systems Analysis (Cambridge, Mass.: Ballinger, 1981), pp. 199–201.
33. WEC, *Oil and Resources,* p. 78.
34. See Richard Nehring, *Giant Oil Fields and World Oil Resources,* prepared for the Central Intelligence Agency, Washington, D.C., pp. 14–22. The cutoff between giant and super giant fields has been proposed as 4 billion to 10 billion barrels of recoverable oil, Nehring notes. Similarly, the giant category has not been uniquely assigned a cutoff of 500 million barrels of recoverable oil but this cutoff has become the accepted cutoff in the last decade (Nehring, pp. 6–7).
35. "Gasification Plant Carbon Dioxide Eyed for Oil Field Use," *Oil and Gas Journal,* October 27, 1980.
36. See DOE, *1980 Annual Report to Congress,* p. 317. Direct communication with a scientist at the Institute for Energy Analysis confirmed the presence of carbonate rock in the Four Corners region. Other oil shales are technically shales, do not contain the carbonate rock, and therefore, CO_2 is not produced during the shale oil extraction process. The occurrence of the carbonate oil "shale" worldwide is unknown. Therefore, regions with large carbonate oil "shale" resources may have a large, relatively cheap source of CO_2 produced as a by-product of shale oil production.
37. Robert Stobaugh and Daniel Yergin, *Energy Future: Report of the Energy Project at the Harvard Business School* (New York: Random House, 1979), p. 43.
38. J. H. Gary, "Foreword to the Seventh Oil Shale Symposium" *Quarterly of the Colorado School of Mines,* vol. 69, no. 2, April 1974, p. v, cited in Congress of the United States, Office of Technology Assessment, *An Assessment of Oil Shale Technologies,* Report No. OTA-M-18 (Washington, D.C.: U.S. Government Printing Office, 1980), p. 108.
39. Stobaugh and Yergin, *Energy Future,* p. 43.
40. Guy Elliot Mitchell, "Billions of Barrels of Oil Locked Up in Rocks," *National Geographic Magazine,* February 1981, p. 205, cited in Stobaugh and Yergin, ibid., p. 43.
41. WEC, *World Energy Resources,* p. 67.
42. OTA, *Assessment of Oil Shale Technologies,* p. 13.
43. For example, see Robert V. Jelinek, *Costs of Synthetic Fuels in Relation to Oil Prices,* Congressional Research Service Report (Washington, D.C.: U.S. Government Printing Office, 1981); John Sterman, "The Transition to Synfuels: A Receding Horizon?" in *Plexus: Systems Dynamics News,* vol. 1, no. 2, November 1980, MIT Energy Laboratory Newsletter; and OTA *Assessment of Oil Shale Technologies,* pp. 180–190.
44. One can explore the likely increase necessary for oil prices to catch up to estimated costs of shale oil, given that energy is an input into the shale mining and refining process. Even when the problem is designed as favorable to the case of parallel movement, one finds this explanation insufficient, as a sole cause, in explaining the observed relations during the 1970s. The problem is set up as

$$(1) \quad P_s = P_E g + h$$

where P_s is the price of shale oil, P_E the price of the energy input, g the amount of direct energy input used per unit of shale oil output, and the aggregated cost of all other inputs.

We wish to formally recognize embodied energy in all other inputs, and resulting increases in their prices, as a contributor to increasing shale oil costs as oil prices rise. Thus, h is decomposed as

$$(2) \quad h = P_E h_E + P_m h_m$$

where P_E is defined as before, h_E is the energy input per unit of h, and P_m is the price and h_m the amount of nonenergy input per unit of the nondirect energy aggregate into shale oil production. Substituting (2) in (1) yields

$$(3) \quad P_s = P_E(g + h_E) + P_m h_m$$

where both the direct (g) and indirect (h_E) effects of energy are explicitly taken into account. Given that the price of energy is something below the estimated cost of shale oil, we wish to know what rise in energy prices will result in equal costs for both oil and shale oil. We know that

$$(4) \quad \frac{dP_E}{dP_E} = 1$$

and, from (3), that

$$(5) \quad \frac{dP_s}{dP_E} = g + h_E$$

Given an initial shale oil cost estimate of P_{so} and an initial price of energy of P_{eo}, we want to calculate a change in P_E such that $P_E^* = P_s^*$.

(6) $P_s^* = \Delta P_E (g + h_E) + P_{so}$

and

(7) $P_E = \Delta P_E(1) + P_{Eo}$

Setting (6) and (7) equal and solving for the change in the price of energy gives

(8) $\Delta P_E = \dfrac{P_{so} - P_{Eo}}{1 - (g + h_E)}$

The term $1/1 - (g + h_E)$ is a multiplier factor.

The following table is from Richard H. Shachem and H. Jones Leach, *Maintaining Automotive Mobility Using Fuel Economy and Synthetic Fuels to Compete with OPEC Oil* (Arlington, Va.: The Energy Productivity Center, Mellon Institute, 1980), p. 74. Using this data and the estimated life of the plant (25 years), the number of operating days per year (328), and an assumed required rate of return of 15 percent, it is possible to calculate the shale cost attributed to direct energy (assuming utilities are all energy costs) and nonenergy costs, using a straight-line depreciation.

Capital and Operating Costs: Oil Shale Retorting and Upgrading (88,000 barrels per operating day)

	Facility Costs ($million, Jan. 1980)
Capital costs	
Construction	919.3
Land and land right	0.9
Base year value of escalation and interest during construction	63.6
Working capital	122.2
Total	1105.8
Annual operation costs	
Manpower	19.0
Materials and parts	27.8
Utilities	16.0
Ad valorem taxes	13.8
Insurance	8.3
Rents, royalties, leases	1.5
Miscellaneous	2.2
Feedstock:	
Shale at $3.88 per ton	209.9
By-products credits	
Sulfur at $26.85 per ton	−2.1
Ammonia at $152.50 per ton	−16.0
Total	208.4

Making the necessary calculations gives

$$P_s = 16.99$$
$$P_{EG} = 0.55$$
$$h = 16.44$$

Assuming P_E was roughly 25 dollars per barrel in 1980, implies

$$g = 0.022$$

As a rough estimate one could argue that embodied energy in the input factors is likely to be roughly an economy average. Energy costs accounts for no more than 10 percent of the economy, thus

$$P_E h_E = 1.644$$

and

$$h_E = 0.06576$$

Based on these estimates, the multiplier factor is approximately 1.1. Thus, for a given difference in the estimated cost of shale oil and crude oil prices, crude oil prices need only increase by the original difference plus 10 percent before shale oil becomes competitive. Even if one argued that the shale oil production inputs embodied far more energy than the average for the economy—20 percent, for example—the multiplier is only 1.18.

To put these estimates in historical perspective, a reasonable relation between shale oil cost estimates and oil prices in 1972 (in 1979 dollars) was $7 per barrel for shale and $3 per barrel for oil. Thus, oil prices should have had to increase by only $4.40 before shale oil became competitive (under the more reasonable, lower multiplier estimates) or $4.70 under the higher multiplier estimates. By 1974, the price of oil had increased by more than $10 (1979 constant dollars) from the 1972 price.

Further, the problem formulation will tend to overstate the multiplier effect for two reasons. First, it assumes fixed energy inputs. A more general formulation would allow for price response to increasing oil prices in the shale oil production function. Second, all energy is assumed to be oil, or equivalently, all energy prices move with the price of conventional oil. Since scarcity of oil has been the driving force behind energy price increases, the sympathetic price responses for other energy types will be *at most* equal to the initial oil price change. In addition, this formulation allows no substitution away from oil to other types of energy.

45. See footnote 43.
46. OTA, *Assessment of Oil Shale Technologies.* The increases in various industrial cost indices during the 1970s is also documented in Robert V. Jelinek, *Costs of Synthetic Fuels.*
47. OTA, *Assessment of Oil Shale Technologies,* p. 189.
48. Ibid.
49. O. E. Williamson, "Predatory Pricing: A Strategic and Welfare Analysis," *Yale Law Journal,* vol. 87 (December 1977), pp. 284–340.

50. OTA, *Assessment of Oil Shale Technologies,* p. 189.
51. Ibid.
52. Donald C. Duncan and Vernon E. Swansen, "Organic Rich Shales of the United States and World Land Areas," Geological Survey Circular 523, Washington, D.C. (1965), p. 2.
53. D. C. Gibson, "Oil Shale in Australia—Its Occurrence and Resources" in Duncan and Swansen, ibid.
54. Ibid., p. 19.
55. D. L. Gibson, "Oil Shale in Australia."
56. Duncan and Swansen, "Organic Rich Shales," p. 17.
57. "Moroccans Pick Davy McKee for Oil Shale Project," in *Energy Daily,* May 19, 1980, p. 3.
58. "Soviets Use Shale Oil to Fire Power Station Boilers," in *Oil and Gas Journal,* April 14, 1980, p. 65.
59. Assuming 20 gallons per metric ton, 42 gallons per barrel, 5.8×10^6 Btus per barrel of oil.
60. See Duncan and Swansen, "Organic Rich Shales," and Anonymous (1980), "Neglected Ukranian Oil Shale Deposits," in *Energy in Countries with Centrally Planned Economies,* January 1980, p. 14.
61. "Yugoslavia to Start Shale Oil Production," in *Energy in Countries with Centrally Planned Economies,* December 1979, p. 17.
62. "Jordan: Scientists are Studying Shale Oil," in *World Business Weekly,* December 17, 1979, p. 14.
63. Gibson, "Oil Shale in Australia."
64. "Shale Project in Australia Put on Hold," in *The Energy Daily,* Thursday, April 16, 1981, pp. 3–4.
65. See Alvaro Franco, "Petrobas Proposes Oil Shale Plant," *Oil and Gas Journal,* April 11, 1977, p. 85; and Vincente T. Padula, "Oil Shale of Permian Irati Formation, Brazil," in *Journal of American Association of Petroleum Geologists,* vol. 53 (March 1969), pp. 591–602.
66. WEC, *Oil and Gas Resources,* p. 70.
67. WEC, *World Energy Resources,* p. 47.
68. From personal communication with an analyst at Pace Consulting, Denver, Colorado, concerning cost rankings of unconventional oil sources appearing in Cameron Engineers, "Overview of Synthetic Fuels Potential to 1990," in U.S. Congress, Senate, Committee on the Budget, *Synthetic Fuels,* Report by the Subcommittee on Synthetic Fuels, 96th Cong., 1st sess. (Washington, D.C.: U.S. Government Printing Office, 1979).
69. Cameron Engineers, "Overview of Synthetic Fuels," p. 170.
70. Ibid., p. 171.
71. Harold Wilson, "Interest Building in Coal-Based Synfuels," *Oil and Gas Journal,* March 10, 1980, pp. 50–52.
72. Text of joint communique appearing in *Energy Clearinghouse,* vol. 2, no. 26, July 2, 1981, p. 2.
73. DOE news release, "Successful Run at Synfuels Plant Marks Milestone in Readying Technology for Private Sector, Edwards Says," *Energy Clearinghouse,* vol. 2, no. 24, June 18, 1981, p. 2.
74. Cameron Engineers, "Overview of Synthetic Fuels," p. 166.
75. Ibid.
76. Ibid., p. 156.
77. Sydney Katell, *Economic Analysis of Coal Mining Costs for Underground and Strip Mining Operation,* prepared for the U.S. Department of Energy Information Administration, No. HCP/I760/8-01, October 1978, p. 2.
78. Estimated yield from the EXXON Solvent process. See DOE news release, "Successful Run at Synfuels Plant."
79. EXXON, *The Offshore Search for Oil and Gas,* EXXON Background Series, 1980, p. 2.
80. Ibid., p. 9.
81. Ibid.
82. Ibid., p. 13.
83. Ibid., p. 18.
84. See, for example, National Petroleum Council, *Enhanced Oil Recovery* or Lewin and Associates, "Potential and Economics of Enhanced Oil Recovery."
85. Joy Dunkerley et al., *Energy Strategies for Developing Nations* (Baltimore, Md.: Johns Hopkins University Press for Resources for the Future, 1981), p. 195.
86. Ibid., p. 196.
87. Ibid.
88. This tradeoff has been examined by a large number of authors spanning the full range of potential biases, including Dunkerley et al., *Energy Strategies;* Denis Hayes, *Rays of Hope: The Transition to a Post-Petroleum World* (New York: W. W. Norton, 1977); and the National Research Council, *Energy in Transition 1985–2010,* Final Report of the Committee on Nuclear and Alternative Energy Systems (San Francisco: W. H. Freeman, 1979).
89. Dunkerley et al., *Energy Strategies,* p. 200.

Conventional Gas 9

Natural Gas is a more convenient fuel than oil in many ways. It requires little or no processing or storage before being delivered for end use. It is a cleaner burning fuel and is associated with low capital and operating costs. Since the end of World War II, its use has grown rapidly, first in the United States and then in Europe and the Soviet Union. In the past, most marketed natural gas was associated with oil production and the marketing was largely an effort to make a profit on a by-product that had previously been vented or flared. However, the natural gas market has expanded so rapidly in the developed regions of the world that nonassociated gas fields have been developed. The oil price hikes of 1973 accelerated the search for nonassociated natural gas and have encouraged transportation and distribution pipelines in the developing countries, as well.

Nevertheless, as recently as 1975 the National Academy of Sciences reported that an estimated 80 percent of gross gas production was flared or vented in the oil-producing developing countries.[1]

This chapter reviews estimates of global and regional gas resources, historical gas discovery and production data, and trends in gas flaring and venting. Finally, the problem of transporting gas from areas of high resource/low demand to areas of low resource/high demand are reviewed. The discussion is restricted to "conventional" natural gas. Thus, synthetic gas from coal and biomass, blast furnace, coke oven and still gas, geopressured gas resources, and gas from tight gas formations, coal seams, and shale are excluded. (See chapter 10 for a discussion of these unconventional gas resources.) Further, we follow the definition of conventional oil and discuss only gas that is exploitable with existing technologies at 1980 prices.

CONVENTIONAL RESOURCES

Global Estimates

In most regions of the world (the primary exception being the United States and more recently Europe and the USSR), either gas was abundantly available as a by-product of oil production or imported oil was the cheaper fuel. As a result, there has been relatively little exploration for nonassociated natural gas. This has changed considerably since 1973, but the relatively short history of demand for natural gas, beyond that available as associated gas, make data on the occurrence of nonassociated gas relatively scarce. As the one study notes, "it was not long ago that finding a dry gas field somewhere outside the U.S. was considered a catastrophe."[2] Hubbert argues that, with the exception of the United States and Canada, where independent estimates of gas resources are available, "about the best means available for estimating the ultimate amount of natural gas that a given region will produce is to compute the gas ultimate from the prior crude oil estimate by means of a gas/oil ratio."[3] Using this method, he obtains an estimate of 10,000 trillion cubic feet (tcf) of gas. Obviously, this provides a very crude estimate of gas resources.

Table 9-1 summarizes the findings of several researchers on the size of global natural gas resources. It indicates the fairly wide range of estimates—between 6,000 and 16,370 tcf. In fact, the range of estimates is considerably broader than indicated in the table. Odell reported that gas resource estimates discussed at the 1975 World Energy Conference ranged from 5,000 tcf to possibly six times that amount.[4]

Parent concludes that total gas resources are about 10,000 tcf and, based on the standard deviation of various estimates around this central value, argues that the uncertainty is of the order of 2,500 tcf.[5] Yet, even this level of uncertainty, given the early stage and short history of natural gas production, appears overly optimistic with regard to the quality of gas resource estimates. Ion lists four reasons why "all estimates of gas resources are even more nebulous than those of crude oil."[6] In short, he argues that, first, both associated and nonassociated gas may have the same origin as crude oil, but why gas or oil was formed remains controversial. Second, some gas has an origin from material that could not lead to the formation of crude oil. Third, utilization of associated gas depends on crude oil utilization. Fourth, many natural gases contain nonhydrocarbon elements.

TABLE 9-1. World Natural Gas Resources (10^{12} ft^3)

		As Given
1967	Ryman	12,000
1967	Shell	10,200
1968	Weeks	6,900
1969	Hubbert	8,000–12,000
1970	F.P.C.	16,370[a]
1971	Weeks	7,200
1973	Coppack	7,500
1973	Hubbert	12,000
1974	Hubbert	12,800
1975	Kirby and Adams	6,000
1975	Moody and Geiger	8,150
1975	National Academy of Sciences	7,120
1977	American Gas Association	10,510

Source: Joseph A. Parent, *A Survey of United States and Total World Production, Proved Reserves, and Remaining Resources of Fossil Fuels and Uranium* (Chicago: Institute of Gas Technology, 1979), p. 57.

[a]Hendricks' figure revised by Hendricks and Pepper of U.S. Geological Survey.

Regional Estimates

Parent assembled past estimates of global gas resources by region, noting that such estimates were extremely scarce.[7] Based on these estimates and other data, he arrives at regional estimates of natural gas resources. Table 9-2 reports Parent's (IGT), the National Academy of Sciences' (NAS), and IIASA's regional estimates of natural gas resources. The regional definitions are not completely consistent, as noted in the table. The IGT and NAS figures are for total resource, produced, proved, and speculative while the IIASA study figures are for proved reserves and undiscovered resources.

As a base case, estimates of total resources for natural gas by region as developed by the World Energy Conference have been accepted and are reported in table 9-3. It has been necessary to change the groupings to conform to those used throughout our study. In most cases this involved summing over more disaggregated categories. The more recent WEC study (1981) contains estimates of gas resources by country; however, the data are incomplete; many countries did not report estimates for additional resources beyond proved and inferred or speculative reserves.

TABLE 9-2. Regional Natural Gas Resources (10^{12} ft^3)

	IGT[a]	NAS[b]	IIASA[c] Reserves Jan. 1, 1977	IIASA[c] Undiscovered Resource
North America	1,615–2,025	1,669[d]	274	1,536
United States	1,290–1,700	—	—	—
Canada	325	—	—	—
Latin America	800	615[d]	95	530
Africa and South East Asia			126[j]	424[j]
Africa	800	471		
South East Asia	—	—	—	—
Western Europe, Australia, New Zealand, Japan	850[f,g]	1,010[f,i]	179	512
Western Europe	700[f]	493[f]	—	—
Oceania	150[g]	517[i]	—	—
Centrally Planned Countries	3,400[h]	2,658[e]	821	2,437
USSR	—	—	800	2,084
Eastern Europe	—	—	—	—
China and Centrally Planned Asia	—	—	21	353
Middle East	2,500	1,377	747[j]	2,755[j]
Total	9,965–10,375	7,800	Total resource	9,888

[a]Joseph D. Parent, *A Survey of United States and Total World Production, Proved Reserves and Remaining Resources of Fossil Fuels and Uranium* (Chicago: Institute of Gas Technology, 1979), p. 59.

[b]National Academy of Sciences, *Mineral Resources and the Environment* (Washington, D.C., 1975), p. 87. Based on data reported in figures 10, 13, 14, 16 on cumulative production, reserves, and undiscovered resources. Converted from cubic meters to cubic feet using a conversion factor of 0.028317.

[c]International Institute for Applied Systems Analysis, *Energy in a Finite World, A Global Energy Systems Analysis.* Wolf Häfele, Project Leader (Cambridge, Mass.: Ballinger, 1981), p. 65.

[d]Mexico and Caribbean included in North America, excluded from Latin America.

[e]Asia, excluding Indonesia, Brueni, Malasia, etc.

[f]Includes Eastern Europe.

[g]Excludes Japan.

[h]Includes non-communist Asia, including Indonesia, Brunei, Malaysia, but excludes Eastern Europe.

[i]Includes Indonesia, Brunei, and Malaysia.

[j]Includes North Africa as part of the Middle East; North Africa excluded from Africa and South East Asia.

A comparison of tables 9-2 and 9-3 gives an idea of the variety of estimates in this area. However, the level of uncertainty associated with regional gas estimates is considerably greater than is suggested by the dispersion of available estimates, for at least two reasons. First, the global totals based on the regional estimates represent a far narrower range than the range of global gas resource estimates reported in table 9-1. Second, only the NAS and WEC regional estimates have any claim to independence from other estimates. The IGT estimates are based completely on a review of available regional estimates, while the IIASA estimates are based primarily on the WEC estimates.

One should expect gas resources to be more evenly distributed than oil resources because the conditions under which gas was formed have a much broader temperature and pressure range than oil.[8] Such a phenomenon is observed within the United States. This is only partially reflected in the regional gas estimates; among the developed, centrally planned, and Middle East regions, gas resources are considerably more evenly distributed than oil. However, the entire developing country region, including Latin America, Africa, and

TABLE 9-3. Natural Gas Resources, Reserves, and Cumulative Production (10^{12} ft^3 at 1,000 Btu/ft^3)

Region	*Resource*[a]	*Proved Reserves*[b] *(January 1, 1980)*	*Cumulative*[c] *Production (through 1979)*	*Cumulative Share Produced*	*Cumulative Share Discovered*
1. US	2,014	199.885	593.7	0.295	0.374
2. OECD WEST	1,031	223.376	110.9	0.108	0.324
3. JANZ	150	38.250	1.7	0.011	0.266
4–5. Centrally planned total	3,375	963.985	191.0	0.057	0.342
EUSSR	2,993	938.210	186.0[e]	0.062[e]	0.376[e]
ACENP	382	25.775	5.0[e]	0.013[e]	0.081[e]
6. MIDEST	2,315	796.595	80.1	0.035	0.379
7. AFR	755	186.160	22.0	0.029	0.276
8. LA	855	173.395	78.7	0.092	0.295
9. SEASIA	355	88.757	16.6	0.047	0.299
Total	10,850	2,670.403	1,094.5[d]	0.101	0.247

[a]Based on estimates presented by the World Energy Conference (WEC). The resource estimate for OPEC group 1 (non-Arab OPEC) was allocated to Africa, Latin America, and South East Asia after comparing Parent's estimates (p. 59) for these continential regions with WEC's.

[b]*Source: Oil and Gas Journal,* "World Reserves Holding Up Despite Record Production" (December 31, 1979), pp. 70–71. Standardized to 1,000 Btu/ft^3.

[c]Includes marketed, flared, and vented gas and gas used for repressuring. Cumulative production through 1977 from Parent (1979, p. 60). 1978 and 1979 production from *Oil and Gas Journal,* December 31, 1979, with adjustments for nonmarketed gas production. Standardized to 1000 Btu/ft^3.

[d]Compare with E. N. Tiratsoo, *Natural Gas,* 3rd ed. (Beaconsfield, England: Scientific Press, 1979), p. 13; estimate for cumulative world gross production through 1978 of 1,014.9 tcf.

[e]Estimated.

South and East Asia, is estimated to have only between 12 and 18 percent of the world's gas resources, while, even in the case of "conventional wisdom" oil resource estimates of tables 7-3 and 7-4, this region is estimated to have nearly 23 percent of the oil resources of the world. The estimates of oil resources accepted as a base case (table 7-4) show the developing countries accounting for nearly 33 percent of the world's estimated conventional oil.

These considerations suggest that the regional estimates for gas resources in the developing countries may be somewhat low. However, as is the case for global oil estimates, little detail is given, so that it is impossible to determine the factors that were or were not included in any estimate.

HISTORICAL PRODUCTION AND DISCOVERY

Figure 9-1 illustrates the historical rate of production of natural gas. Only since 1955 has production outside the United States begun to expand rapidly; by 1979 the United States accounted for 37 percent of world production, down from 92 percent in 1950. At present, the largest producer of gas outside the United States is the Soviet Union, accounting for approximately 23 percent of world production in 1979. Beyond the United States and the Soviet Union, the major gas producers are Canada, 7 percent of total world gas production in 1979; the Netherlands, 6 percent; China, 4 percent; and the United Kingdom, 2.6 percent. With the exception of the United States, all major gas producers have shown fairly rapid expansion of production through the 1970s.

Together, the six major producers accounted for 80 percent of world gas production in 1979. However, since the increase in world oil prices, the developing countries have shown more interest in developing domestic energy reserves, including natural gas. The development of petro-

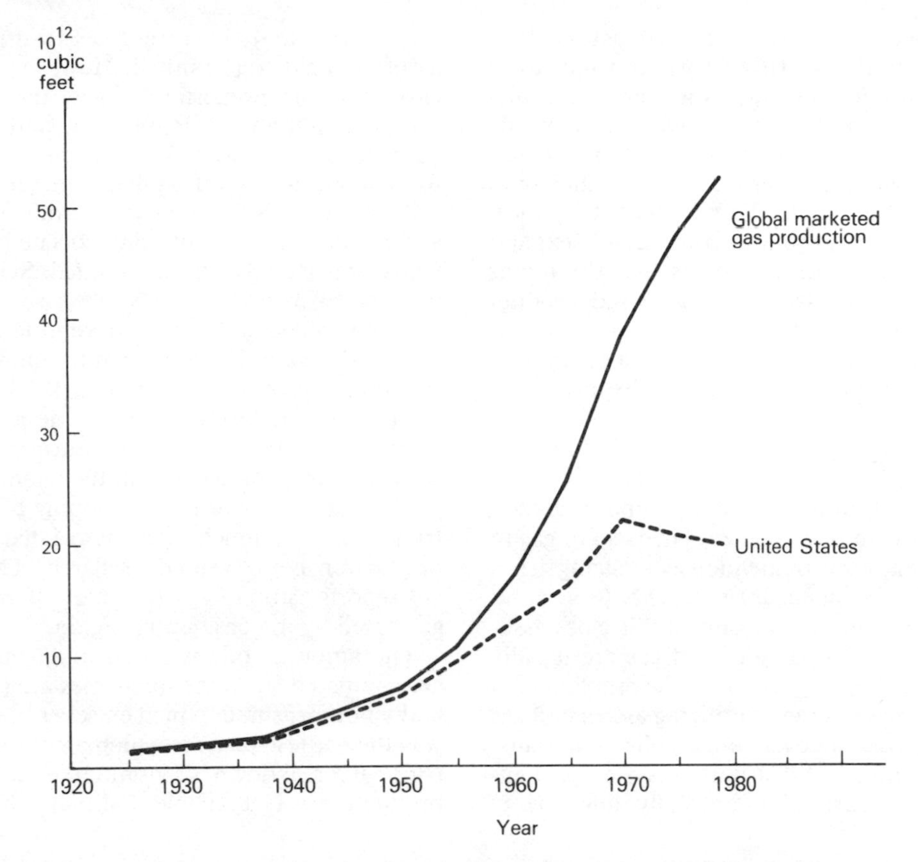

FIGURE 9-1. Historical production of marketed natural gas. (Historical consumption is drawn from Joel Darmstadter, *Energy in the World Economy,* Baltimore, Md., Johns Hopkins University Press for Resources for the Future, 1971; and Energy Economics Research Ltd., *Oil and Energy Trend: 1980 Statistical Review,* Berkshire, United Kingdom, 1980.)

leum resources was aided considerably when, in 1977, the World Bank began lending for gas and oil exploration. Previously, private financing for risky exploratory work was largely unavailable unless the project was undertaken by one of the major oil companies. In most cases, the oil companies were interested only in areas that were believed to have the potential to be significant oil exporters. Countries with the potential for gas deposits or relatively small oil deposits were of little interest. The scope of the Bank program is to finance geological or geophysical surveys to provide accurate information which will either enhance the country's ability to attract foreign capital or assist national petroleum companies to mount well-prepared drilling programs.[9] In 1979, twenty developing countries were producing 0.1×10^9 ft^3 of gas per day or more, up from only twelve countries in 1970.[10] This trend reflects both the increased use of associated gas and the development of nonassociated gas fields.

Despite these developments, a comparison of the 1980 gas reserves by region (table 9-3) and country shares of global gas production in 1979, indicates the relatively underdeveloped gas potential. U.S. gas production is far out of proportion compared to the region's share of reserves. The reserves of the EUSSR, principally the Soviet Union, indicate the tremendous potential for increased gas production. Production in the MIDEST will continue to be constrained by domestic demand limits and transportation considerations. While the reserves of the other developing country regions are considerably smaller than either the MIDEST or EUSSR, they indicate there is room to considerably expand production. Ehrlich notes that the major constraining factor on increased production of gas in several countries is the development of local distribution systems or facilities for liquefaction and export.[11]

Gas Flaring

Gas and oil are normally found in association with one another. The ratio of gas to oil can vary tremendously, ranging from primarily oil to dry gas fields (gas found without oil). As a result of this close association, some amounts of gas are usually produced along with oil. The problem usually encountered in utilizing associated gas is the cost of developing a collection, transportation, and distribution system to handle the gas. With generally low energy prices prior to 1973, incentives to utilize associated gas were limited. However, the clean burning properties of gas, the low capital requirements to burn it, and the convenience of pipeline delivery combined to generate a market sufficient to justify utilization of associated gas production and the development of nonassociated gas fields, particularly in the United States, prior to 1973–74.

As a result, gas flaring and venting had been reduced to a practical minimum in the developed countries prior to 1973–74. In contrast, the lack of sufficient domestic markets and the expense of transporting gas to the areas of demand in the major oil producing regions of the developing countries resulted in much of the associated gas production being vented or flared. Table 9-4 reports estimates of the share of gross gas flared by major country region.

The amounts of gas flared and vented are estimated since the quantities are generally not measured. Nineteen seventy-two was the earliest year for which a complete regional breakdown of nonmarketed gas production was available. Estimates have

TABLE 9-4. Natural Gas Vented, Flared, or Used for Repressuring as a Share of Gross Gas Production

Region	*1972*[a]	*1974*[b]	*1978*[b]
1. US	0.074[c]	0.055	0.063
2. OECD West	0.023[d]	0.070	0.079
3. JANZ	—	0.005	0.021
4–5. Centrally Planned	0.060[e]	0.050	0.055
EUSSR	N.A.	0.049	0.053
ACENP	N.A.	0.107	0.094
6. MIDEST	0.774	0.727	0.693
7. AFR	0.837	0.736	0.608
8. LA	0.653[f]	0.523	0.404
9. SEASIA	0.511[g]	0.349	0.248

N.A. = Not available; — = negligible.

[a]NAS, *Mineral Resources and the Environment,* 1975, p. 111.

[b]Computed from data in U.S. Department of Energy, *Energy Data Report: World Natural Gas* (Washington, D.C.: U.S. Government Printing Office, 1980), pp. 6–11.

[c]Includes Mexico and Central America and Canada.

[d]Excludes Canada.

[e]Includes all of Centrally Planned Asia and Eastern Europe.

[f]Excludes Central America and Mexico.

[g]Indonesia, Malaysia, Brunei, Australia, New Zealand.

been made of cumulative gross gas production and, therefore, account for gas vented and flared prior to 1972. The estimates reported in table 9-3 are an example. Such estimates are inferred from historical oil production and marketed gas production figures and are subject to error. For some countries, current data on flaring and venting are obtained in a similar fashion.

Two cautionary notes should be made concerning the interpretation of the nonmarketed gas shares in table 9-4. First, gas used to repressure oil wells is included. The amount of gas used for this purpose is likely to have increased as rising oil prices have justified a wider application of such enhanced recovery oil techniques. This may partially counter a trend toward less gas flaring and venting. Gas used for repressuring has become the major nonmarket use in the United States; flared and vented gas comprised less than 1 percent of gross gas production while gas used for repressuring was 4.5 percent of gross production.[12]

A second problem is regional differences in dry gas production. Before Australian oil finds began producing, nearly all gas production in this region came from dry gas wells; none of this was flared, vented, or used for repressuring oil wells. The low share of nonmarketed gas registered for this region does not necessarily represent greater success in utilizing associated gas than in the United States, for example, but only indicates that very little associated gas was produced relative to the amount of nonassociated gas.

Together, these considerations suggest some problems in interpreting the differences in nonmarket shares between regions and across time. However, these problems are only secondary. The basic conclusion stands; the developed and centrally planned regions have reduced gas flaring and venting to a practical minimum while the developing regions have shown considerable progress during the 1970s in reducing the amount of gas flared and vented. This point is supported by observed trends toward development of energy-intensive industries in the Middle East,[13] the development of transport and distribution systems within the developing countries,[14] and the development of liquefied natural gas handling facilities at both the shipping and receiving end.[15]

TRANSPORTATION OF NATURAL GAS

Technologies

The first natural gas delivery system is believed to have been built centuries ago by the Chinese, using hollowed-out bamboo for pipes and natural reservoir pressure to move the gas.[16] Since that time, natural gas transportation has grown into a worldwide network of pipelines, tankers, and processing facilities. Today there are two basic types of international gas transport technologies—pipelines and liquefied natural gas (LNG) systems.

Pipelines

The basic principle of using pressure to move gas through pipes is still in operation today. Natural gas is now moved through pipelines under pressure through the use of compressor stations. The pipes are made of steel and are various diameters in width. They can be built above or underground and under water.

Technological improvements in the pipeline system have generated new international trade and have resulted in more than one-half million miles of large diameter pipeline laid throughout major producing areas.[17] Recent improvements in the pipeline system include the use of larger diameter pipelines to handle large volumes of gas. The Soviet Union has 56-inch diameter pipes in operation and has plans to use 64-inch pipes. Underwater lines of 32-inch and 36-inch diameter and reaching a depth of 150 meters are in operation in the North Sea. Other technological improvements include the use of higher pressures and the reduction of fuel gas consumption in compressor stations. The use of sophisticated computers is also enhancing efficiency.[18]

LNG

The transportation of gas long distances overseas requires numerous special facilities and tankers. After producing, gathering, processing, and treating the natural gas, it is moved to a liquefaction plant where it is cooled to −162 degrees centigrade. At that temperature, natural gas turns into a liquid which occupies 1/600th of the volume of the gas.[19] It can then be stored for later loading onto tankers for overseas shipment.

Storing and moving the low-temperature LNG requires specially designed and insulated storage tanks and ships. The storage facilities usually consist of several tanks, with the typical tank holding 95,000 m^3. The standard size is 125,000 m^3. It has been estimated that this capacity tanker carries enough LNG to heat a city of 100,000 for a month.[20]

After the gas is shipped to the receiving terminal, it is unloaded and stored until regasified. The regasification plant usually uses either treated ocean water, river water, or fired heaters to vaporize the liquid gas for distribution to consumers.[21]

The Development of Gas Trade

The volume of gas traded internationally has increased rapidly in recent years. Between 1974 and 1978, growth in trade increased at an average annual rate of 10 percent a year. The amount of gas traded, as a share of total marketed production, increased from 8.9 percent in 1974 to 11.6 percent in 1978. In 1978 the volume of gas reached 6,011 billion cubic feet (bcf) compared to 4,111 bcf in 1974. Approximately 75 percent of the increase was accounted for by exports from the Soviet Union (37.6 percent), Norway (26.5 percent) and Indonesia (10.4 percent).

Most of the trade in 1978 occurred between a few countries. Exports from the Netherlands, the Soviet Union, and Canada accounted for 30.6 percent, 20.1 percent, and 14.7 percent of the total respectively. Shipments from the Netherlands went to Belgium, France, Germany, and Italy. Over half of the Soviet Union's exports went to Western Europe, with the other half going to Eastern Europe. All of Canada's exports went to the United States. The largest importers were Germany, the United States, Italy, and Japan. Together they accounted for approximately 60 percent of the total world imports. The flows of trade are illustrated in figure 9-2.

Pipelines

Most of the gas traded internationally is moved by pipelines to neighboring countries. The largest share of this trade occurs among countries in Eastern and Western Europe and the Soviet Union. Figure 9-3 shows a map of the 1978 flow of pipeline trade among these countries. Of the 5,055 bcf traded in 1978, Western Europe imported 3,166 bcf. Of this total, 2,530 bcf came from countries within region 2 (OECD West), primarily from the Netherlands and Norway's North Sea operations. The remainder of Western Europe's imports came from the Soviet Union.

Region 4 (EUSSR) is the second most active area of natural gas trade. In 1978 the Soviet Union exported 1,210 bcf, of which 580 bcf were consumed by countries within the region. In region 1 (US) the United Stated imported 881 bcf of natural gas from Canada and exported 4 bcf to Mexico. Trade throughout the rest of the world was minimal. The Soviet Union imported 256 bcf from Iran and 82 bcf from Afghanistan. And in Latin America, 80 bcf of gas from Bolivia and Chile went to Mexico.

Major Trends and Recent Pipeline Activity

U.S./Canada

The largest pipeline currently under construction is the Alaskan Highway project. This 4,800-mile system is designed to carry

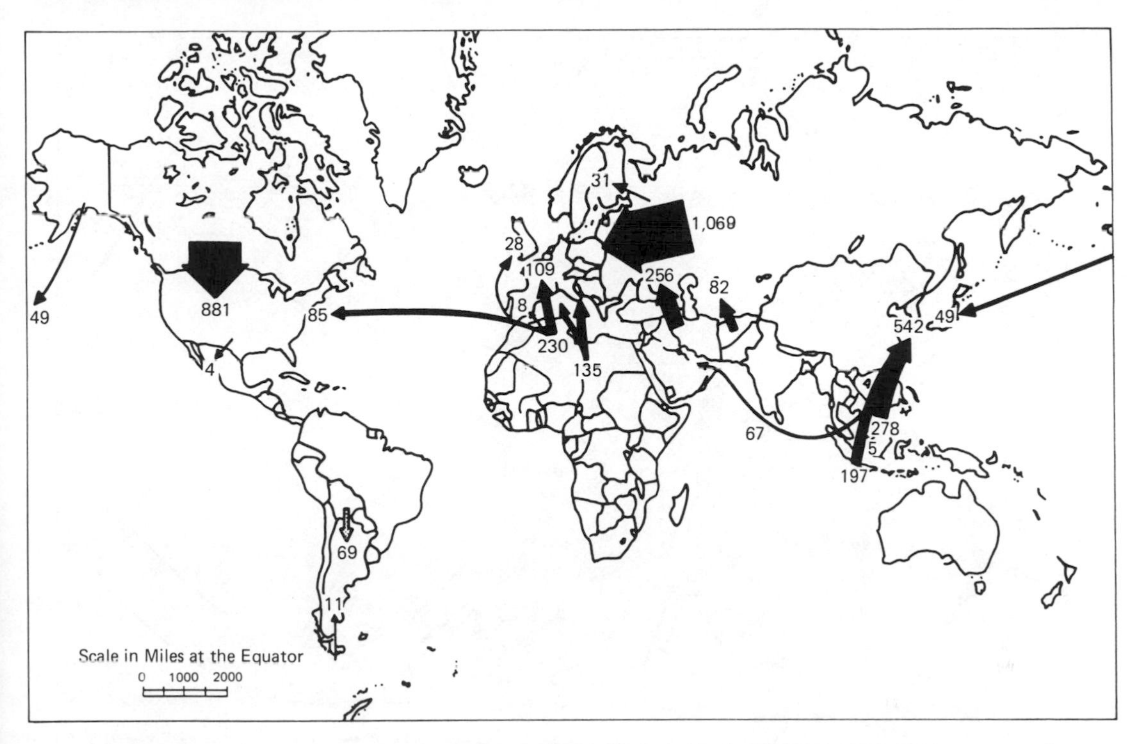

FIGURE 9-2. International natural gas flow, 1978 (billion cubic feet). (From U.S. Department of Energy, Energy Information Administration, *1980 Annual Report to Congress,* vol. II, no. DOE/EIA-0173(80)/2, Washington, D.C.: U.S. Government Printing Office, April 13, 1981, p. 114.)

about 2.4 bcf/day of Prudhoe Bay gas to the U.S. lower 48 states, beginning in 1985–86. It will cross Alaska and Canada, splitting north of Calgary. The western leg will run to San Francisco, the eastern leg to Chicago.[22] The amount of gas to be carried by this system represents about 5 percent of current U.S. consumption. However, it is estimated that by the turn of the century, the capacity of the pipeline could triple.[23] In the meantime, Canadian exports to the United States are expected to remain stable throughout the 1980s, but as Canadian domestic consumption increases, exports are expected to decline until phased out by 1995.[24]

Soviet Union/Western Europe and Other Regions

The Soviet Union is currently constructing a 3,000-mile pipeline system to carry Siberian gas to Western Europe. The pipeline's projected capacity of 1.4 trillion to 2.4 trillion cubic feet of gas a year (tcf/y)[25] would supply additional gas to ten Western European countries: West Germany, Italy, Austria, France, Belgium, Switzerland, the Netherlands, Sweden, Finland, and Greece.[26] The pipeline is expected to be built largely on the basis of western financing, with West Germany providing the largest share of credit.[27] A tentative financing agreement was reached in July 1981 between West Germany and the Soviet Union, so preparatory work could begin in the winter of 1981.[28] Completion is not expected before 1986.[29] A major implication of this trade would be the increasing dependence of Western Europe on Soviet gas supplies. It is estimated that by the 1990s, West Germany would be dependent upon the Soviet Union for up to 40 percent of its gas supplies.[30]

During 1978 the Soviet Union imported

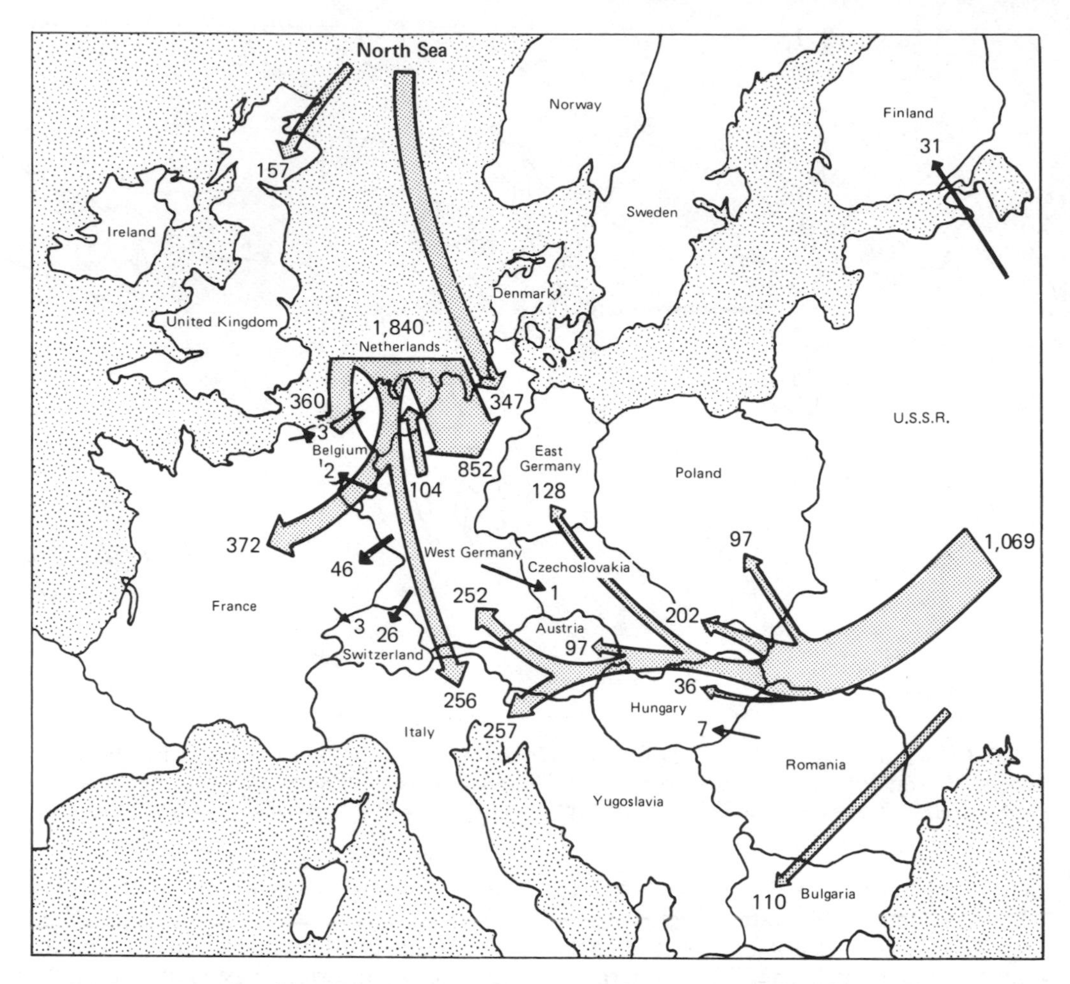

FIGURE 9-3. Natural gas trade by pipeline in Eastern and Western Europe and the Soviet Union, 1978 (billion cubic feet). (From U.S. Department of Energy, Energy Information Administration, *1980 Annual Report to Congress,* vol. II, no. DOE/EIA-0173(80)/2, Washington, D.C.: U.S. Government Printing Office, April 13, 1981, p. 114.)

256 bcf of gas by pipeline from Iran. Since the fall of the Shah in 1979, Iranian deliveries of natural gas to the Soviet Union under their old agreement (IGAT I) have been interrupted and possibly terminated and deliveries from the projected line (IGAT II), discussed in 1975–79, are not believed likely under the current Iranian government.[31]

Afghanistan exports an estimated 84–91 bcf of natural gas to the Soviet Union from its Hodja-Gugerdag field.[32] According to Moscow reports, Afghanistan's newly developed field of Djar-Kuduk near the Soviet border could produce as much as 70 bcf a year. It is believed that most of the initial flow from this field will be transmitted to the Soviet Union.[33]

Africa/Western Europe

The construction of the trans-Mediterranean line from Algeria to Italy is on schedule and due to be completed by 1982. The system involves 342 miles of gas line in Algeria; 230 miles in Tunisia; a three-line, 99-mile Mediterranean crossing from Cape Bon, Tunisia, to Mazara del Vallo, Sicily; 219 miles in Sicily; 9 miles and 4 lines across the Strait of Messina; and 653 miles in continental Italy.[34] The system was originally planned to carry 1.2 bcf/day, but with the addition of a fourth line across the Strait of Sicily, deliveries could reach 1.7 bcf/day. Plans for another trans-Mediterranean pipeline from Algeria to Spain are underway. It is believed that the proposed system could deliver nearly 1.0 bcf/day by 1985.

Western Europe
A variety of projects are underway in Europe, with much of the activity centered around North Sea production. The largest project is a proposed 490-mile line designed to connect two North Sea fields to a third line where the gas would then run to a terminal in Scotland.[35] Elsewhere in Europe, construction activity is primarily within national boundaries. With the exception of the projects discussed earlier, a few smaller international projects are also proposed or under construction.

The Netherlands currently supplies over half of Europe's pipeline imports, but increasing domestic demand has led the Netherlands to caution that it may not renew its current export contracts when they expire at the end of the 1980s and in the 1990s.[36]

Latin America
There is very little international trade of natural gas in Latin America. Small quantities are currently exported from Bolivia and Chile to Argentina. Recent activity includes the conclusion of a feasibility study by Brazil and Argentina in July 1981 for 1,429 miles of pipeline to initially deliver 350 million cubic feet (mcf) of gas to Brazil.[37] The plans are awaiting final political approval.

Natural gas trade between the United States and Mexico was never large and has declined rapidly in recent years. In 1974 the United States exported 14 bcf of gas to Mexico, while Mexico exported less than 500 million cf to the United States. By 1978 the United States was exporting only 4 bcf to Mexico, and Mexico was not exporting any natural gas to the United States.[38] Since this time, Mexico's policy has been to use as much gas as possible domestically instead of exporting large volumes to the United States.[39]

China
China is not a participant in the international trade of natural gas, and nationally it lacks a comprehensive gas distribution system. No long-distance pipelines are known to exist in China. Aside from the 623 miles in Szechuan Province, pipelines elsewhere are minimal. Even the gas from the Takang and Sheng-Li fields is not piped to large nearby cities, such as Peking, but rather is used in the manufacture of petrochemicals or is reinjected into the producing reservoirs to maintain pressure.[40]

Liquefied Natural Gas

Movement of gas by LNG tanker accounts for a small but rapidly growing share of the world gas trade. In 1974 LNG shipments accounted for 398 bcf or 9.7 percent of total trade. By 1978 the volume had increased to 956 bcf and accounted for 16 percent of all trade. The largest importer of natural gas was Japan. It imported 591 bcf from Brunei, Indonesia, the United Arab Emirates, and the United States. Western Europe was the second largest consumer of LNG, with 280 bcf from Algeria and Libya. The United States imported 85 bcf from Algeria.

Since the first delivery of LNG from Algeria to the United Kingdom in 1964, trade has expanded to include six exporting countries operating thirteen projects with contract volumes totaling over 2 tcf by July 1979. (Since some of these projects have not reached their full plateau rates, actual deliveries are less than contracted volumes.) In addition to the operating

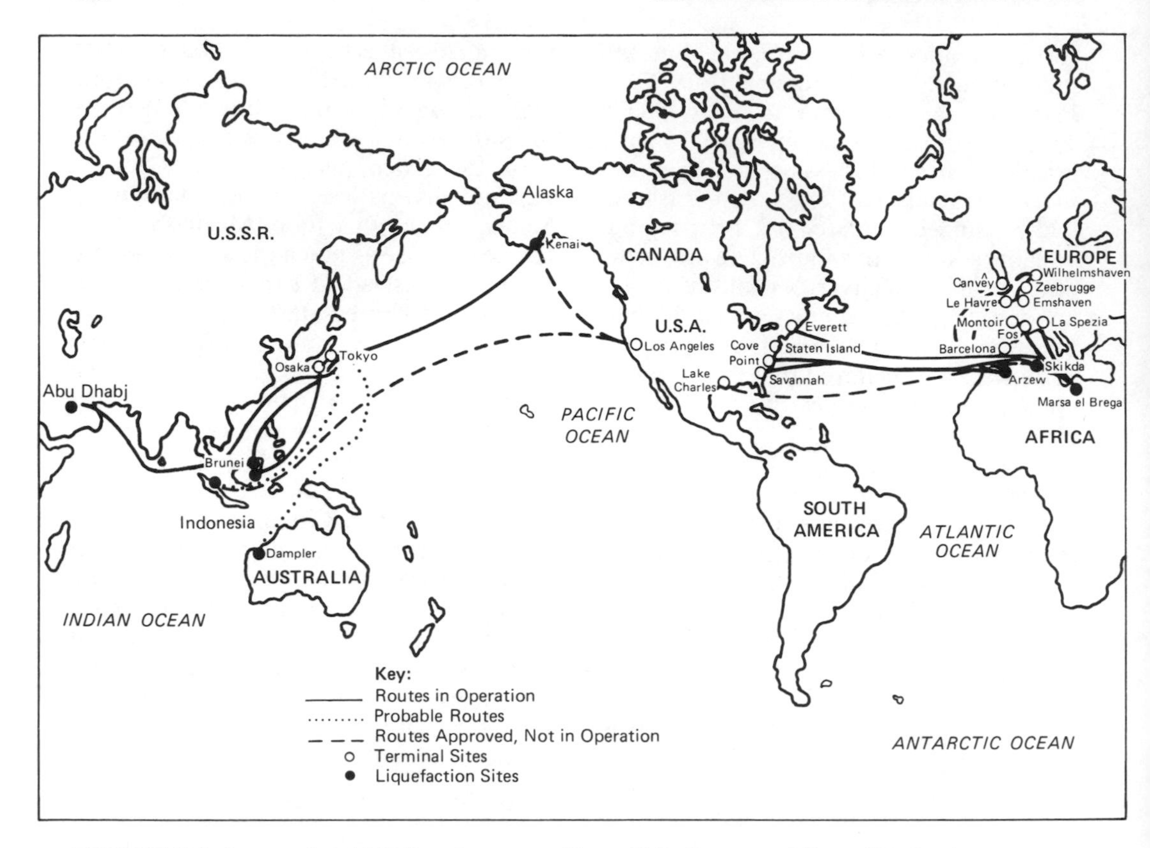

FIGURE 9-4. International LNG trade routes. (From U.S. Congress, Office of Technology Assessment, *Alternative Energy Futures,* Part 1, Report No. OTA-E-110, Washington, D.C.: U.S. Government Printing Office, March 1980, p. 2.)

projects, at least thirty other projects have been suggested, including projects in Indonesia, Australia, Malaysia, Chile, and Bolivia.[41]

These trade routes are illustrated on the map in figure 9-4. Algeria has approved projects with four European countries and the United States, which would add another 1,150 bcf of gas to international trade in the 1980s. If possible Algerian projects are added to this volume, the total could reach close to 2 tcf of additional gas. Possible exports from the Soviet Union to Japan and the United States could amount to 1,500 bcf. Other potentially large projects include 600 bcf from Nigeria to the United States and Europe, 480 bcf from Indonesia to the United States and Japan, and an equal amount from Australia to the United States and Japan. If all the approved, probable, and possible projects are realized during the 1980s, then additional contract capacity could reach more than 7 tcf, with ten new exporting countries accounting for approximately 60 percent of the new capacity. Further in the future, countries such as China, Thailand, and New Zealand could become LNG exporters.

Pipeline Capital Costs

The cost of transporting gas by pipeline varies enormously. The most important factors affecting the final cost include the diameter and length of the pipeline, the location (either on or offshore), and such variables as the nature of the terrain and the number of river and road crossings. In addition, offshore construction tends to be more expensive. The average U.S. onshore pipeline costs range from about $150,000 (1979 dollars) per mile for 8-inch pipelines up to about $700,000 per mile for 36-inch pipelines In contrast, offshore costs average about $375,000 per mile for 8-inch pipelines and up to $800,000 for 36-inch pipelines.[42]

The estimates outlined above are only indicative; costs vary tremendously from project to project, depending on the terrain and other factors. Examples of this range include two of the most ambitious international gas pipeline projects—the proposed Soviet-European pipeline and the Alaskan U.S.–Canadian pipeline. The Soviet-European pipeline, running 3,000 miles from western Siberia, is expected to cost at least 10 billion, or 3.3 million dollars per mile.[43] The U.S.–Canadian Alaskan natural gas system of 4,800 miles has escalated to a cost of 30 billion, or 6.3 million per mile.[44] Both these costs are considerably higher than those outlined above.

The major investment costs for pipelines are the line pipe and construction of the pipeline. Together they account for over two-thirds of the total cost. Other major costs include conditioning plants and metering stations. In addition, hidden costs, such as inflation and regulatory delays, can increase the initial cost estimate by as much as 30 to 100 percent.[45]

LNG Capital Costs

LNG projects are expensive, long, and complex undertakings. The chain of supply from producer to consumer requires the construction of liquefaction and regasification facilities, LNG ships, marine terminals, and storage facilities. The initial investment cost for these facilities and ships runs into billions of dollars. For a world-scale project involving 1 bcf/day beginning in the early to mid-1980s, the total capital required is estimated to be about 6 billion (1981). Gas production and liquefaction facilities account for approximately 40 percent of this cost, another 40 percent is applied to the cost of LNG ships, and the remaining 20 percent is for import terminals and regasification facilities.[46]

These shares are by no means exact. An important variable influencing the total investment cost is the length of the voyage. A typical LNG ship costs between 150 and 200 million dollars. For a relatively short voyage from, for example, the Caribbean to North America, six ships costing between 900 million and 1.3 billion would be sufficient to assume service for an average-sized project. However, for a long voyage, such as from the Middle East to Japan, two to three times as many ships would be needed, bringing the cost to between 1.8 and 3.6 billion.[47]

The financing of an LNG project requires a complex consortium of several partners, including a seller, a buyer, and a transporter, each with their associated affiliates. All financing and pricing contracts as well as site approval for plants and pipelines and government reviews usually have to be agreed upon before the project can begin.[48] Furthermore, the development and implementation of the project can easily extend over 15 years.[49] For these reasons, contracts between buyers and sellers are long term, typically involving 20 to 25 years.[50]

Prices

There is a vast range of prices for internationally traded gas. Prices generally reflect the value to the buyer and are therefore likely to reflect the price of competing fuels. Since the cost of transporting gas can differ significantly, depending upon the distances involved, there is no uniform f.o.b. or c.i.f. price. For example, Algeria receives a significantly lower price from the United States for its gas than Canada does. The f.o.b. price of Algerian LNG paid by the United States is $4.11/mbtu (July 1981). Ocean transport adds $2.00 to this price and storage, terminalling and regasification add another $1.25, so the final delivered price is $7.37/mbtu.[51] At the same time, Canada receives $6.41/mbtu for its gas to the United States; however, after trucking and regasification costs are added, the final delivered price of $7.62/mbtu[52] (April 1981) is roughly equivalent to Algeria's gas. In addition, Algeria's f.o.b. price differs in each market. Belgium, for example, has agreed to pay Algeria almost $1 more per mbtu than the United States (April 1981).[53]

Despite these price disparities, a major trend in gas pricing can be perceived. The price of gas is moving closer to oil prices. The export price of gas has risen from a small fraction of the equivalent oil price to about two-thirds. Producing nations have argued that the export price should be on a par with oil exports. In the case of Algeria, this would mean about $7/mbtu. Consuming nations reject this argument on the basis of the added high costs of transporting, storing, and regasifying LNG.

Comparison of Internationally Traded Oil and Gas Costs

For short distances up to 8,000 kilometers, pipeline transport is the most economical way to move gas because of the liquefaction cost saving. Beyond this distance, LNG is the less expensive alternative.[54] In comparison with oil transport, both modes of gas transport are more expensive. The cost of transporting natural gas by pipeline is roughly twice that for crude oil[55] and sea transport for LNG is roughly 5 to 7 times as expensive.[56] Furthermore, for producing countries, the cost of liquefying gas relative to the energy content of oil products is 1½ to 2 times the cost of refining crude oil in a fuel-type refinery and the energy consumed in the process is twice as much as petroleum refining.[57] Other major differences between oil and gas costs include the large front-end investment required for gas projects and the lengthy contract terms. In contrast, crude oil projects usually do not require such large initial investments and the contracts seldom cover more than one or two years.[58]

These added costs have limited the attractiveness of gas trade to many potential producers, particularly in the Middle East, where oil attracts a much higher price for less cost than gas. On the other hand, it is clear that LNG and pipeline projects can be very profitable and are certainly preferable to flaring gas.

ENVIRONMENTAL AND SAFETY ISSUES

Liquefied Natural Gas

The major environmental concern associated with liquefied natural gas is the likely effects of an LNG spill. The consequences of a major spill, particularly in a densely populated area, where LNG is typically stored and transported, could be catastrophic. The resulting gas cloud could spread into underground sewers or subways, creating explosions and fires across a very wide area. The explosions and widespread fires would be beyond the capability of firefighting methods.[59]

The results of previous smaller spills provide some insight into the possible consequences of a large spill. For example, in June 1977 a spill of 15 cubic meters of naptha (which is much less volatile than LNG) into the sewers of Akron, Ohio, created violent explosions more than 8 miles from the point of the spill.[60] The only major LNG spill in the United States occurred in 1944 when a storage tank in Cleveland ruptured, spilling 6,200 cubic meters into the streets and sewers. The resulting fires and explosions killed 128 people and injured 300.[61]

Storage

Until recently, natural gas was liquefied primarily for convenience of storage, with the gas from these storage facilities used during peak demand. These plants have existed for decades with very few accidents. However, the expansion of gas use and international LNG trade has led to a parallel increase in both the number and size of storage facilities, as well as LNG ships and trucks. The quantities stored in tanks today are about 15 times as large as the tank that ruptured in Cleveland and the probability of a major accident increases with the number of facilities and the years of operation.[62]

On the other hand, improvements in safety standards have reduced some of the risks. For example, the U.S. Bureau of Mines concluded that the Cleveland tank failed because it was made of 3.5 percent nickel steel, which becomes brittle on contact with the extreme cold of LNG. It has since become standard practice in the LNG industry to use 9 percent nickel steel, aluminum, or concrete and to surround storage facilities with dikes capable of containing the contents of the tank.[63] These improvements, however, do not eliminate the risks associated with natural forces, such as earthquakes or floods, nor do they address the possibility of sabotage. Furthermore, some of the improvements provide only a limited safety margin. For example, the dikes are designed to contain the spill of LNG from relatively slow leaks. They cannot contain the surge of LNG from a massive rupture or the collapse of a tank wall.[64]

Transport/Ships

The most likely cause of a major spill from an LNG tanker is a ship collision and the most likely place for a collision to occur is in a busy port shipping channel. Since most ports are adjacent to large urban areas, the magnitude and extent of the damage from a ship collision could be very severe. Due to the large quantities of LNG carried by tankers and the fact that a water spill would spread much farther and evaporate much more quickly than a land spill, an LNG ship collision is believed to be the most likely event that could trigger the most serious type of LNG accident.[65]

However, of all the LNG transportation and storage systems, LNG ships are believed to be the least vulnerable. These ships are specifically designed to withstand major damage. The standard 125,000 cubic meter tanker usually has five cargo tanks made of aluminum alloy or 9 percent nickel steel, with layers of insulation on the outside. These tanks are either welded to cylindrical skirts or tied to supporters which are welded to the ship structure. The ship's hull becomes, in effect, the outer tank so that LNG ships are double hulled, unlike the single-hulled oil tankers.[66] A demonstration of the protection provided by this double hull occurred in June 1979 when the tanker *El Paso Paul Kaiser* ran aground while on a loaded voyage off Gibralter. The ship remained on the rocks for four days, sustaining extensive damange over the entire cargo area but no LNG was released.[67] If the ship had been an oil tanker, there would have been a massive spill. However, it is possible that if the grounding had occurred in bad weather rather than the calm seas which existed at the time, the results could have been much different.[68]

Transport/Trucks

The possibility of an LNG accident occurring to a truck is much greater than for a ship, although the quantities of LNG carried by truck are only a small fraction of the amount contained in ships. A typical truck contains about 40 cubic meters of LNG. If spilled on an urban street, however, this quantity is enough to fill more than 110 miles of a 6-foot diameter sewer line, or 15 miles of a 16-foot diameter subway system.[69]

LNG Risk Reduction

Numerous actions can be taken to reduce the likelihood of a major LNG spill. For storage facilities, the risk of a massive rupture or collapse of a tank wall can be virtually eliminated by building tanks underground. Japan uses many underground storage tanks which are operating satisfactorily and cost about the same as aboveground tanks with dike installations. Risks can be further reduced by building all new facilities in remote areas and requiring more stringent building and operating standards, such as those used for nuclear power plants. Sabotage could be deterred by requiring guards at LNG facilities to carry weapons and use them if necessary. In the transportation of LNG, the most important preventive measure is highly trained personnel in the ships, ports, and terminals, as well as measures to divert LNG trucks from routes which are the most susceptible to accidents.[70]

International LNG Safety Requirements

Unlike other energy industries, such as coal, whose practices, facilities, and equipment can vary enormously among different countries, LNG transport and storage technology is standardized throughout the world. Therefore, the risks associated with its use are roughly comparable from country to country. Added safety precautions can and have been implemented to various degrees in different countries, although it is impossible to predict which countries will adopt what measures in the future or to what extent these measures will prevent a major spill.

The Odds of a Major LNG Spill

Table 9-5 compares the risks of LNG operations with known risks. It shows that the likelihood of a fatality from an LNG operation is equal to about 1 in 10 million to 1 in 100 million. In comparison, the likelihood of a fatality from lightning is 1 in 50 million. Nonetheless, it is important to note that only one major spill could set back the LNG industry decades. The Cleveland accident halted U.S. use of LNG for 20 years, and in comparison with what could happen today in much larger facilities and ships, this was a minor accident.[71]

TABLE 9-5. Comparison of LNG Operations Risk Assessment with Known Risks

Accident Type	*Fatality Probability per Exposed Person per Year*
Motor vehicles	1 in 27,000
Falls in public places	1 in 240,000
Fires in homes	1 in 260,000
Falls in homes	1 in 460,000
Fires in public places	1 in 3,000,000
Air transport	1 in 5,000,000
Poisoning by gases	1 in 5,000,000
LNG operations	1 in 10,000,000 to 1 in 100,000,000
Lightning	1 in 50,000,000

Sources: Supply of Liquefied Natural Gas to the Northeast (Upton, N.Y.: National Center for Analysis of Energy Systems, Brookhaven National Laobratory for U.S. Energy Research and Development Administration, April 1976), p. 68. *Accidents and Unscheduled Events Associated With Non-Nuclear Energy Resources and Technology* (Washington, D.C.: U.S. Environmental Protection Agency, February 1977), p. 174. Figures shown for LNG were derived from assessments performed by Science Applications Inc. and El Paso Alaska Co. Range reflects different sites and methodologies. Cited in American Gas Association, *LNG Fact Book* (Arlington, Va.: American Gas Association, December 1977), p. 26.

Pipeline Construction

The environmental effects of pipeline construction are minimal compared with LNG safety hazards. In the United States, concern has been expressed over the possible adverse effects of the Alaskan pipeline. The Department of the Interior's report on the Alaska natural gas transportation system lists the following outstanding impacts the pipeline would have on the environment:[72]

Permafrost. Thawing will occur in the ice-rich, fine-grained material. The extent of the thawing will depend on the care that the applicants take during construction.

Geology. Construction of the pipeline trench may trigger mud flows, soil falls, and landslides.

Soils. Stripping of the vegetation along the right-of-way will expose the soil to erosion by wind and water along the entire length of the route.

Water Resources. Large amounts of water will be required that may be difficult to obtain during winter construction periods.

Vegetation. Vegetation will be destroyed along the entire length of the route.

Wildlife. Pipeline construction will have a detrimental effect on various mammalian species.

Wilderness and Historical Sites. Other impacts include aesthetic degradation of the wilderness and possible damage to archeological and historical sites.

Many of these impacts apply only to the Alaskan pipeline; others can apply to pipeline construction elsewhere in the world. The fact remains that these environmental factors have not been a binding constraint in the construction of pipelines in the past. It is doubtful that they will inhibit growth in the future.

NOTES

1. National Academy of Sciences, *Mineral Resources and the Environment* (Washington, D.C.: 1975), p. 87.
2. Wolf Häfele, Project Leader, *Energy in a Finite World: A Global Systems Analysis,* Report by the Energy Systems Program Group of the International Institute for Applied Systems Analysis (Cambridge, Mass.: Ballinger, 1981), p. 210.
3. M. King Hubbert, "World Oil and Natural Gas Reserves and Resources," in *Project Interdependence: U.S. and World Energy Outlook through 1990* (Washington, D.C.: Congressional Research Service, 1977), p. 635.
4. Peter R. Odell, "Energy Resources and Energy Demand," *Energy Policy* (September 1975), p. 251.
5. Joseph D. Parent, *A Survey of United States and Total World Production, Proved Reserves and Remaining Resources of Fossil Fuels and Uranium* (Chicago: Institute of Gas Technology, 1979), p. 59.
6. D.C. Ion, *Availability of World Energy Resources,* 2nd ed. (London: Graham and Trotman, 1979), p. 36.
7. Parent, *A survey of United States and Total World Production.*
8. Oral communication with spokesman from the American Gas Association.
9. World Bank, *Energy in Developing Countries* (Washington, D.C.: World Bank, 1980).
10. Energy Economics Research Ltd., *Oil and Energy Trend: 1980 Statistical Review* (Berkshire, United Kingdom, 1980).
11. Everett Ehrlich, "Oil and Gas Prospects in Non-OPEC Less Developed Countries." Paper presented at the Workshop on Energy and the Developing Countries (Palo Alto, Calif., 1980), pp. 5, 9, 13.
12. U.S. Department of Energy, Energy Data Reports, *Natural Gas Production and Consumption: 1977* (Washington, D.C.: U.S. Government Printing Office, October 18, 1978).
13. Al Janabi Adnan, "Estimating Energy Demand in OPEC Countries," in *Energy Economics* (April 1979).
14. Ehrlich, "Oil and Gas Prospects."
15. Ibid.
16. John Kean, "China's Gas Industry, A Sleeping Giant," *Pipeline and Gas Journal* (October 1979), p. 40.
17. United Nations, Economic and Social Council, Committee on Natural Resources, 6th Session, *Reassessment of Natural Gas Prospects* (E/C.7/106), April 20, 1979, p.4.
18. 14th World Gas Conference, Toronto, 1979, *Report of Committee C, Transmission of Gases* (Toronto: Bruce Handerson, 1979), p. 13.
19. United Nations, *Reassessment of Natural Gas Prospects,* p. 35.
20. U.S. Congress, Office of Technology Assessment, *Transportation of Liquefied Natural Gas* (Washington, D.C.: U.S. Government Printing Office, September 1977), p. 15.
21. U.S. Department of Energy, *Leading Trends in Environmental Regulations That Affect Energy Development,* Final Report, prepared by Flow Resources Corp. and International Research and Technology Corp., Contract #DE-AC03-78EV-01682 (Washington, D.C.: Government Printing Office, January 1980), p. 124.
22. "Early Building Okayed for Alaska Line Segment," *Oil and Gas Journal,* January 7, 1980, p. 53.
23. Robert J. Samuelson, "Here Comes the World Gas Market—More Energy and New Uncertainties," *National Journal,* January 31, 1981, p. 188.
24. "GAO: Don't Pin Energy Hopes on Alaska, Canada, Mexico," *Oil and Gas Journal,* September 22, 1980, p. 53.
25. John F. Burns, "Soviet Speeding Pipeline Deal," *New York Times,* August 5, 1981, p. D15.

26. U.S. Congress, Joint Economic Committee, *Energy in Soviet Policy,* 97th Cong., 1st sess. (Washington, D.C.: Government Printing Office, June 11, 1981), p. 96.
27. Ibid., p. 97.
28. John F. Burns, "Soviet Speeding Pipeline Deal," p. D15.
29. "Pipelines Hold Key to Soviet Gas Production," *Oil and Gas Journal,* June 29, 1981, p. 42.
30. John F. Burns, "Soviet Speeding Pipeline Deal," p. D15.
31. U.S. Congress, *Energy in Soviet Policy,* p. 96.
32. "Afghanistan Gas Production Capacity Rises 65%," *Oil and Gas Journal,* February 18, 1980, p. 66.
33. Ibid.
34. "Pipelines Set Sights on 51,401 Miles of Construction, *Oil and Gas Journal,* January 19, 1981, p. 28.
35. Ibid.
36. Robert J. Samuelson, "Here Comes the World Gas Market," p. 185.
37. "Argentine-Brazil Gasline 'Feasible'," *World Gas Report,* July 6, 1981, p. 9.
38. U.S. Department of Energy, *World Natural Gas 1978,* Report No. DOE/EIA-0133(78) (Washington, D.C.: U.S. Government Printing Office, July 1980), pp. 32–33.
39. "World Pipeline Projects Aiming For 62,376 Miles," *Oil and Gas Journal,* January 21, 1980, p. 30.
40. John Kean, "China's Gas Industry," p. 44.
41. Op. cit. footnote 24 and Everett M. Ehrlich, "Oil and Gas Prospects in Non-OPEC Developing Countries," in Peter Auer, ed., *Energy and Developing Nations* (New York: Pergamon, 1981), pp. 13–33.
42. Earl Seaton, "Pipeline Building Costs Continue Steep Climb," *Oil and Gas Journal,* August 13, 1979, p. 69.
43. Robert J. Samuelson, "Here Comes the World Gas Market," p. 186.
44. "Alaska Gasline Pushes for Producer Equity," *Oil and Gas Journal,* June 1, 1981, p. 54.
45. Earl Seaton, "Pipeline Building Costs," p. 82.
46. U.S. Congress, Office of Technology Assessment, *Alternative Energy Futures, Part 1, The Future of Liquefied Natural Gas Imports* (Washington, D.C.: U.S. Governemnt Printing Office, March 1980), p. 69.
47. United Nations, *Reassessment of Natural Gas Prospects,* p. 36.
48. Robert J. Samuelson, "Here Comes the World Gas Market," pp. 185.
49. P. Takis Veliotis, "Marketing LNG Vessels to U.S. Import Projects—The Shipbuilders' Experiences and Risks," *Sixth International Conference on Liquefied Natural Gas,* Kyoto, Japan, April 4–10, 1980, vol. 2 of 2, sessions III & IV (Chicago: Institute of Gas Technology, 1980), p. 4.
50. Robert J. Samuelson, "Here Comes the World Gas Market," p. 185.
51. "Algeria—U.S. Gas Trade Restarts Next Month," *World Gas Report,* July 6, 1981, p. 6.
52. "ERA Okays LNG Imports at 6.41/MMBtu," *Oil and Gas Journal,* April 27, 1981, p. 115.
53. Roger Vielvoye, "Parity for LNG?" *Oil and Gas Journal,* April 20, 1981, p. 41.
54. M. J. Prior and M. Teper, "The Future Supply of Gas," *Energy Policy* (December 1980), p. 311.
55. P. J. Anderson, "Steady Growth Seen in Next Decade for World Trade in LNG," *Pipeline and Gas Journal,* March 1980, p. 44.
56. Robert J. Samuelson, "Here Comes the World Gas Market," p. 187.
57. T. Mossadeghi, "The Prospects of LNG Export & Its Limitations," *Fifth International Conference on Liquefied Natural Gas,* Dusseldorf, Germany, August–September 1977, vol. 1 of 2, sessions I & II (Chicago: Institute of Gas Technology, 1977), p. 8.
58. United Nations, *Reassessment of Natural Gas Prospects,* p. 37.
59. OTA, *Transportation of Liquefied Natural Gas,* p. 8.
60. U.S. General Accounting Office, Report to the Congress, *Liquefied Energy Gases Safety,* vol. 1, no. EMD-78-28 (Washington, D.C.: U.S. Government Printing Office, July 31, 1978), p. 2.
61. OTA, *Transportation of Liquefied Natural Gas,* p. 9.
62. GAO, *Report to the Congress, LNG Safety,* p. 24.
63. OTA, *Transportation of Liquefied Natural Gas,* p. 9.
64. GAO, *Report to the Congress, LNG Safety,* p. 9.
65. OTA, *Transportation of Liquefied Natural Gas,* p. 18.
66. Ibid., pp. 15–16.
67. Roger Flooks, *Natural Gas by Sea* (London: Gentry Books, 1979), p. 208; and Dean Hale, ed., "Developments Proceed Slowly in World LNG Industry," *Pipeline and Gas Journal,* March 1980, p. 18.
68. Roger Flooks, Natural Gas by Sea, p. 210.
69. GAO, *Report to the Congress, LNG Safety,* p. 17.
70. Ibid., pp. 10–23.
71. Ibid., p. 25.
72. U.S. Department of the Interior, *Alaska Natural Gas Transportation System, Final Environmental Impact Statement* (Washington, D.C.: U.S. Government Printing Office, 1976), pp. 136–138.

Unconventional Gas 10

Among the fossil fuels, perhaps the least is known about the magnitude of global unconventional gas resources and cost of recovery. Only in the United States has much effort been invested in exploring for and developing technologies to exploit unconventional gas resources. This term is used to cover a broad category of gas resources, from tight gas formations to geopressured gas to gas produced from coal and biomass. It is possible to upgrade gas produced from any of these sources to pipeline quality (roughly 1,000 Btu/ft^3), allowing it to be introduced directly into existing conventional natural gas pipeline and distribution systems. Once upgraded, the end-use qualities of each are, for practical purposes, identical to conventional natural gas. However, the supply technologies differ tremendously across these sources. Each has problems or critical uncertainties associated with it, ranging from immature technology to cost.

RESOURCE BASE

Table 10-1 lists the various unconventional gas resources and the estimated existing resource base. In general, the figures for non-U.S. unconventional gas, if available at all, are highly speculative. In the case of gas from biomass, the resource base is, appropriately, the land or global surface area and the energy input is renewable solar energy. For biogas, the table includes some estimates of theoretical contributions of biomass to the energy flow. It should be noted that no one methodology for computing reasonable estimates for the theoretical contribution of renewables is firmly established. Global insolation is many orders of magnitude larger than commercial energy consumption but biomass must also compete for land area with food crops and other land uses. On the other hand, it may be possible to farm the sea for biomass crops. The estimates in table 10-1 show that gas production based on gas hydrates, coal, geopressured aquifers, or shale would have a resource base that was effectively inexhaustible given a time horizon of 2050.

The energy content of the world's coal resources in terms of cubic feet of gas is close to 30 times the total estimated conventional resources of natural gas; estimated gas in place as hydrates is many thousands of times conventional gas resource estimates. However, the resource base is a very broad term without any presumption that much or any of the gas will ever be produced.

Table 10-1 also describes the prospects and problems likely to be encountered in producing gas from these sources and gives a rough idea of the stage of development of each source. While the resource base is large, tapping even the most accessible will be relatively costly compared with conventional gas production, given existing technology. Large shares of the resource base are judged technically unrecoverable. While the technologies are immature in most cases, there are unlikely to be technological breakthroughs that will drop the average costs of production below that of conventional gas. Obtaining gas from tight gas formations, coal seams, geopressured resources, shale, and hydrates in permafrost regions requires drilling, as does conventional gas, but each requires additional or special production or processing steps before the gas can be marketed. Tight gas formations and shale gas require special efforts to fracture the rock holding the gas. Attempts at fracturing through detonation of underground nuclear devices was tried in the United States and in the Soviet Union, beginning in 1967. However, the results did not meet expectations, implying the need for a fairly large number of nuclear explosions to achieve satisfactory gas flow rates. The poor results, combined with concern over such a widespread use of nuclear devices, resulted in termination of these tests in 1973. The most promising fracturing technique being explored today is known as massive hydraulic fracturing (MHF) and involves pumping sand and liquids into the bore hole.

Gas from biomass often appears economically attractive on a small scale and in pilot plants, but often these plants (1) use wastes that are of uniform size and quality, (2) the waste is the output of a single operation, making collection costs minimal, and (3) the gas output is used in the waste-producing operation so transportation and distribution costs are nonexistent. Under these circumstances, the production costs can be very competitive with conventional natural gas. However, there is a considerable jump in extrapolating the economies of small scale gas-from-waste to conversion of the waste output of a society to gas. Even if the theoretical waste output of the United States was fully utilized (which is highly unlikely and would be fairly costly in terms of collecting waste from the more dispersed waste generators), the contribution to energy supplies would be relatively small—some 20 percent of present gas consumption in the United States.[1]

As a result, a biomass-gas future would have to rely on energy farms. As noted in the table, theoretical estimates suggest that 4.5 percent of the land area of the United States could yield enough organic material to produce 20–30 tcf of gas. However, 4.5 percent of the land area of the United States is equivalent to roughly a quarter of the land presently under cultivation.[2] On such a scale, a biomass operation is likely to compete significantly with the food and natural fiber industry for prime agricultural land. While any plant can be used for biomass production, thus broadening the category of usable land beyond presently tilled acreage, the crops producing highest organic matter (sorghum, for example), do best under generally the same conditions as traditional agricultural crops. (See chapter 15 for a development of biomass production potential.)

Marine-based farming could augment or substitute for land-based biomass production, but, again, the surface area required, while small in relation to the ocean area, would be massive, and realistic costs of producing and harvesting the organic matter on a large scale are unknown.

Without a doubt, biomass will contribute something but most likely this contribution will be from small-scale, user-operated waste management programs. However, the amount of biomass waste that meets this criterion is nearly insignificant in terms of likely total energy use or in terms of the calculated maximum total contribution of biomass reported in table 10-1. In terms of specific industries, a notable exception is the pulp and paper industry, in which a large share of the energy requirements are supplied by waste produced by the industry. But, in these cases, the waste can often be used more directly, thus bypassing the added cost of conversion to high-quality gas.

TABLE 10-1. Unconventional Gas Resource Base (all figures for gas are 10^{12} ft^3)

Gas Resource	Description	Prospects/Problems	Development Stage	Resource Base (10^{12} ft^3)
Geopressured gas	Salt brines in high pressure aquifiers saturated with methane.	Recovery factors are largely speculative. Potential technical and environmental problems in brine disposal. Likely occurrence of economic reservoirs is unknown. Inability to locate fault blocks of high permeability is an obstacle to reasonable production rates. Potential to use heat and/or pressure to produce electricity as a by-product of gas production.	First commercial production expected in 1990. Gas dissolved in lower pressure brines has been produced in the U.S. and Japan as a by-product of other processes.	(U.S.) 3,000–50,000[a] (Global) unknown, but overpressure regions exist worldwide.
Gas from tight gas formations	Conventional gas in sands of low permeability.	Production rate (volume per year given the field size) and recoverable fraction tends to be low. Improvements in fracturing technology could increase production rate and recoverable fraction. Price rises would stimulate production.	Presently, 0.09 tcf/year produced in the United States.	(U.S.) 400[b]–924[c]–1,300[d] (Global) 2,700–3,600[b]
Coal seams	Methane existing in association with coal deposits.	Technological improvements in drilling technology possible. The cost of collecting coal seam gas from relatively dispersed occurrences has discouraged production where conventional gas has been widely available.	Commercial production was less than 0.1 tcf in 1975.[g]	(U.S.) 398[e]–550[f] (Global) 2,280[e]

TABLE 10-1. Unconventional Gas Resource Base (all figures for gas are 10^{12} ft^3) (*Continued*)

Gas Resource	*Description*	*Prospects/Problems*	*Development Stage*	*Resource Base (10^{12} ft^3)*
Hydrate zones	Gas in a semisolid form in permafrost regions or on the ocean floor.	Onshore production technology is largely experimental. Requires some heating to produce the gas. No known technologies to produce ocean bed hydrates.	Small commercial production in the Messoyala field in the USSR	(Onshore) 500–1,400,000[h] (Offshore) 270,000,000[h]–350,00,000[i]
Shale gas	Gas is associated with orgnic shale, either as free gas or absorbed in the shale matrix.	Free gas from naturally fractured shale is relatively easy to produce. Fracturing technologies are being explored that would increase the production rate of free gas. However, the larger share of shale gas is absorbed in the shale matrix. It is a subject of debate as to whether fracturing causes gas desorption from the matrix and, if so, the rate of desorption.	0.1 tcf is produced commercially from naturally fractured shale in the U.S.[l]	(U.S.) 2,300–3,900[j] (Global) The energy content of organic shales is huge (some 4×10^{21} Btu in higher grade—10–65% organic matter—shale) but the share that may be present as gas is unknown.[k]
Oil shale gasification	Where organic matter in shale has not naturally been converted to gas, production processes can be designed to make the conversion. In addition, the production and refining of shale oil will produce refined fractions amenable to production of SNG.	No problems exist in converting refined fractions of shale oil to gas. The procedure is identical to the conversion of naptha fractions of crude oil to SNG. Direct production of SNG from oil shale would require demonstration of technical and commercial feasibility.	Not commercial. Largely dependent on development of the shale oil industry.	The shale oil resource base is known to be very large both in the United States and globally (see above).

Biomass-organic waste	Conversion of human and animal waste and solid waste to methane.	Collecting and processing waste on a large scale presents a problem. Most promising conversion technique is anaerobic digestion. Cold weather and the introduction of synthetic detergents and heavy metals inhibit the process. Natural decomposition in landfills produces methane which can be mined.	About 0.002 tcf[n] produced in 1980 in the United States. A large number of small-scale gas plants utilizing animal waste exist in India, China, and Korea. (An estimated 2 million such plants in China.)[m]	(See chapter 15.)
Biomass-energy farms	Grow land, freshwater, or marine crops specifically for conversion to gas.	Problem is finding plants that produce a large amount of organic matter per acre per year. For land crops the focus is on sorghum, sugarcane, kenaf, and eucalyptus and silviculture. The water hyacinth is a freshwater plant with potential. The giant kelp is the fastest growing marine plant. The land operations compete with agriculture.	No commercial amounts produced.	(U.S.) It has been estimated that 4.5% of U.S. land area devoted to these crops could provide U.S. gas requirement in 2000 (20–30 tcf).[o] (Global) Total maximum available biomass energy would be about 180 tcf per year.[p] (See chapter 15.)

TABLE 10-1. Unconventional Gas Resource Base (all figures for gas are 10^{12} ft^3) (*Continued*)

Gas Resource	*Description*	*Prospects/Problems*	*Development Stage*	*Resource Base (10^{12} ft^3)*
Synthetic gas from coal	Basic process involves heating coal while restricting available oxygen. Major alternatives are *in situ* or aboveground processes.	Basic technology has existed and been used for a long time. Town gas and gas produced as a by-product of coking are examples. Problem has been cost of producing pipeline quality (1,000 Btu/ft^3) gas. *In situ* technologies have several advantages in scaled up industries; *in situ* avoids most of the labor-intensive mining effort and reduces waste disposal and reclamation problems. Further *in situ* technology can recover gas from coal that is too deep or in seams too thin for conventional mining.	Development and full-scale testing of alternative production methods for pipeline quality gas is in the works. Production expected by 1985 in the United States. Town and coking gas have been utilized in Europe and the United States for many years.	(U.S.) /5,000[q] (Global) 295,000[q]

Sources: Sources as noted; the technological problems and prospects summarize discussions appearing in sources listed in notes a–q and U.S. Department of Energy, *Assessment of Industrial Activity in the Utilization of Biomass for Energy,* prepared by Booz, Hamilton, and Allen, Inc. (Springfield, Va.: National Technical Information Service, 1980); U.S. Department of Energy, *Environmental and Economic Evaluation of Energy Recoverable from Agriculture and Forestry Residues,* prepared by Argonne National Laboratory (Springfield, Va.: National Technical Information Service, 1980); and U.S. Department of Energy, *Environmental Residuals and Capital Cost of Energy Recovery from Municipal Sludge and Feedlot Manure,* prepared by Argonne National Laboratory (Springfield, Va.: National Technical Information Service, 1980).

TABLE 10-1. Notes

[a]Vello A. Kuuskraa, and Richard F. Meyer, *Review of World Resources of Unconventional Gas,* paper presented at the International Institute for Applied Systems Analysis (IIASA) Conference on Unconventional World Natural Gas Resources (Laxenburg, Austria: IIASA, 1980), p. 55. The American Gas Association (AGA), *The Gas Energy Supply Outlook: 1980–2000* (Arlington, Va.: American Gas Association, 1980), p. 40 lists the range as 860–100,000 tcf, but does not cite the references for these estimates.

[b]Low estimate for U.S. in Kuuskraa and Meyer, *Review of World Resources,* p. 17.

[c]National Petroleum Council (NPC), *Unconventional Gas Sources,* Executive Summary (Washington, D.C.: National Petroleum Council, 1980), p. 7.

[d]World Energy Conference (WEC), *World Energy Resources 1985–2000* (New York: IPC Science and Technology Press, 1978), p. 200.

[e]NPC, *Unconventional Gas Sources,* p. 11.

[f]Kuuskraa and Meyer, *Review of World Resources,* p. 41.

[g]Tiratsoo, E. N., *Natural Gas* (Beaconsfield, England: Scientific Press, 3rd ed., 1979), p. 335.

[h]Kuuskraa and Meyer, *Review of World Resources,* p. 41. Inconsistencies appearing in the text have been resolved through telephone conversation with Meyer.

[i]Tiratsoo, *Natural Gas,* p. 336.

[j]Kuuskraa and Meyer, *Review of World Resources,* p. 23.

[k]Ibid., p. 25

[l]Ibid., p. 21.

[m]Tiratsoo, *Natural Gas,* pp. 338–339.

[n]AGA, *Gas Energy Supply Outlook,* p. 39.

[o]Tiratsoo, *Natural Gas,* p. 340.

[p]Converted from terawatt years/year (twyr/yr) as reported in Wolf Häfele, Project Leader, *Energy in a Finite World: A Global Systems Analysis,* Report by the Energy Systems Program Group of the International Institute for Applied Systems Analysis (Cambridge, Mass.: Ballinger, 1981), p. 351. Assumes all biomass used for gas production. The figure reported is amounts in excess of food consumed by humans and animals and amounts needed to support a lumber industry (population assumed to be 8×10^9). It requires "cultivation of virtually all of the productive land in the world" and implies "sophisticated, very careful management of the 'photosphere'"; it includes the assumption that all waste from food and grain crops practicably recoverable is recovered. It does account for losses due to conversion efficiencies; however, the conversion efficiencies are not necessarily those applicable to conversion to gas. The focus is on liquid fuels.

[q]Energy value of estimated total remaining coal resources in tcf. No account is taken of conversion efficiencies in coal to gas transformation. Based on coal resource estimates of chapter 11. Assumes no coal used for any other purposes.

RECOVERABLE AMOUNTS OF UNCONVENTIONAL GAS

Table 10-2 lists various estimates of technically recoverable resources for unconventional gas sources. At this point, gas produced from biomass and coal is separated because of the different considerations important to their future as sources of gas; biomass, because it is a renewable resource and gas from coal because it could be, essentially, an unlimited gas source through 2050, given the size of the coal resource.

A comparison of technically recoverable resources with estimates of the full re-

TABLE 10-2. Technically Recoverable Resources of Unconventional Gas (10^{12} ft^3)[a]

Resource Type	Estimate by	Resource Amount U.S.	Resource Amount Global
Geopressured gas	Various authors, reported in NPC (1980, vol. V, p. 18)	44–5000[b]	N.A.
	Potential Gas Agency, reported in Ion (1979, p. 78)	53–246	N.A.
	USGS Bulletin #790 onshore and offshore	150–1500	N.A.
	USGS Bulletin #790 onshore only	97	N.A.
Gas from tight gas formations	Lewin Associates, reported in Kuuskraa & Meyer (1980, p. 17)	110–200	370–860
	Potential Gas Agency, reported in Ion (1979, p. 78)	70–300	N.A.
	NPC (1980, vol. I, p. 27)	608	N.A.
Gas from coal seams	Kuuskraa and Meyer (1980, p. 43)	40–60	170–260
	Potential Gas Agency, reported in Ion (1979, p. 78)	200–250	N.A.
	AGA (1980, p. 39)	10–60	N.A.
	NPC (1980, vol. II)	44.7[c]	N.A.
Hydrate zones	—	N.A.	N.A.
Gas from shales[d]	Kuuskraa and Meyer (1980, p. 26)	10–50	10–350
	Potential Gas Agency, reported in Ion (1979, p. 78)	2–900	N.A.
	AGA (1980, p. 33)	60–600	N.A.
	NPC (1980, vol. III, p. 4–5)	25.3–49.9	N.A.
Total[e]		126–6758	594–7020

N.A. = Not available or no estimates made.

Sources: National Petroleum Council (NPC), *Unconventional Gas Sources,* Executive Summary (Washington, D.C.: National Petroleum Council, 1980); D. C. Ion, *Availability of World Energy Resources,* 2nd ed. (London: Graham and Trotman Ltd., 1980); U.S. Geological Survey (USGS), *Assessment of Geothermal Resources of the United States,* Circular 790 (Washington, D.C.: U.S. Government Printing Office, 1978); Vello H. Kuuskraa and Richard F. Meyer, *Review of World Resources of Unconventional Gas,* paper presented at the International Institute for Applied Systems Analysis (IIASA) Conference on Unconventional World Natural Gas Resources (Laxenburg, Austria: IIASA, 1980); American Gas Association (AGA), *The Gas Energy Supply Outlook: 1980–2000* (Arlington, Va.: American Gas Association, 1980).

[a]"Technically recoverable" is best interpreted as an operationalized definition of resource as defined in Chapter 6. It has been operationalized by examining specific technologies. No explicit economic criteria were applied in most cases. However, the fact that the technologies have been seriously explored implies some belief that production would occur at some plausible price. The sense of the literature is that the upper range of plausible prices is at most 2 to 4 times present natural gas prices of $3–$5/mcf.

[b]Most authors based estimates on aquifiers in Texas and Louisiana only; these are viewed as by far the most promising. Kuuskraa and Meyer (1980, p. 55) note that reinjection could increase recoverable amounts by five- to sixfold.

[c]Highest estimated recovery at $9/mcf (the highest price considered).

[d]Refers only to production of naturally occurring gas in shales; potential for synthetic gas from shale is excluded.

[e]Sum of the lowest and highest estimates for each resource. For the global total, U.S. figures are used when no global estimates are available or when the upper range value of the U.S. estimate is higher than the reported upper range value for the entire world.

source base shows clearly the limits of available extraction technologies. Notably, gas from tight gas formations, with one of the smaller estimates of gas in-place, appears to be one of the more important sources in terms of technically recoverable amounts of gas.

As is evident from the table, little effort has been aimed at examining unconventional gas resources outside the United States. The Kuuskra and Meyer study is the only one, to our knowledge, that has attempted to look at global unconventional gas resources. Their analysis is based almost entirely on extrapolation of conditions in the United States to other areas of the world. They repeatedly note the speculative nature of their estimates. However, at this stage in the development of these resources, more extensive geological analysis would probably have little, if any, payoff in increased certainty in the estimates of global recoverable resources of unconventional gas. Rather, intensive analysis of the known aquifiers in Texas and Louisiana is likely to considerably narrow the range of values for such technological parameters as likely flow rates, potential for repressurization of wells, likely costs, degree of saturation of the brine with methane, and land subsidence problems, among others. A narrowing of the range for these values would lead to a clearer idea of how much geopressured gas might be available in the future than carefully locating, delineating, and assessing the world's geopressured aquifiers. Similarly, focusing shale gas research on the technological questions of gas flow rates and gas desorption from the shale matrix is likely to result in a narrowing of estimates of gas recoverable from shales.

The one area where we know of no major research effort in the United States is gas hydrates. The only onshore U.S. hydrate resources would be in Alaska and, at present, facilities for transporting potential conventional gas production to consuming regions are insufficient, so the dearth of research effort on gas hydrates is not surprising. Research on production of gas hydrates is being done in the USSR but relatively little is known about this effort.

Overall, geopressured gas and gas hydrates present the biggest uncertainties because of the huge resource base and the speculative nature of recovery estimates. Should initial tests prove out and the knowledge and technology of locating and producing these sources develop beyond present reasoned expectations, the resource base could support a gas economy. On the other hand, these sources may never provide a significant share of gas supplies.

The most comprehensive study of unconventional gas in the United States to date is the 1980 study by the National Petroleum Council.[3] It benefits from studies that have gone before and also from its detailed analysis of individual fields and potential producing regions. The NPC's estimates of technically recoverable resources of gas from coal seams and shale fall well within the range of other estimates. However, their estimates do not support the fairly optimistic upper end values that others have reported. In contrast, the NPC estimates are considerably more optimistic about gas from tight gas formations and very pessimistic concerning geopressured gas. No estimates of technically recoverable reserves were given for geopressured gas, but they estimated production to rise to only 0.008 tcf by 1995 and then to fall to 0 by 2010 with gas prices of $5.00/mcf.[4] They noted four reasons why geopressured gas was unlikely to be a major contributor to gas supplies by 2000 including: high capital and operating costs, yielding little profit allowance to counter the possibility of dry holes or low productivity holes; low solubility of brine with gas, requiring high flow rates for profitable recovery; low recoverability of gas in place, and generally high risk involved in the effort.[5]

Further, they discounted the possibility of reinjecting brine to maintain the well pressure because the energy needed to do so would use 50 percent or more of the recovered gas. They concluded that using either the heat or pressure from the operation to generate electricity would be unprofitable because of the short useful life of the well for these purposes compared with the life of the electricity-generating equipment.

The AGA foresees a much larger role for geopressured gas, arguing that it could supply from 0.02 to as much as 1 tcf per year by 2000.[6] However, any projections are highly speculative; much more will be known at the conclusion of U.S. government-funded research and development currently underway. These results are not expected until 1984 or 1985 and even the conclusion of these initial studies is likely to raise as many new questions as are answered.[7]

Tight Gas, Coal Seams, Devonian Shale

Table 10-3 reports some attempts to add an economic dimension to estimates of the three unconventional gas resources that will certainly contribute some amount to future gas supplies. All are presently commercialized but together accounted for less than 1 percent of gas production in the United States during 1980.

The two sources for this data were the National Petroleum Council (NPC) and the American Gas Association (AGA).[8] In both cases, the range of estimates for a given price is generated by examining a conventional and an advanced technology case. In general, these estimates suggest that coal seam gas and gas from Devonian shale (the most promising shale for gas recovery because it is naturally factured) will be relatively insignificant gas sources in terms of a time horizon of 2050. The total economically recoverable resource available at the highest prices examined would satisfy the U.S. gas demand for a few years at best. Gas from tight gas formations is likely to make a much more significant contribution, especially under the scenario envisioned by the NPC. The upper end estimate for recoverable gas from tight formations is equivalent to 25–30 years of U.S. consumption at present rates.

No estimates of economically recoverable amounts of gas hydrates or geopressured gas resources, similar to those reported in table 10-3, have been made. Production of gas from hydrates is very speculative and even if the cost of production were comparable to the costs of gas from other sources, the additional costs of transporting gas from arctic regions to

TABLE 10-3. Economically Recoverable Amounts of Unconventional Gas in the United States (trillion cubic feet)

Gas Source	*Price (1979 $/mcf)*					
	2.50	*3.00*	*4.50*	*5.00*	*6.00*	*9.00*
Tight gas						
NPC (1980, Vol. I, p. 27	192–331			365–503		404–574
AGA (1980, p. 29)		30–100[a]	45–120		60–150	
Coal seams						
NPC (1980), Vol. II, pp. 4–5)	2–5			16.7–25.4		33.2–44.7
Devonian shale						
NPC (1980, Vol. III, pp. 4–5)	3.3–118			11.4–27.2		16.6–38.9
AGA (1980, p. 33)		10–20	15–30		25–45	

Sources: American Gas Association (AGA), *The Gas Energy Supply Outlook: 1980–2000* (Arlington, Va.: American Gas Association, 1980); and National Petroleum Council (NPC), *Unconventional Gas Sources,* vols. I–V: vol. I, Executive Summary; vol. II, Coal Seams; vol. III, Devonian Shale; vol. IV, Geopressured Brines; vol. V, Tight Gas Reservoirs (Washington, D.C.: National Petroleum Council, 1980).

[a]Reported as amount available below $3.12/mcf.

areas of demand would have to be factored in. Production of geopressured gas appears to be marginally profitable in the $6–10/mcf range, depending on the quality of the reservoir. However, after reviewing both the optimistic and pessimistic estimates, one finds the fairly simple arguments of the NPC logically convincing. The argument is, basically, that a fairly small amount of gas is present in each barrel of brine produced and falling pressure as production occurs is a severe constraint on the potential producible amount of brine, without repressurization. But repressurization requires some 60 percent or more of the energy content of the gas brought to the surface.

These characteristics suggest a basic breakthough price in the $5–10/mcf range; however, amounts available at this price would be fairly limited. In order to get substantial amounts of gas, repressuring would be necessary. The price would have to reflect the fact that only 40 percent or less of the gas recovered from each barrel of brine could be marketed. As a result, a price capable of supporting production with repressurization would have to be about three times the cost of producing gas without repressurization.

Under this simple assumption, the price for geopressured gas with repressurization would be anywhere from $15 to $30/mcf. Maintenance of pressure over a longer period might make it feasible to install electric generating equipment to utilize heat from the well. This, in turn, could be a cheaper source of power for pressure maintenance than using the gas as the energy source. Nevertheless, likely production costs would probably be uncompetitive with gas from coal. Estimates of costs of gas from coal tend to be less than $\$10/10^6$ Btus.

Gas from Coal and Biomass

Table 10-4 reports some estimates of the costs of gas production from coal and biomass. There are several problems in attempting to compare directly the costs reported in table 10-4 with natural gas prices. First, the costs reported in the table are undelivered costs. The additional costs appropriate for delivery to end users would vary considerably, depending on the gas source and the type of production/use system envisioned. The estimates for biogas are derived from demonstration or actual "commercially" operating plants. However, most of the estimates implicitly assume small systems that are oriented toward situations in which the waste producer is also the gas consumer. Under such a situation, transportation and distribution charges are essentially zero.

In a dynamic scenario of fuel switching from fossil fuels to renewables such as biomass, utilization of waste is usually envisioned as the first step in the switch.[9] In addition, many advocates of the renewables path view decentralized energy production as a natural and desirable energy system structure. However, the amount of gas that could be supplied through such fortunate matches of waste availability and energy requirements is severely limited. For biomass to play a major role as a source of gas, transport costs will have to be incurred, either before or after conversion to gas. Further, it would be necessary to go beyond waste utilization to energy farming.

A second problem with the estimates in table 10-4 is that for all of them, with the exception of the high Btu coal gasification, the gas produced is low or medium Btu gas. For industrial users who combust the gas at high temperature, low or medium Btu gas can be easily substituted for high Btu gas or other energy carriers. However, for lower temperature combustion, a high Btu gas is required. The general pipeline system is designed to serve all types of customers, and as a result, the introduction of a low or medium Btu gas will compromise the performance of gas customers' equipment. Conventional gas produced in the United States is on average somewhat above pipeline quality (1,000 Btu/ft^3) gas, thus some medium Btu gas can be mixed with conventional gas. However, the share would be fairly limited. For example, a mixture of 9 percent medium Btu gas (700 Btu's/ft^3) and 91 percent conventional gas, assuming 1030 Btu/ft^3 for conventional gas, which is roughly the average for the United States, would yield a 1,000 Btu/ft^3 gas.[10] The tolerance limit for high Btu gas equipment is probably as low as 900 Btu/ft^3, giving the possibility of a larger share of low Btu gas. But, eventually, if gas is to remain the high quality fuel it is today, higher Btu gas must be produced. It is possible to upgrade produced gas to high Btu gas by removing noncombustible gases, primarily carbon dioxide. Such upgrading would add to the cost per million Btus of the gas and a major cost is the energy required. For example, removal of 35 percent of the carbon dioxide to upgrade the gas produced by anaerobic digestion to 710 Btu/ft^3 of gas requires 10 percent of the gross gas produced by the process.[11] While medium Btu gas is an imperfect substitute for natural gas, a considerable market for the gas could exist as a substitute for other industrial energy fuels where gas-producing plants could be located near enough major energy-consuming industries to justify designated pipelines.

A third consideration is the cost of the basic feedstock. Several of the estimates

TABLE 10-4. Biogas and Coal Gas Production Costs (1979 dollars except as noted)

Gas Source	*Reference*	*Cost ($/10⁶ Btus)*[a]
Biogas		
From waste in developing countries, e.g., manure, wood, agricultural residues.	Cited in Cecelski et al., p. 54	0.07–3.17[b]
Corn residue pyrolysis (U.S.) low Btu (150 Btu/ft^3)	Estimate based on data in U.S. DOE (a), pp. 67–68	4.40–4.76[c]
Wheat residue pyrolysis (U.S.) low Btu (150 Btu/ft^3)	Estimate based on data in U.S. DOE (a), pp. 86–87	4.60–4.96[c]
Feedlot manure, anaerobic digestion (U.S.) medium Btu (710 Btu/ft^3)	Estimate based on data in U.S. DOE (b), pp. 23–25	3.64–8.34[d] (capital cost only)
Feedlot manure, pyrolysis (U.S.) medium Btu (710 Btu/ft^3)	Estimate based on data in U.S. DOE (b), pp. 30–31	0.62–1.53[e] (capital cost only)
Municipal sludge, anaerobic digestion (U.S.) medium Btu (600 Btu/ft^3)	Estimate based on data in U.S. DOE (b), pp. 7–11	1.27–3.16[f] (capital cost only)
Municipal sludge, pyrolysis (U.S.) medium Btu (600 Btu/ft^3)	Estimate based on data in U.S. DOE (b), pp. 17–20	2.00–4.98[f] (capital cost only)
Coal gas		
Aboveground process, high Btu	Reported by Booz-Allen, p. 102	5.80–8.00
Aboveground process, medium Btu	Reported by Booz-Allen, p. 103	4.00–6.25
Aboveground	Reported by IGT, p. 2	3.45
Aboveground	Reported by EPRI	4.00
Aboveground	Reported by AGA	4.00
Aboveground first generation	Reported by Cameron Engineers, p. 177	5.80–7.70
Aboveground second generation	Reported by Cameron Engineers, p. 177	5.80–8.80
Aboveground	Reported by CRS, p. 739	4.00–6.00
Aboveground	Reported by Bechtel, p. 24	9.80

for biogas are for capital costs only. In the case of conversion of municipal sludge, it is probably appropriate to impute a zero cost to the feedstock, and labor costs are likely to be only a very small addition to the unit cost of the gas. But, it is necessary to impute a fertilizer and soil conditioning value of crop residues and manure, when one considers exploiting the energy potential of these agricultural wastes. Once biomass farms are established, either terrestrial or marine, there will be obvious production costs that must be considered as part of the cost of biogas.

A fourth consideration is the potential for using useful by-products of gas production. Biogasification of agricultural wastes preserves its fertilizer value. The by-product residue may be used as a fertilizer and soil conditioner or may even be used as feed for livestock (refeed material). In contrast, pyrolysis destroys most of the fertilizer value. However, pyrolysis of urban solid waste provides the useful by-product service of reducing the volume of waste, thereby reducing the size of the waste disposal effort. A large biogasification industry based on marine farming could significantly augment fertilizer supplies and livestock feed for land-based agriculture. A U.S. Department of Energy report noted that many companies presently involved

Sources:

U.S. Department of Energy, (a) *Environmental and Economic Evaluation of Energy Recoverable from Agriculture and Forestry Residues,* prepared by Argonne National Laboratory, National Technical Information Service, Springfield, Va. (1980).

U.S. Department of Energy, (b) *Environmental Residuals and Capital Cost of Energy Recovery from Municipal Sludge and Feedlot Manure,* prepared by Argonne National Laboratory, National Technical Information Service, Springfield, Va. (1980).

Booz-Allen & Hamilton, "Analysis of Economic Incentives to Stimulate a Synthetic Fuels Industry," in U.S. Congress, Senate, Committee on the Budget, *Synthetic Fuels,* Report by the Subcommittee on Synthetic Fuels, 96th Cong., 1st sess. (Washington, D.C.: U.S. Government Printing Office, 1979), p. 102.

U.S. Congress , House, Committee on Interstate and Foreign Commerce, *The Energy Factbook,* by the Congressional Research Service, Library of Congress, 96th Cong., 2d sess. (Washington, D.C.: U.S. Governemnt Printing Office, November 1980), p. 739.

Cameron Engineers, "Overview of Synthetic Fuels Potential to 1990," in U.S. Congress, Senate, Committee on the Budget, *Synthetic Fuels,* Report by the Subcommittee on Synthetic Fuels, 96th Cong. 1st sess. (Washington, D.C.: U.S. Government Printing Office, 1979), p. 155.

D. F. Spencer, M. J. Gluckman, and B. N. Looks, *A Comparative Analysis of Implication of Various Incentives for Mature Commercial Synthetic Fuel Plants* (Palo Alto, Calif: Electric Power Research Institute, July 1979).

H. F. Brush, Bechtel Corp., "Readiness: Where Do We Really Stand?," paper presented at the Government Research Corporation Conference, "Synthetic Fuels: Worldwide Outlook for the 80s," San Francisco, California, 19-20 February, 1981, pp. 1 and 11.

IGT, *Energy Topics,* June 19, 1978, p. 2.

American Gas Association, *Fact Book* (Washington, D.C.: American Gas Association, September 1979), p. 2.

Elizabeth Cecelski, Joy Dunkerley, and William Ramsey, *Household Energy and the Poor in the Third World,* Resources for the Future, Washington, D.C., 1979.

[a]All costs were converted to 1979 constant dollar prices using a GNP deflator from the *Survey of Current Business,* December 1980, p. 18, except as noted.

[b]Converted from dollars/GJ. Range based on estimates from various sources. No information given on constant/current dollar prices so it was not possible to convert to constant 1979 dollars.

[c]Based on data reported on plant lifetime, capital cost, operating expenses. Low estimates assumes 5 percent real return, high estimate assumes 20 percent rate of return. Excludes cost of marketing and transporting gas.

[d]See note c except no operating cost estimates were available. System produces medium Btu gas (710 Btu/ft^3); the cost reported is per 1000 Btu/ft^3 gas equivalent. Actually upgrading gas to 1,000 Btu/ft^3 would add to the cost. The system size would handle waste from 25,000 cattle. No operating costs or costs of manure included but, since anaerobic digestion produces fertilizer as a by-product, a zero cost manure input may be a reasonable assumption.

[e]See note c except no operating costs available. System would handle waste from 145,000 head of cattle. System size dictated by economies of scale. However, this size would mean much higher operating costs for transporting manure from several feedlots to the plant. In addition, it would be necessary to impute a positive price for manure because of its lost value as a fertilizer.

[f]See note c except no data on operating costs available. System would handle waste from 3 million people. Gas used primarily for city water treatment plant, so no upgrading is necessary; this also minimizes transport and distribution cost element of gas production. Labor costs and maintenance costs are the major omitted operating costs since the feedstock, sewage, has no alternative uses.

in anaerobic digestion projects indicated that the refeed material and fertilizer produced by the system had a higher economic value than the gas.[12] Thus, it may be more appropriate to consider the energy and refeed/fertilizer outputs as joint products of the anaerobic digestion process, with some part of the costs imputed to the refeed/fertilizer output and some part to the gas output.

Finally, there are differential environmental impacts among the various gas-producing technologies. Anaerobic digestion is probably the most environmentally benign. Pyrolysis has air pollution and other residue disposal problems associated with the process. The costs reported in table 10-4 generally include air pollution equipment; char residue can be used, like solid coal, to supply energy. Similarly, the coal conversion technologies have air pollution implications, but the costs reported include stack scrubbers to remove polluting agents. However, the level of pollution control included in these costs may be more or less than will be required for actual operation, depending on the region or the area within a region where the plant is located and the pollution control regulations in effect in the future.

Illustrative Costs of Biogas and Coal Gas

The previous section noted several problems in arriving at a single figure that might be considered a breakthrough price for gas from coal or biomass. In this section, an attempt is made to distill the relevant data on capital and operating costs for biogas and coal gas production and relate these to the cost of the basic feedstock. Combining supplies of the basic feedstock at various costs (from chapter 15 for biomass and chapter 11 for coal) makes it possible to begin to characterize long-run supply for coal gas and biogas.

Biogas

Table 10-5 reports illustrative costs of biogas for various feedstock costs. It is assumed that anaerobic digestion is the most

TABLE 10-5. Illustrative Costs of Biogas Using an Anaerobic Digestion Process

Biomass Input Cost[a]		*"Barebones" Selling Price (Dollars/10⁶ Btus)*	
Dollars/metric ton	*Dollars/10^6 Btus*	*Medium Btu Gas*[b] *(600 Btu/ft^3)*	*High Btu Gas*[c] *(900 Btus/ft^3)*
0	0	4.73	5.16
9.37	0.50	5.98	6.72
18.75	1.00	7.23	8.29
28.13	1.50	8.48	9.85
37.50	2.00	9.73	11.41
46.87	2.50	10.98	12.97
56.25	3.00	12.23	14.54
93.75	5.00	17.23	20.79

Note: See Doan L. Phung, "The Discounted Cash Flow (DCF) and Revenue Requirement (RR) Methodologies in Energy Cost Analysis," ORAU/IEA-78-18(R) (Oak Ridge, Tenn.: Oak Ridge Associated Universities, 1978). The "barebones" selling price is calculated using the revenue requirement method. However, assumptions have been made to extremely simplify the estimates. No distinction is made between debt and equity financing, which is equivalent to assuming that the required rates of return are identical for each (a 15 percent real return was assumed). In addition, no estimates for taxes are included. The 15 percent real return is assumed to be sufficient to cover taxes and risk. A straight-line depreciation method was used to calculate depreciation charges.

[a]Dollars per dry ton of biomass at 18.75×10^3 Btus/kg of dry matter.

[b]Assuming 40 percent net conversion efficiency. Net conversion efficiency is the real-life operating efficiency of the process less energy debits associated with running the plant.

[c]Additional costs include the capital costs of adding a gas purifier to remove noncombustible gases (primarily CO_2) plus a reduction in net efficiency to 32 percent based on the additional energy required to purify the gas.

likely biogas process to achieve widespread adoption. Further, it is assumed that the municipal sludge system, referenced in table 10-4, represents a standard size plant—one where most of the economies of scale have been exhausted. Labor and maintenance costs are assumed to be 2.03/10^6 Btus in 1979 dollars.[13] Depending on the rate of return required on the capital investment, capital costs for a medium Btu plant are between $1.90/$10^6$ Btus (10 percent real return), $2.53/$10^6$ Btus (15 percent real return), and $3.17/$10^6$ Btus (20 percent real return). The reported lifetime of the plant is 20 years; the capital costs per cubic foot assume recovery of capital expenditure plus the real return on the initial investment. Taxes, compensation for risk, and profits are assumed to come out of the real return on capital. For the high Btu plant, the capital costs include the cost of a gas purifier capable of removing noncombustible gases, primarily CO_2. Under the 10, 15, and 20 percent real return benchmarks, the capital costs are respectively $2.22, $2.96, and $3.70 per 10^6 Btus with the addition of the purifier.

By relating the cost of the basic biomass feedstock and the energy efficiency of conversion for the complete process, it is possible to construct a schedule of feedstock costs versus the final gas cost. The efficiency value is a net efficiency of the plant. Net efficiency is used here to mean the ratio of marketable gas production (gross gas production minus energy requirements of plant operations) to the energy content of the basic biomass input. It does not account for energy requirements of fertilizer for growing the biomass and energy required for planting, harvesting, and transporting the biomass to market. The net efficiency of the process, using this all-encompassing definition, is a subject of considerable debate, with some arguing that more energy is put in than is gotten out of the process. Other calculations suggest that this is not true or that the energy cost is justified because the quality of the liquid or gaseous fuel derived is much improved over the quality of the input energy.

The selling price computed is "barebones" in the sense that it does not include any marketing costs and other miscellaneous costs beyond the production process. The calculation for the high Btu gas is probably somewhat low relative to the calculation for the low Btu gas because the purification costs were linearly extrapolated from actually estimated costs of upgrading gas from 550 to 710 Btus per cubic foot. In these types of processes, the cost of removing additional marginal increments usually increases.

Finally, it is important to bear in mind that the value of important fertilizer or refeed by-products has not been imputed as an offset to the gas price. These are very difficult to assess, as was noted earlier; companies considering biogas production have suggested that the value of fertilizer and refeed products may be higher than the gas. But a large-scale biogas industry would undoubtedly put on the market quantities of the fertilizer and refeed material that could significantly depress prices for these products. In order to fully assess such a joint process, a joint production function would have to be specified and integrated into a model specifying demands for both the gas output and the fertilizer and refeed outputs.

Coal Gas

Table 10-6 lists illustrative costs of coal gas for different costs of the coal input. The coal input is relatively cheap per 10^6 Btus compared with biomass. As developed in chapter 15, the larger share of potentially available biomass would be available only at costs exceeding $1.50 to $2.00 per 10^6 Btus. Thus, high Btu gas from biomass would cost $10 or more. Coal delivered to end use consumers is likely to be available at $30 per ton in regions with large amounts of high grade coal—North America, Australia, South Africa, China, the USSR. Coal gasification plants located near sources of coal should be able to obtain coal at mine mouth prices—possibly as low as $5 to $10 but probably under $20. Thus, gas from coal should become widely

TABLE 10-6. Illustrative Costs of Coal Gas (1979 dollars)

Coal Input Cost		
Dollars/Metric Ton Coal Equivalent	*Dollars/10^6 Btus*	*"Barebones" Selling Price (dollars/10^6 Btus)*
5	0.18	6.07–8.21
10	0.36	6.55–8.68
15	0.54	7.02–9.16
20	0.72	7.49–9.63
30	1.08	8.44–10.58
60	2.16	11.28–13.42

Note: Calculated using revenue requirement method assuming an 8 and 15 percent real return straight-line depreciation, total capital cost of $2,984 million plant producing 0.295×10^{12} Btus per day, 330 operating days per year, 25-year life, and operating costs (minus feedstock) of $186.8 million per year. Based on data in Shackson and Leach, *Using Fuel Economy and Synethetic Fuels to Compete with OPEC Oil* (Mellon Institute, Arlington, Va., 1980). The plant description allows for production of methanol or methane or a combination. Capital costs per 10^6 Btus output are 5.82 (15 percent real return) and 3.68 (8 percent real return), operating and maintenance costs are $1.92 per 10^6 Btus of output. Coal is converted into gas at a 38 percent overall thermal efficiency.

available at costs in the range of $6.50/$10^6$ Btus to $9.50/$10^6$ Btus, depending on the rate of return and local coal costs.

CONCLUSIONS

Unconventional sources of gas fall into one of two broad categories; naturally occurring and produced gas. Naturally occurring gas—gas from tight gas formations and coal seams, geopressured gas, shale gas, and gas found in hydrate zones—represents a very uncertain group of sources. Technologies to produce tight gas, gas from coal seams, and gas from shale are known but will be relatively costly. There is considerable uncertainty as to the likely extent of resources. However, the likely magnitudes of available resources suggest that these sources will not play a large role in the future. The resource base for the remaining sources of naturally occurring gas (hydrates and geopressured gas) appears huge but the technologies for developing it are immature. The present focus of development efforts is to determine the feasibility of recovery. As a result, the cost of commercially exploiting these resources is highly speculative. Present indications are that the costs of exploiting a significant amount of either of these are likely to be prohibitive.

The technologies for producing gas from coal or biomass are more advanced in most cases. Gas from coal has an apparent advantage over biogas in terms of cost. While biomass represents a flow of energy available indefinitely into the future, the size of the flow and costs associated with devoting the biomass flow to energy production suggest a severe constraint on the global biomass energy potential.

NOTES

1. The theoretical waste output does not include solid waste.
2. Calculated from data in U.S. Department of Agriculture (USDA), *Agricultural Statistics* (Washington, D.C.: U.S. Government Printing Office, 1980, p. 420.)
3. National Petroleum Council (NPC), *Unconventional Gas Sources,* vols. I–V: vol. I, Executive Summary; vol. II, Coal Seams; vol. III, Devonian Shale; vol. IV, Geopressured Brines; vol. V, Tight Gas Reservoirs (Washington, D.C.: National Petroleum Council, 1980).
4. Ibid., vol. V,
5. Ibid., vol. IV, pp. 3–4.
6. American Gas Association, *The Gas Energy Supply Outlook: 1980–2000* (Arlington, Va.: American Gas Association, 1980).

7. Personal communication with spokesman from American Gas Association, May 1981.
8. NPC, *Unconventional Gas,* vols. I–V.
9. See for example, Denis Hayes, *Repairs, Reuse, Recycling—First Steps Toward a Sustainable Society* (Washington, D.C.: Worldwatch Institute, 1978). Denis Hayes, *Rays of Hope: The Transition to a Post-Petroleum World* (New York: Norton, 1977).
10. Federal Energy Administration, *Energy Interrelationships: A Handbook of Tables and Conversion Factors for Combining and Comparing International Energy Data* (Washington, D.C.: U.S. Government Printing Office, 1977).
11. U.S. Department of Energy *Environmental Residuals and Capital Costs of Energy Recovery from Municipal Sludge and Feedlot Manure,* prepared by Argonne National Laboratory (Springfield, Va.: National Technical Information Service, 1980).
12. U.S. Department of Energy, *Gas Resources RD&D Plan* (Washington, D.C.: U.S. Government Printing Office, 1980).
13. Reported operating and management costs for methanol from wood converted from dollars per gallon. From Congressional Research Service, *The Energy Factbook* (Washington, D.C.: U.S. Government Printing Office, 1980), p. 761.

Coal 11

Coal is by far the most plentiful conventional fossil fuel resource. However, no studies published to date have seriously attempted to estimate the resource base or coal resources in a way comparable to the conceptual definitions of these terms as set out in chapter 6. The major attempts to examine coal resources are those of the United States Geological Service (USGS) and the World Energy Conference (WEC). The methodology used is to estimate resources in place in known deposits. In practice, the estimates rely heavily on reports of widely varying quality by individual countries and indicate a tremendously uneven distribution of coal. The United States, the Soviet Union, and China possess over 80 percent of the world's known coal resources. This fact may, in part, reflect more extensive exploration efforts in the Northern Hemisphere; however, many regions in the Southern Hemisphere are considered geologically less favorable for large coal deposits.[1] While no careful documentation of the actual magnitude of coal resources currently exists, experts cite estimates of as high as 50 trillion short tons of coal in place, perhaps 30–35 trillion metric tons of coal equivalent (mtce), compared with WEC estimates on the order of 11 trillion mtce.

The above estimates refer to coal in place. A 50 percent recovery of in-place reserves is a standard rule of thumb. Assuming a like fraction for resources would indicate that recoverable coal resources are many times larger than indicated by reserve estimates. Such a resource size makes coal a practicably inexhaustible source of energy over at least the next 75 to 100 years.

The following sections discuss the size of the coal resource, costs of mining and transportation, international coal trade, and non-CO_2 environmental impacts of coal production and use.

RESOURCE CONSIDERATIONS

The quality of estimates of the coal resource base is relatively poor. Moreover, the concepts of reserves and resources tend to be used with somewhat different meanings as applied to coal than when applied to other energy resources. These problems arise largely because coal is practicably inexhaustible over the horizon of most energy studies. As a result, it has been enough to know that coal resources are big. However, as we push out the time horizon of energy studies and imagine heavy use of synfuel technologies, coal use can expand rapidly and resource estimates which are "big" in a shorter time frame may become a constraint.

The most often quoted estimates of coal resources and reserves refer to amounts of coal in place.[2] Thus, there is no presumption that the full amount of coal identified as falling in these categories can ever be mined or can be mined at costs less than a given level. Some fraction of the coal will remain, for practical purposes, unminable. On the other hand, the fraction itself is dependent on the assumed economic conditions and technological regime.

A more serious deficiency in coal "resource" estimates is that deposits referred to as coal resources are really comparable to the concept of inferred reserves as applied to oil. Specifically, coal "resource" estimates are based on rough estimates of amounts of coal in known deposits; a zero probability is implicitly assigned to finding new deposits. No estimates of coal "resources" conceptually comparable to oil or gas resources have been published. The magnitudes discussed for coal resources cannot be considered as indicating an ultimate constraint on coal available under any technological/cost regime. The discovery process can uncover more, but no one has begun to think about how much more coal might exist after the amounts that are now labeled resources are exhausted.

Given this inadequacy, it is meaningless to assign grade categories to coal reserves or deposits labeled resources and argue that the exhaustion of higher grade categories at some future date will lead to increased coal prices. Additional high-grade resources could easily be discovered in the intervening period. At this time, there is absolutely no basis on which to approximate within an order of magnitude how much more of this high-grade coal might exist.

It is not clear why *no* such attempts have been made even though it is apparent that simple estimation methods cannot be applied to coal. For example, in estimating oil and uranium resources (see chapters 7 and 12), researchers have argued that over suitably large areas (petroleum basins) one might expect to find the mineral in more or less equal amounts per unit of land area (basin). While such methods can be criticized and at best offer very crude estimates, there is no compelling reason to discount them. In contrast, the uneven distribution of coal is marked and it appears that the unevenness is due to more than differential exploratory effort.

Despite these problems, one can hope to answer the question: On the basis of admittedly conservative resource estimates, can resources be ruled out as a constraint on production over a specific time? If the answer is "yes," the result is fairly conclusive. If the answer is "no," then the only conclusive statement possible is that the resource estimate represents a lower bound constraint on consumption (given prices and technologies) over the forecast period.

Figure 11-1 shows global coal resources, including the standard World Energy Conference (WEC) estimates.[3] The high range estimate in the unproved category is based on speculation by experts.[4] The box areas have been scaled to represent the amount of the resource in each category. (Already produced amounts have been excluded from the relevant entries.) The diagram illustrates the huge amounts which remain as speculative or hypothetical resources.

Estimates of the world's resources have been revised upward considerably in the past few years. Table 11-1 shows the revisions made by the WEC since 1974. This in and of itself is indicative of the "inferred

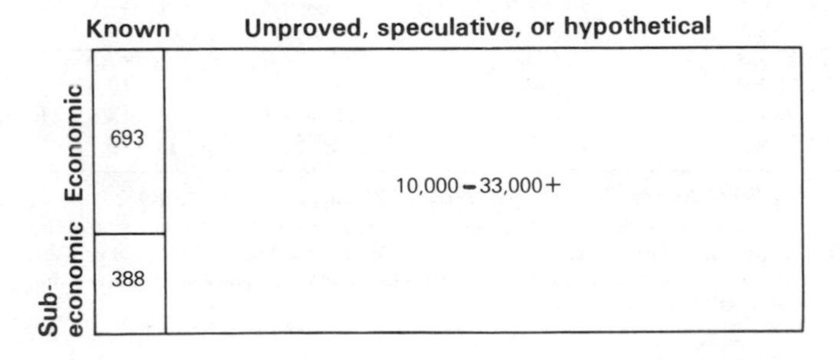

FIGURE 11-1. Classification of global coal resources (10^9 tons of coal equivalent). (From table 11-2; also see note 4.)

TABLE 11-1. Increases in WEC Estimates of Coal "Resources" and Reserves Since 1974 (10^9 tons of coal equivalent)

Year	*"Resources"*	*Technical and Economically Recoverable Reserves*
1974	8,603	473
1976	9,045	560
1978	10,125	636
1980	11,184[a]	693[b]

Source: World Energy Conference, *Survey of Energy Resources 1980,* by the Federal Institute for Geosciences and Natural Resources, Hanover, Fed. Republic of Germany (London: World Energy Conference, 1980), p. 54.

[a]Includes 122 billion tce of peat resources.

[b]Includes 6 billion tce of recoverable peat reserves.

resource" definition implicit in the coal resource concept. While resource estimates might be revised, an appropriate methodology should not yield increases in resources estimates as exploration proceeds, and changes in estimates caused by changes in economic/technological assumptions should be explicitly distinguished from those caused by improved estimating methodologies.

Regional Resources

Coal resources and reserves are widely but unevenly distributed throughout the world. Table 11-2 shows the world's total

TABLE 11-2. Global Resources and Reserves (10^6 tce)

Regions	*Proved Reserves in Place*	*Proved Recoverable Reserves*	*Additional Resources*	*Total Resources*
Region 1 (US)	*344,195.4*	*190,890.0*	*2,529,090.0*	*2,873,285.4*
U.S.	344,195.4	190,890.0	2,529,090.0	2,873,285.4
Region 2 (OECD West)	*139,505.2*	*89,112.4*	*747,724.7*	*887,229.9*
Canada	12,579.0	4,586.8	403,925.0	416,504.0
Germany	63,574.6	34,977.0	186,300.0	249,874.6
U.K.	45,000.0[a]	45,000.0	146,892.0	191,892.0
Other OECD West	18,351.6	4,548.5	10,607.7	28,959.3
Region 3 (JANZ)	*68,249.8*	*37,521.8*	*614,036.3*	*682,286.1*
Australia	59,496.2	36,302.0	611,600.0	671,096.2
Other JANZ	8,753.6	1,219.8	2,436.3	11,189.9
Region 4 (EUSSR)	*307,827.4*	*224,157.7*	*4,575,404.1*	*4,883,231.5*
Poland	65,598.9	30,600.0	92,558.8	158,157.7
USSR	211,390.8	169,063.7	4,469,845.0	4,681,235.8
Other EUSSR	30,837.7	24,494.0	13,000.3	43,838.0
Region 5 (ACENP)	*117,737.4*	*99,684.0*	*1,344,954.0*	*1,462,691.4*
China	99,000.0[a]	99,000.0	1,339,838.0	1,438,838.0
Other ACENP	18,737.4	684.0	5,116.0	23,853.4
Region 6 (MIDEAST)	*385.4*	*193.0*	*215.0*	*600.4*
Region 7 (AFR)	*72,745.1*	*32,659.0*	*146,400.4*	*219,145.5*
Botswana	7,000.0	3,500.0	100,000.0	107,000.0
South Africa	58,749.0	25,290.0	33,762.0	92,511.0
Other AFR	6,996.1	3,869.0	12,638.4	19,634.5
Region 8 (LA)	*6,833.8*	*4,656.4*	*36,831.9*	*43,665.7*
Region 9 (SEASIA)	*23,876.8*	*14,395.7*	*107,701.0*	*131,577.8*
India	21,392.0	13,134.0	91,231.7	112,623.7
Other SEASIA	2,484.8	1,261.7	16,469.3	18,954.1
World	1,081,356.3	693,270.0	10,102,357.4	11,183,713.7

Source: Calculated from tables presented in the WEC, *Survey of Energy Resources 1980,* Part B, tables 1.1.–1.4.

[a]According to the WEC (p. 55, *Survey of Energy Resources*), the proved in-place reserves of the United Kingdom, China, and some other countries in their survey are unknown and should really have been set higher than the estimates of proved recoverable reserves. This probably occurred because these countries state quantities for reserves in place after most of the various losses have been subtracted (losses due to safety pillars, slopes, etc.).

coal resources as estimated by WEC, proved in-place reserves, economically and technically recoverable proved reserves, and additional resources by major countries and regions. As can be seen in this table, the Soviet Union, the United States, and the People's Republic of China possess about 80 percent of the world's coal resources and over 66 percent of the recoverable proved reserves. Another 14 percent of the resources and 17.6 percent of the recoverable proved reserves are found in the other non-U.S. OECD countries, primarily West Germany, Australia, the United Kingdom, and Canada. Only 3.5 percent of the resources and 7.5 percent of the recoverable reserves are found in Africa, Latin America, and Asia and over half of these resources and three-fourths of these reserves are accounted for by South Africa and India. The absence of significant quantities of coal throughout most of the less developed countries is as striking as the large quantities known to exist in the industrialized countries. At least some of this disparity is due to differential exploration effort; however, a large amount may derive from an uneven distribution of the resource. Until there is some sound effort to use probabilistic techniques to assess unexplored or inadequately explored regions, it is impossible to say much more about the distribution of speculative amounts, i.e., amounts above the 11 trillion tons in table 11-2.

A rough way of judging the adequacy of coal reserves is to suppose continued growth in annual production and determine the number of years known amounts would last. When the historical average annual growth rate of coal production (2.6 percent) is used as a yardstick to determine the adequacy of regional availabilities of *proved reserves,* only Centrally Planned Asia (region 5) and South and East Asia (region 9) are shown to exhaust their recoverable *proved reserves* before the end of the projection period (2050; see table 11-3). This does not begin to account for resources and while these calculations pro-

TABLE 11-3. Economically and Technically Recoverable Reserves by Region: Years to Exhaustion at 2.6 Average Annual Growth in 1978 Production

Region	*Number of Years*	*Year of Exhaustion*
1. OECD US	89	2067
2. OECD West	81	2059
3. OECD JANZ	90	2068
4. Centrally Planned Europe	76	2054
5. Centrally Planned Asia	61	2039
6. The Middle East	72	2050
7. Africa	89	2067
8. Latin America	81	2059
9. South and East Asia	62	2040
Total world	78	2056

Source: Same as table 11-2 and the United Nations, *World Energy Supplies 1973–1978,* Statistical Papers, Series J, No. 22 (New York: United Nations Publishing Service, 1979), pp. 2–43.

vide some insights into the adequacy of regional coal reserves, they also point up apparent problems in the data and the reserve concepts when applied in this way. In particular, China is thought of as rich in coal, but by this measure is most in danger of running out, whereas Europe is shown to have relatively plentiful reserves.

A reasonable explanation for this apparent contradiction is as follows: An abundance of coal in relatively accessible deposits gives little incentive to explore and develop other deposits to a point where they can be classified as proved reserves. This is the most likely explanation for the relatively close exhaustion date for the ACENP region (dominated by China). In contrast, the easily accessible coal in Western Europe has, by and large, been recovered already and the location and degree of economic exploitability of most of the coal resources are fairly well known. A relatively small share of Europe's additional resources are likely to be moved to the proved reserve category simply as a result of greater exploration. Higher coal prices or improved mining technology

TABLE 11-4. Research Reserves According to Production Costs

	Country	<15	15–30	30–60	>60
		US $/t			
		Bituminous Coal and Anthracite			
Africa	South Africa	22 090,0	3 200,0		
	Zambia		11,7	3,6	
America	Canada			1 607,0	
	USA	15 627,0	57 614,0	33 942,0	
	Brasilia	123,0	66,0		
	Mexico		1 200,0		
	Venezuela		134,0		
Asia	India	12 610,0			
	Indonesia			10,9	
Europe	Belgium	31,0	62,0	62,0	174,0
	France			550,0	
	Federal Republic of Germany			23 991,0	
	Great Britain			45 000,0	
	Netherlands				130,0
	Norway				18,0
	Span			247,0	151,0
Oceania/Australia	Australia	25 400,0			
		Subbituminous Coal			
America	Canada	2,182,0	neg.		
	USA	91 676,0			
	Argentina			100,0	
	Brasilia	600,0	324,0		
	Mexico		384,0		
	Venezuela		2,5	3,8	
Asia	Indonesia		108,4		
	Taiwan			140,0	
Europe	France			10,0	
	Spain		123,0		
Oceania/Australia	Australia[a]	1 500,0			
		Lignite			
America	Canada	2 117,0			
	USA	24 400,0			
Asia	Indonesia		420,0		
	Thailand	103,0			
Europe	Federal Republic of Germany	10 000,0			
	Italy	6,0	15,0		
	Span	430,0			
Oceania/Australia	Australia	32 440,0			
		Peat			
Europe	Finland	2 340,0			
	Federal Republic of Germany[a]	900,0			
	Ireland	99,0	12,0	77,0	

Source: Reprinted with permission from WEC, *Survey of Energy Resources 1980* (London: World Energy Conference, 1980), table 1.10, p. 72. World Energy Conference.

Note: While not specified in the source, we take the dollars to the current dollars at the publication date minus 1 year to represent a publication lag. Thus an interpretation of 1979 dollar values seems appropriate.

[a]Other sources.

could justify exploiting these deposits but given the abundance of relatively inexpensive coal in several regions of the world, it is unlikely that prices will increase to a level that justifies exploiting coal in Europe's relatively inaccessible deposits. Regional differences in costs of mining coal have and can persist because of the considerable cost of transporting coal. However, these differences have changed as transport costs have varied. As transport technology has improved and as several regions have begun seriously developing export markets, both the intra- and interregional differences in prices become more difficult to maintain.

Table 11-4 shows regional and individual country responses to a WEC questionnaire for production costs of recoverable reserves. As can be seen, abundant supplies of coal can be exploited at relatively inexpensive costs. In general, it is much more expensive to produce coal in Europe than in other regions. As a consequence, as part of their employment programs, West Germany and Great Britain have adopted various measures, such as subsidies and import restrictions, that allow their domestic coal to compete favorably with imported coal. If these countries allow domestic coal industries to be gradually backed out of the market, relatively little of Europe's coal might be exploited.

AN INITIAL ATTEMPT TO DEDUCE AN ORDER-OF-MAGNITUDE SUPPLY SCHEDULE FOR COAL

If one begins to envision high coal use scenarios or the study horizon is pushed out much beyond 2050, the calculations of table 11-3 do not yield a convincing argument for modeling coal as an unconstrained resource. This section attempts to develop an initial order-of-magnitude global supply schedule for coal in the hope that it will stimulate a careful examination of the issue.

A reasonable process for arriving at an order-of-magnitude estimate of one point on the supply curve might be to assume 50 percent recoverability of estimated resources in place at a \$55–\$85/ton "marginal reserve cost."[5] One might then use WEC resource estimates to portray lower bound estimates and the higher estimates cited earlier (34 trillion mtce) to produce an upper end estimate.

One supply point might be contained in range A in figure 11-2. To obtain some ad-

FIGURE 11-2. An order of magnitude, long-range coal supply curve.

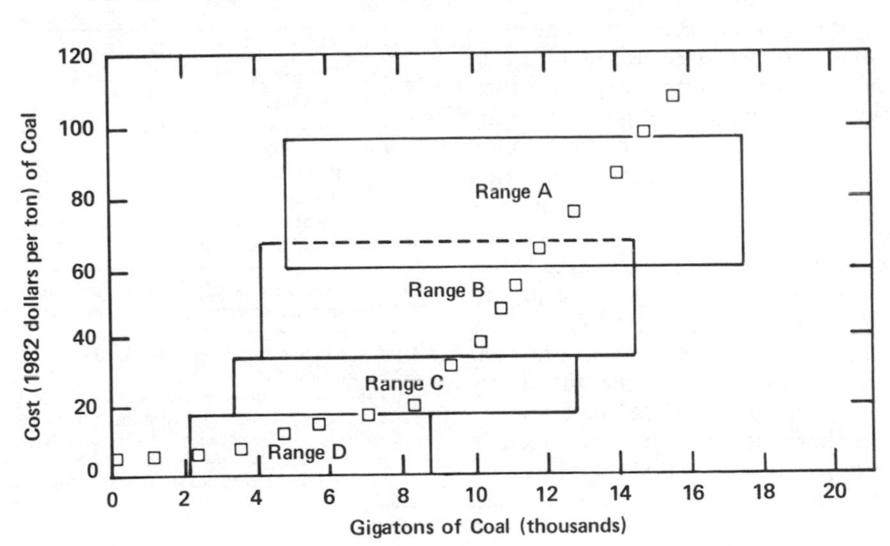

ditional points on the supply curve, we might suppose that the recoverable amounts of unproved resources are distributed by cost category similarly to those reported by the WEC in table 11-4. Doing this yields ranges B, C, and D. WEC estimates put 51 percent of reserves available at less than $15/ton (1982 dollars), 73 percent available at less than $30/ton, and 83 percent available at less than $60/ton. These are conservatively low percentages because about 13 percent of the amount under $30/ton was not broken down into the less than $15 and $15–$30 categories. In computing these percentages we have assumed that the full 13 percent fell in the $15–$30/ton category and none fell in the less than $15 category. Similarly, 65 percent of the amount in ranges A and B was not categorized between $30 to $60/ton and above $60/ton. Again we have assumed that the full 65 percent fell into the higher cost category. This conservatism has the effect of making the supply curve steeper at the lower cost levels. On the other hand, one might expect, if anything, that unproved resources might be somewhat skewed to higher cost categories. Or, one might argue that by the time proven grades are exploited, technological progress in extracting coal could make deposits available which are currently costly compared with present reserves.

Thus, the biases in the various assumptions made to produce the figure are at least not all in the same direction. Moreover, while it is difficult to judge the magnitude of bias error in either direction, on the face of it, the errors in either direction seem of the same order. Since there is widespread feeling that the 10,000 GT resource estimate is too low, one might wish to skew a best guess supply curve toward the high end of ranges A, B, C, and D. Once the ranges A, B, and C are constructed, a supply curve like the heavy dashed line in figure 11-2 can be drawn.

On the basis of the above, we conclude that perhaps 5,000 to 18,000 GT of coal are available for exploitation at costs less than $85/ton (1979 dollars). In addition, a large amount—3,800 to 13,000 GT—might be available at less than $30/ton (1982 dollars). Thus, eventually recoverable amounts below the $30/ton cost category could be as much as 5 to 20 times proved reserves.

PRODUCTION AND TRANSPORT COSTS

Production

There is no one world reference cost for coal. Costs vary enormously not only among different countries but also within a particular country. The differences in the types of coal produced, the methods of production, and environmental, health, and safety restrictions in each region all contribute to the disparity in the estimates. In order to arrive at a best estimate of coal production costs for each region, it was necessary to synthesize a wide variety of information from a number of sources (see table 11-5).

In general, the regional costs were derived from a weighted average of the costs

TABLE 11-5. Global Coal Production Costs by Region ($1979 U.S./tce at mine mouth)

Region	
1. US	$30
2. OECD Europe	67
3. OECD Asia	20
4. Centrally Planned Europe	33
5. Centrally Planned Asia	30
6. Middle East	—
7. Africa	20
8. Latin America	35
9. South and East Asia	35

Sources: Calculated largely from information given by the U.S. Department of Energy, Interagency Coal Export Task Force, *Interim Report of the Interagency Coal Task Force,* Draft for Public Comment, NO. DOE/FE-0012 (Springfield, Va.: National Technical Information Service, January 1981); International Energy Agency, *Steam Coal Prospects to 2000* (Paris: Organisation for Economic Cooperation and Development, 1978); and Report of the World Coal Study, *Coal—Bridge to the Future,* by Carroll L. Wilson, Project Director (Cambridge, Mass.: Ballinger, 1980).

for the largest producers in each region. For example, the estimate of $67/tce at the mine for OECD Europe was derived from production cost estimates for West Germany and the United Kingdom. These two countries account for about 79 percent of OECD Europe's coal production. In particular, the high cost and large quantity of West German coal production strongly influences the average regional cost. The estimate for region 1 (US) of $30/tce takes into consideration the types of mining performed in the United States and the amount produced by each method. The estimate for region 3 (JANZ) of $20/tce is for Australian coal since this country overwhelmingly dominates the production of the region. Estimates of production costs for the Communist countries of regions 4 (EUSSR) and 5 (ACENP) were not available. The costs shown in table 11-5 of $33/tce for Centrally Planned Europe and $30/tce for Centrally Planned Asia were derived by subtracting transportation and handling charges from estimates of Soviet, Polish, and Chinese total delivered costs to Europe and Japan. Estimates for Africa, Latin America, and South and East Asia reflect the production costs of a representative country in each region.

Transport

Coal is very expensive to transport. The mine production costs shown in table 11-5 generally represent only about 60 percent of the final delivered price of coal on the international market. Estimates of the final delivered costs of coal from regions 1 (US), 3 (JANZ), 4 (EUSSR), 5 (ACENP), and 7 (AFR) to markets in Europe, Asia, and Latin America take into consideration a wide variety of regional and individual country factors, such as transportation costs to ports, port handling charges, and ocean freight shipping costs. The final delivered cost of coal for region 1 (US) of $61/tce in Europe, $58/tce in Asia, and $56/tce in Latin America reflects the cost of production for different types of mines in the United States, the location of these mines, the distances to ports, and the likely final consumers. For example, underground mining in the East is generally more expensive than surface mined western coal although transportation charges to ports are more expensive for West Coast coal. Ocean freight shipping costs for West Coast coal to Asian markets are also more expensive than shipping costs for East Coast coal to Europe. The final delivered costs take all these factors into consideration. The cost in Asia is less expensive than in Europe because of the large difference in the cost of production for East and West Coast coal (approximately $29/tce vs. $43/tce). The large difference in estimates of region 3's (JANZ) costs to Europe and Latin America ($57/tce) and Asia ($41/tce) is due to differences in ocean freight shipping costs. Region 3 (JANZ) is much closer to its Asian market. Region 4's (EUSSR) delivered cost to Europe of $49/tce is lower than all its competitors due to its closeness to the market. Region 5's (ACENP) cost of $47/tce in Asia is higher than Region 3's (JANZ) largely because of estimates of higher production costs. Region 7's (AFR) cost of $51/tce to Europe, Latin America, and Asia reflects its central location to these three markets.

INTERNATIONAL COAL TRADE

Types of Coal and Applications

Most of the coal now traded internationally is metallurgical or coking coal (bituminous and anthracite) for use in iron and steel manufacture. In 1979 metallurgical coal accounted for 70 percent of world coal trade.[6] Despite its large share of the market, most recent studies show very slow growth in trade of metallurgical coal in the future.[7] This conclusion is based on the following observations: Steel consumption per unit of gross national product per capita is declining in industrialized countries; the proportion of steel produced by direct reduction from scrap without the use of coke is increasing; and the amount of coke needed to produce a ton of pig iron is declining.[8] Significant expansion of coal trade, if it occurs, will come largely from expansion of steam coal trade resulting from heavy use of coal to generate electricity. Such demand is conditional on the cost and availability of other electricity-generating technologies, most directly coal vis-à-vis nuclear, but, with greater uncertainty and more caveats, solar, hydro, oil and gas, and others.

Coal Markets

In an expanded coal-use scenario, potential interregional coal markets exist. The most immediate market for coal is OECD Europe. Imported coal, despite high transport costs, is generally competitive with Europe's expensive domestic coal. U.S., South African, and Australian coal compete in Europe at present, as does coal from Eastern Europe—primarily Poland. Japan presently imports coal, but the combination of Japan with Australia and New Zealand makes region 3 a net coal exporter. Both China and the Soviet Union have vast, relatively low-cost coal resources.

The developing country regions offer the greatest uncertainty in terms of coal use and import demand. Proved reserves tend to be low, with the exception of southern Africa and India. However, lack of proved reserves of coal may be more indicative of a neglect of coal in favor of cheap oil until 1974. Should a switch to coal occur in these regions, it may be associated with the location and development of indigenous coal resources. Among the three non-OPEC developing country regions, Latin America appears most likely to develop into a major net importer. In addition, South and East Asia is a potentially heavy importer of coal. India's production has not kept pace with consumption despite large coal reserves. Thus, this region could be a likely net coal importer.

Regions 1 (US), 3 (Janz), 4 (EUSSR), 5 (ACENP) and 7 (AFR) are likely net exporter, supplying regions 2 (OECD West), 8 (LA), and 9 (SEASIA). Region 6 (MIDEAST) is unlikely to participate in coal trade. While lower cost coal could substitute for domestic oil use, freeing the oil for export, it is more likely that domestic gas will displace oil since the alternative value of gas in export will be relatively low given the high cost of liquefying the gas.

NON-CO_2 ENVIRONMENTAL IMPACTS OF COAL PRODUCTION AND USE

Coal has always been an inexpensive fuel in terms of its cost per joule of energy. It has also always been an expensive fuel in terms of its associated nonfuel costs, especially its capital costs. In addition, virtually every stage of coal production and use creates by-products, which can adversely affect the environment. These range from gases and particles released into the atmosphere to mine wastes and other discharges. Figure 11-3 illustrates some of the disturbances caused by coal-related activities. Coal production and consumption processes have coincident environmental hazards, which are associated with virtually every class of environmental degradation. The abatement of pollution related to coal use has added significantly to the costs of combustion in the United States. Pollution penalties represent a potential barrier to world coal penetration.

Production

Health and Safety

One of the most serious occupational hazards associated with coal mining is the possibility of contracting pneumoconiosis, or black lung disease, through the inhalation of coal mine dust. Characterized by scarring and deformation of lung tissue, it is often disabling and can even be fatal. The risk of contracting black lung disease has been reduced through improved ventilation and water spraying techniques. For

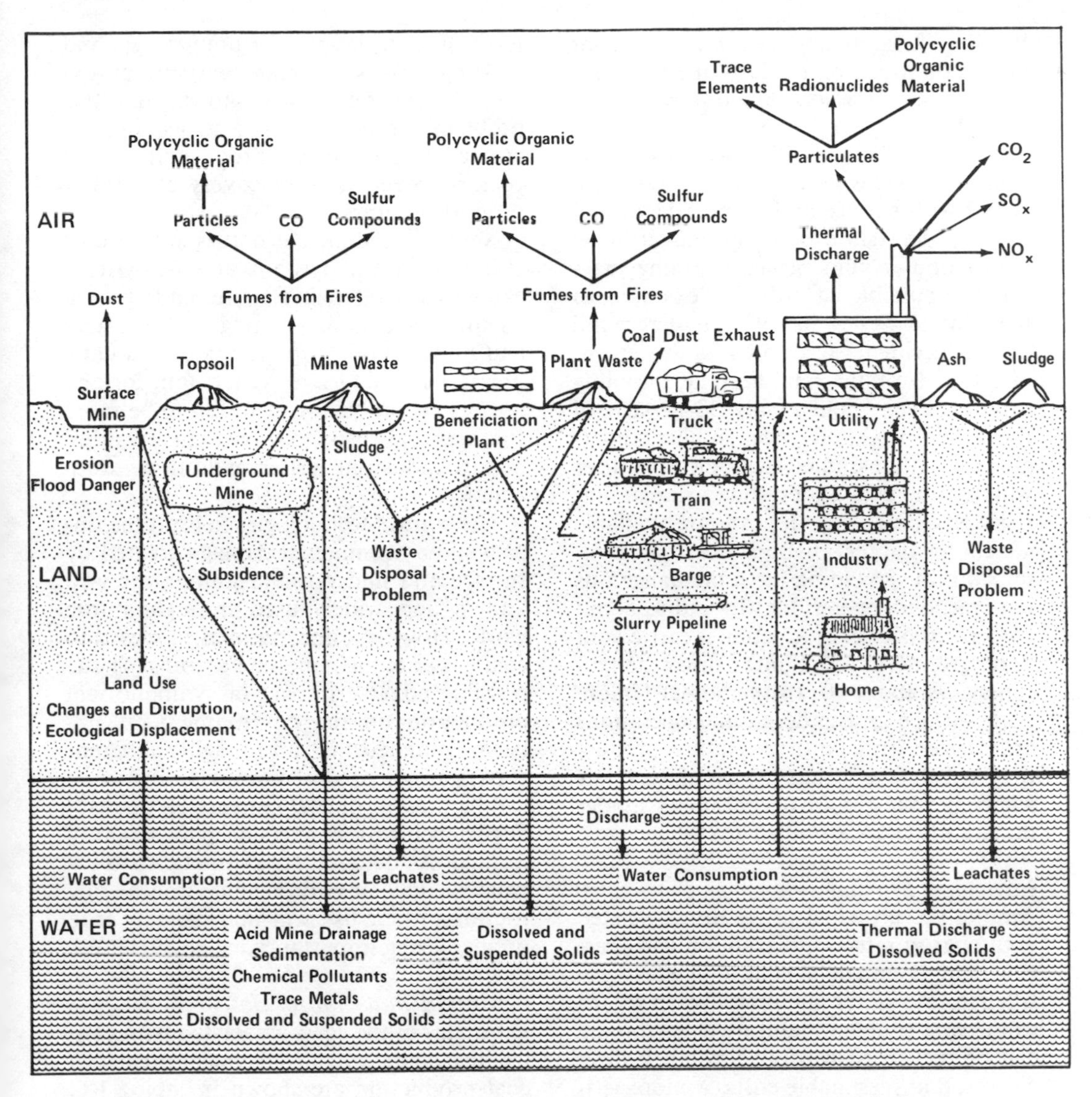

FIGURE 11-3. Environmental disturbances from coal-related activities. (From U.S. Congress, Office of Technology Assessment, *The Direct Use of Coal,* Report No. OTA-E-86, Washington, D.C.: U.S. Government Printing Office, April 1979, p. 184.)

example, the United Kingdom reported that such measures reduced the incidence of black lung disease in miners under the age of 35 years by 40-fold over the past 20 years.[9] Other coal mining hazards, such as accidental injury and death, have been reduced through more stringent safety measures, although the safety improvements have not been as large as the reduction in the incidence of black lung. The United Kingdom has shown a 5-fold improvement since 1952 in the number of deaths per shifts worked. It currently averages about 1 per million shifts worked, which is 3 times better than the averages in the United States, West Germany, France, and

Belgium.[10] Safety and health statistics for other countries, particularly the Communist and less developed countries, are not available.

There is a distinct difference between the occupational health and safety risks associated with surface and with underground mining. The close quarters and artificial ventilation in underground mining magnify the possible hazards. For example, in the United States in 1979, underground mining accidents accounted for 80 percent of all coal mining fatalities and disabling injuries, even though only 38 percent of the coal produced was mined underground.[11] It is assumed that other countries which mine coal underground run similar risks. In China, large deposits are available at shallow depths, enabling open cast mining and thus avoiding the more dangerous underground operations. In India, about 75 percent of coal production is carried out underground and in South Africa the percentage is about 90. In the Soviet Union, both surface and underground operations are carried out; however, future production will shift to locations east of the Urals where permanent frost conditions hinder the use of open-cast mining.[12]

Ecological Damage

Among the principal problems associated with coal mining are disruption and scarring of land, and damage from wastes, subsidence, and acid mine drainage. In regard to the first problem, land used for mining can be reclaimed without too much difficulty and at reasonable costs. A more serious environmental hazard is posed by solid wastes. These are susceptible to spontaneous combustion, are vulnerable to landslides, and can leach impurities into surface and ground waters.[13] These wastes can be controlled by grading, compacting, and covering with nontoxic material.[14]

Damage from subsidence occurs when the support of the mine roof either shifts or collapses, causing ground slopes which can damage roads, water and gas lines, and buildings; change the natural drainage patterns of river flows; interrupt aquifers and existing springs; or create new springs and seeps.[15] Experience has shown that this problem can be controlled by leaving unworked pillars of coal for roof support. The cost, however, is the recovery of a lower proportion of reserves.[16]

Acid mine drainage occurs where coal seams are rich in pyrite. When the pyrite is exposed to water and air, it undergoes a chemical reaction, creating sulfuric acid and metal ions which are toxic to aquatic life; it generally leaves water unfit for any use.[17] This problem is believed to be more complex and costly to solve than any other major industrial water pollution problem.[18]

The control technologies to limit the impact of environmental damage created by surface and underground mining are well known and widely used throughout the world. However, uncertainties in regard to coal production remain. Foremost is the possible adverse impact of synthetic fuel production on the environment. It has been estimated that up to eighty different classes of potentially hazardous substances can be present in coal conversion plants.[19] Experience with conversion plants is very limited. As demand for synthetic fuels increases, the environmental issues surrounding their use will undoubtedly take on increasing importance.

Environmental Control Costs

Estimates of the costs associated with controlling some of the adverse impacts of coal production are shown in table 11-6. They were prepared by the Organisation for Economic Co-operation and Development (OECD) to show the indicative ranges of environmental control costs. The range is only indicative because, as the OECD noted, "It is impossible for the estimates to fully reflect the wide variety of site-specific factors, environmental regulations, and coal types that will influence actual costs."[20] Furthermore, these estimates are directly applicable only to industrialized OECD countries; cost estimates of the

TABLE 11-6. Indicative Cost Estimates for Specific Environmental Coal Mining and Cleaning Measures ($/ton of coal, 1977 U.S. $)

	Surface Mines		
	Contour Surface Mining (thin seams)	*Area Surface Mining*	*Underground Mines*
1. Reclamation of active mines[a] (including prevention of mine subsidence)	2.80–3.00	0.15–0.90	1.00–5.00
2. Fee for reclamation of abandoned mines[b]	0.10 (lignite) 0.35 (coal)		0.15
3. Dust control	0.10–0.20		N.A.
4. Mine drainage control[c]	0.35–0.50	0.15–0.40	0.07–0.60
5. Occupational health and safety requirements	N.E.	N.E.	6.00
6. Coal cleaning—prevention of runoff from storage and wastes[d]	0.9		0.09

Source: Organisations for Economic Co-operation and Development, *Steam Coal Prospects to 2000* (Paris: International Energy Agency, 1978), p. 93.

N.E. = Not estimated.

N.A. = Not applicable.

[a]Higher for surface mining in steep sloped areas.

[b]U.S. legislation.

[c]1985 technology.

[d]Per ton cleaned.

measures taken in the rest of the world are not available. Nonetheless, the estimates do offer a useful indication of the relative costs of each environmental measure. According to this table, occupational health and safety requirements are the most expensive environmental costs, followed by reclamation efforts and prevention of mine subsidence.

Consumption

The environmental problems associated with coal production are, for the most part, site specific and controllable with today's technology. A more serious concern and uncertainty exists over the air pollution emitted during coal combustion. The effects of coal-induced air pollution are more widespread than the localized coal mining disruptions. Furthermore, the air pollution control technologies utilized in one area can actually intensify the level of pollution in another. In order to minimize the effect of emissions on local areas, utilities have used tall stacks to disperse gases and particles. Due to prevailing wind patterns and atmospheric conditions, these pollutants can be transported and deposited on other areas, creating both national and international air pollution problems. For example, prevailing northeast winds in the United States carry pollution from the Ohio Valley and midwestern United States to Canada and the New England states.[21] Similarly, Sweden attributes half of its sulfur pollution level to emissions from central Europe and England.[22]

The emissions from coal combustion include sulfur, nitrogen, and particles of ash. Once released into the atmosphere, these emissions react with other natural and manmade gases and particles, making it difficult to distinguish the effects of one type of gas or particle from another or to determine the ultimate source of the pollutants. Despite these uncertainties, emissions from coal combustion are believed to be major contributors to air pollution. More than half of the world's manmade sulfur emissions are attributed to this cause.[23] These sulfur emissions, as well as nitrogen and particulate emissions, are associated with human illness and death; the creation of acid rain; and damage to plants, animals, and fish.

Health Effects

High levels of air pollution have been known to cause illness and death. This occurred, for example, in Donora, Pennsylvania, in 1948 and London in 1952.[24] Subsequent use of environmental control technologies has eliminated much of the danger of acute, high-concentration episodes in this country and in other industrialized nations. However, although current clean-up methods have been successful in removing larger particles from emissions, there is evidence that smaller particles are potentially even more dangerous.[25] Fine particles are more likely to become lodged deep in the lungs and to travel great distances.[26] It has, for example, been demonstrated that for a given concentration of sulfate compound, toxicity increases with decreasing particle size.[27] The health effects associated with sulfur compounds and other emissions have been linked to common chronic illnesses which may have other causes. Nevertheless, there is evidence that sulfur compounds can initiate or aggravate emphysema, bronchitis, asthma, and lung cancer.[28]

Acid Rain

The acidity of precipitation in many parts of the world has been steadily increasing over the past few decades. The increase in sulfur and nitrogen oxides in the atmosphere is believed to be responsible for this phenomenon.

Acidity can be measured on a pH scale of 0–14. A pH of 7 is neutral, neither acid or alkaline. Each number below 7 represents a tenfold increase in acidity; each number above 7, a tenfold increase in alkalinity.[29] Normally, precipitation has a pH of about 5.7[30] Acid precipitation lowers the pH of the bodies of water it falls into. Water more acidic than about pH 5.5 tends to impair the reproduction function of trout and other fish.[31] If the pH falls below 5.0, the fish will die. In New York's Adirondack Mountains, 51 percent of the lakes above 2000 feet now have a pH below 5.0 and 90 percent of these lakes support no fish populations.[32] Beyond its effect on aquatic life, there is evidence that acid rain may also be harmful to vegetation and soil. Specifically, studies have shown that acid rain may damage foliage, accelerate erosion, affect the germination of seeds, decrease soil respiration, affect the availability of nitrogen in the soil, and increase leaching of nutrient ions from the soil.[33] In turn, wildlife is affected by the damage done to its environment.

The study of the possible harm resulting from acid rain is complicated by the fact that the effects may be cumulative, so that even a constant rate would result in increasing damage to the environment.[34] Furthermore, some areas may be more vulnerable than others. For example, in Scandinavia where most of the forest land is acid podzolic soil and thousands of lakes are underlaid with granite bedrock, the damage from acid rain has been severe.[35] In contrast, in the western United States where most of the lakes and soils are alkaline, the effects of acid precipitation may never become serious.[36]

Since acid precipitation is a relatively recent phenomenon, there is no long-term experience in dealing with it. One method used to reduce the acidity of lakes is to add limestone. This can raise the pH level but additional applications are needed to sustain this higher level. The cost of this method varies with the size of the lake, the tonnage of limestone, and frequency of application. There is, however, evidence that this method can have its own adverse impact. Studies in Scandinavia have shown that lime tends to release toxic mercury compounds normally tied up in plant tissues.[37] Therefore, outside of reducing the overall level of sulfur and nitrogen emissions, there is no definitive solution to this problem.

Environmental Control Costs

The control of sulfur compounds is perhaps the most serious non-CO_2 coal-related environmental challenge. Flue-gas desulfurization (FGD) units are able to lower the level of sulfur emissions but these units are relatively costly. For countries whose coal is naturally low in sulfur, this expense is unnecessary. This is the case in Australia, Poland, South Africa, and India. For most of the other major coal producers, the quality of coal is mixed, containing both low and high sulfur content. For those countries concerned with levels of sulfur emissions, the use of tall stacks is still the primary control technology, but as discussed earlier, this method can intensify levels of pollution in other areas. Regulations to use low-sulfur coal are also in effect in many of these coal-producing countries.

The control of sulfur emissions is very expensive, averaging between \$7–\$12 (U.S. 1977 dollars) per ton of coal using a lime/limestone FGC system. In contrast, the control of particulate emissions is estimated to average \$1.05–\$2.20 per ton of coal and the control of nitrogen emissions only \$0.20–\$0.30.

Regional Environmental Attitudes and Policies

There is a wide spectrum of views concerning the environmental impacts of coal production and use. On the one extreme are industrialized OECD countries, such as the United States, where numerous laws and regulations governing coal are in effect. In contrast, most developing countries have taken few steps to control environmental damage. China's limited initiatives include domestically produced devices that eliminate soot cheaply but fail to eliminate hazardous fine particulates. To eliminate these would require expensive electrostatic precipitators imported from the West.[38]

In India—the largest Non-Communist LDC coal producer and user—there is no concern for environmental damage resulting from industrial processes. For India, industrialization is seen as "the primary means of improving the environment for living, of providing food, water, sanitation and shelter, of making the deserts green and the mountains habitable."[39] The attitude of other less developed countries toward environmental problems is similar to India's. Maurice Strong, the executive director of the UN Environment Programs aptly summarized this attitude by pointing out that a man facing starvation and other diseases of poverty views risks from contamination of the seas or the atmosphere as so remote as to be irrelevant. Factory smoke means money, jobs, and needed consumer goods.[40]

In contrast to the LDCs, the industrialized countries have devoted an enormous amount of attention to environmental problems. This attention is a logical extension of the fact that the western industrialized world was one of the first regions to experience the impact of industrial pollution, could afford to limit its effects, and was willing to forgo some economic profit to improve or at least maintain the environment. The issues, laws, and policies of these countries are well documented and have been the subject of continuous debate, particularly during the past decade.

The wealth of environmental information available in the industrialized OECD countries contrasts sharply with the virtual absence of published material in the Soviet Union. While the Soviet Union shares the same environmental problems as the rest of the industrialized world, the extent and the pace of their control efforts lag behind. Environmental issues are not a top priority, although they have received increasing attention in the past few years. In June 1980 the Soviet Union issued new air quality and wildlife laws as part of the 1981–86 Five-Year Plan. Soviet environmental laws are generally somewhat vague and impose minor penalties; however, a new marine pollution decree is notable for its stiff and explicit fines.[41] This could indicate the beginning of a shift toward stricter environmental regulations.

The foregoing discussion demonstrates the global diversity of attitudes and approaches to environmental problems. Within this context, attempts to control the adverse impacts of coal production and use are only one small part of the global goal to improve the quality of life. The means to achieve this end are as diverse among countries as they are within a particular country, and the costs and tradeoffs involved are difficult to measure. For example, the production of coal includes an implied cost in terms of human life and health. While the risks associated with this occupation can be reduced, it is unlikely that they will be eliminated. The cost of coal will continue to include accidental deaths, injuries, and illnesses. In the past, most countries were willing to pay these social costs in return for an improved standard of living. It is only in recent years and primarily in developed countries that the costs and benefits of industrialization have come into question. The extent of the environmental program initiated in the future will depend upon each country's perception of what means can improve the quality of life and what costs and tradeoffs they imply.

CONCLUSIONS

Coal resources will not constrain the use of coal in the future. They are capable of supplying the world's total energy needs for another 1,300 years at present consumption rates. In terms of cost per joule, coal is generally cheaper than other energy fuels, but associated costs, including capital charges and handling costs, tend to raise the cost of delivering final energy services from coal to levels comparable with other fuels.

Finally, the production and use of coal is associated with the emission of relatively large amounts of pollutants of known and suspected toxicity to human, plant, and animal life. Control of these pollutants varies from country to country, with considerably more efforts being made in the developed western countries than in other regions. Regulations requiring pollution control equipment and taxes or fees have resulted in higher costs of using and producing coal. It is impossible to predict how global environmental regulations will evolve over the long term.

NOTES

1. Carroll L. Wilson, Project Director, *Coal—Bridge to the Future,* Report of the World Coal Study (Cambridge, Mass.: Ballinger, 1980), p. 36.
2. The definition of resources used by the World Energy Conference (WEC) differs somewhat from the definitions set out in chapter 6. The known reserves are differentiated between reserves in place and the amount that could be recovered. Total resources are the sum of the proved reserves in place and additional resources. The definition for each is as follows:

Proved reserves represent the fraction of total resources that has not only been carefully measured but has also been assessed as being exploitable in a particular nation or region under present and expected local economic conditions (or at specified costs) with existing available technology.

Proved recoverable reserves are the fraction of proved reserves in place that can be recovered (extracted from the earth in raw form) under the above economic and technological limits.

Additional resources embrace all resources, in addition to proved reserves, that are of at least foreseeable economic interest. The estimates provided for additional resources reflect, if not certainty about the existence of the entire quantities reported, at least a reasonable level of confidence. Resources whose existence is entirely speculative are not included.

See the Federal Institute for Geosciences and Natural Resources, Hanover, Fed. Rep. of Germany, *Survey of Energy Resources 1980* (London: World Energy Conference, 1980), p. 48. The important difference between these definitions and those set out in chapter 6 is the exclusion of resources "whose existence is entirely speculative" under additional resources. Such an exclusion puts a definite conservative (downward) bias in the total amount which might eventually be available.

3. Ibid., pp. 50–51. The WEC used the United Nations' conversion factors "as far as possible" when converting tons of different types of coal into tons of coal equivalents.
4. A seminar was held on August 5, 1982 to address the issue of coal resources and costs. Participants included Chester L. Cooper, Institute for Energy Analysis (IEA): Louis DeMouy, U.S. DOE; James A. Edmonds, IEA; Richard Gordon, Pennsylvania State University; Kurt Nagle, National Coal Association; Walter Hibbard, Virginia Polytechnic Institute; Frederick A. Komanoff, U.S. DOE; Gregg Marland, IEA; John M. Reilly, IEA; Alvin M. Weinberg, IEA; and Gordon Wood, U.S. Geological Service. There was general agreement that WEC resource estimates were too low. An estimate of 50 trillion short tons of coal was offered as a higher but perhaps still conservative estimate. The 33 trillion mtce is based on the 50 trillion short ton figure assuming that conversion of metric tons to mtce was simply the average conversion factor resulting from individual assessment of deposits in the WEC estimates. The seminar is documented in "Summary of Dinner Seminar on Global, Long-Term, Coal Resources and Costs," August 5, 1982, Institute for Energy Analysis, unpublished manuscript.
5. These estimates are based on coal seminar proceedings, ibid.
6. U.S. Department of Energy, Interagency Coal Export Task Force, *Interim Report of the Interagency Coal Task Force,* Draft for Public Comment, No. DOE/FE-0012 (Springfield, Va.: National Technical Information Service, January 1981), p. 2–12.
7. See, for example, ibid., p. 2–12; and Wilson, *Coal,* p. 120.
8. International Energy Agency, *Steam Coal Prospects to 2000* (Paris: Organisation for Economic Co-operation and Development, 1978), p. 24.
9. Wilson, *Coal,* p. 139.
10. Ibid.
11. The President's Commission on Coal, *Staff Findings* (Washington, D.C.: U.S. Government Printing Office, 1980), p. 35.
12. Peters and Schilling, *Coal Resources,* p. 80.
13. U.S. Congress, Office of Technology Assessment, *The Direct Use of Coal,* Report No. OTA-E-86 (Washington, D.C.: U.S. Government Printing Office, April 1979), p. 252.
14. Ibid., p. 255.
15. Ibid., p. 249.
16. International Energy Agency, *Steam Coal Prospects,* p. 85.
17. OTA, *The Direct Use of Coal,* p. 234.
18. International Energy Agency, *Steam Coal Prospects,* p. 85.
19. Ibid., p. 90.
20. Ibid., p. 92.
21. OTA, *The Direct Use of Coal,* p. 200.
22. United Nations, "Air Pollution Across National Boundaries: The Impact on the Environment of Sulfur in Air and Precipitation," Royal Ministry for Foreign Affairs, Royal Ministry of Agriculture, U.N. Conference on the Human Environment, Stockholm, Sweden, cited by National Academy of Sciences, *Mineral Resources and the Environment* (Washington, D.C.: National Academy of Sciences, 1975), p. 237.
23. OTA, *The Direct Use of Coal,* p. 188.
24. Ibid., p. 202.
25. National Academy of Sciences, *Mineral Resources,* p. 10.
26. OTA, *The Direct Use of Coal,* p. 191.
27. Ibid., p. 213.
28. Ibid.
29. Ralph Blumenthal, "Acid Rain in Adirondacks Disrupts the Chain of Life," *New York Times,* June 7, 1981, p. B14.
30. J. N. Galloway, G. E. Likens, and E. S. Edgerton, *Science,* no. 194, 1976, p. 722, cited by OTA, *The Direct Use of Coal,* p. 222.
31. Ralph Blumenthal, "Acid Rain," p. B14.
32. G. E. Likens, *Science,* no. 22, 1976, p. 720, cited by OTA, *The Direct Use of Coal,* p. 223.
33. L. S. Dochinger, paper presented at the *Third Annual Energy/Environment R&D Conference,* June 1978, cited by OTA, *The Direct Use of Coal,* p. 224.
34. OTA, *The Direct Use of Coal,* p. 225.
35. L. S. Dochinger and T. A. Seliga, "Acid Precipitation and the Forest Ecosystem," *Journal of the Air Pollution Control Association,* no. 25, 1975, p. 1104, cited by OTA, *The Direct Use of Coal,* p. 224.
36. OTA, *The Direct Use of Coal,* p. 224.
37. Ralph Blumenthal, "Acid Rain," p. B14.
38. Leo A. Orleans, "China's Environomics: Backing into Ecological Leadership," in U.S. Congress, Joint Economic Committee, *China: A Reassessment of the Economy,* 94th Cong., 1st sess., 1975, p. 119.
39. Indira Gandhi, "The Unfinished Revolution," *Bulletin of the Atomic Scientist,* September 1972, cited by Leo A. Orleans, "China's Environomics," p. 117.
40. Maurice F. Strong, "One Year After Stockholm," *Foreign Affairs,* July 1973, p. 691, cited by Leo A. Orleans, "China's Environomics," p. 117.
41. Personal communication with State Department official.

Nuclear Energy* 12

Nuclear energy is a by-product of weapons research. The first controlled fission reaction was realized by Fermi in 1942. In 1953, the United States launched the Atoms for Peace program with the intention of harnessing the destructiveness of nuclear power and dedicating it to supplying energy to the world. The first unit of electricity from nuclear energy was experimentally generated in 1953. From commercial introduction in 1960, nuclear power expanded rapidly through the early 1970s.

In the United States, three main types of power reactors were developed: the boiling water reactor (BWR), the pressurized water reactor (PWR), and the high temperature gas-cooled reactor (HTGR). The first two are commonly called light water reactors (LWR). In 1955, the Shippingport PWR started generating 60 MW of electricity. In 1960, the Dresden BWR started up with 207 MW. By 1980, the United States had 56,000 MW of nuclear capacity in operation and another 100,000 MW under advanced stages of construction or planning.

Power reactor developments in the USSR were somewhat similar to those in the United States in that enriched uranium was utilized. Two main types were put into operation: the light-water-cooled, graphite-moderated reactor (LGR) and the PWR. In 1980, the USSR had over 11,000 MW of installed capacity and another 13,000 in advanced stages of planning. Furthermore, the USSR supplies reactors for most of the centrally planned countries.[1]

In the 1950s, power reactor developments in Canada, the United Kingdom, and France were based on natural uranium because the United States and the Soviet Union held a monopoly on enrichment capability. Canada developed the CANDU system (Canadium deuterium-moderated, natural uranium-fueled reactor). Although penalized by a large front-end investment (heavy water cost about $120/kg in 1980), the CANDU has many attractive features. The United Kingdom and France developed gas-cooled, graphite-moderated, natural uranium-fueled reactors. In the beginning, this development was also weapons oriented (production of plutonium), hence the heavy investment costs were not of prime concern. When both countries obtained enrichment capabilities, gas-cooled reactors were discarded for the advanced gas-cooled reactors (AGR) in the United Kingdom, and for the LWR in France. In 1980, France had one of the most ambitious nuclear power programs in the world.

*Sections of this chapter are based heavily on a draft report by Doan Phung, "Nuclear Energy," unpublished, Institute for Energy Analysis, 1981. While the credit for these sections is due Doan Phung, we accept full responsibility for any errors or misrepresentations that have crept in.

Installed nuclear capacity in 1980 included 21,000 MW (30 percent of electric capacity) and plans called for another 30,000 MW, increasing the nuclear share in the electric sector to 60 percent by 1990.

Developments in Japan and West Germany followed the American leadership, largely through licenses with American LWR manufacturers. The rapid penetration of nuclear is indicated by the share of electricity in the OECD countries of nuclear origin: 25 percent in France, 23 percent in Belgium, 13 percent in Britain, and 11 percent in the United States.[2] In all, a total of 22 countries had nuclear plants operating in 1980.[3]

However, new orders for nuclear plants began slowing in the United States after 1970 due to a shutoff of government subsidies, combined with the increasing costs for additional safety requirements, a lengthening of the time needed to approve and construct a nuclear plant, and higher than expected operating costs. Moreover, rapidly escalating electricity prices and a stagnant economy combined to slow electricity demand growth. As a result, utilities found they had over-ordered new electricity generating capacity (of all types). The Three Mile Island (TMI) core accident raised additional public concern and doubt about the full cost of nuclear power and the desirability of nuclear power at any cost.

In general, experience abroad has paralleled the U.S. experience. However, government responses have varied, depending on the institutional structures governing nuclear power plant construction and licensing and the perceived need for nuclear power as a "domestic" energy source. France has maintained an ambitious program of nuclear development; Sweden has placed a moratorium on new plant construction.

While the environmental effects and safety aspects of nuclear power have probably been more heavily researched than any other societal issue and "technological fixes" have been devised for problems surrounding nuclear power, public policy has moved from a position of directing the spread of nuclear power to a position of reacting as necessary to events associated with nuclear power production and its spread.

The production of nuclear power, like other forms of energy production, has associated effects on society that may not be incorporated in the private energy producer's decision function. It becomes the domain of public decision makers to regulate safeguards or force private decision makers to incorporate externalities of production into the cost of the delivered fuel. Rather than pass judgment on the appropriateness of present nuclear policies or suggested changes in policy, the intent of this chapter is to outline the issues surrounding nuclear power as they relate to its future supply potential. It remains a question of public policy what specific set of safeguards will be required in the future and who will pay for them. The issues of reactor safety, waste disposal, low-level radiation, and weapons proliferation are too complex to be discussed in one chapter, or indeed, in one book, and we make no attempt to do so here.

URANIUM RESOURCE CONSIDERATIONS

The Resource Base

Uranium occurs with an average crustal abundance of 2 ppm.[4] As such, it is a potentially huge energy resource even if the energy is exploited with conventional light water reactors. But, as with other energy resources, the exploitable potential for nuclear resources is considerably less than indicated by the crustal abundance of the resources.

Unlike petroleum resources, uranium is found in many different types of deposits and most of it is diffuse. Exploitable amounts of uranium have been found in sandstone, conglomerates, vein deposits, and other mineral deposits, but these tend to be small and dispersed through the native rock. As a result, mining operations can become fairly extensive even though the tonnage of uranium finally extracted is small. Large amounts of uranium exist in granites and other magmatic rocks, marine black shales, and in sea water, but these are very dispersed deposits, making the cost of recovery several times more than current costs.[5] Rough estimates of the cost of recovery for these deposits go up to a few thousand dollars per kilogram compared with forward costs of $80 kilogram for the grade of resource typically exploited today. Moreover, the problems of disposing of the spent rock and overburden associated with the activity may increase the cost further or simply be unacceptable to society. These concerns are generally considered a limiting factor in the use of light water reactors.[6]

Many geologists believe that uranium mineralization is unique in that deposits tend to be either a rather high grade (greater than 700 ppm) or a relatively low grade (less than 100 ppm). This has been the experience with uranium finds in the United States.[7] Other minerals tend to be deposited in more continuous grade variations.

Resources

Estimates of uranium resources are plagued by the relatively short history of exploratory effort and the failure, to date, of geologists to identify a limited set of geological conditions in which exploitable uranium deposits occur with some certainty. There are two main approaches to estimating the amount of mineral resources available under given economic conditions: geologic identification and statistical extrapolation. In order for geologic identification to provide an unbiased estimate of resources (as defined in chapter 6) enumeration of all deposits in a region is required. Complete enumeration of deposits is a limiting case which occurs only when the region has been fully explored. It is hard to imagine uses for resource-size information that would justify costly exploration. Thus, resource estimates are a by-product of exploration aimed simply at finding enough high-grade deposits to assure that production can meet demands over the next 5 to 10 years. Using only information generated by exploration of this type will always tend to bias resource estimates downward since the methodology implicitly attributes a zero probability to finding resource deposits outside known fields.

A second approach is to identify geological formations in which the mineral occurs, assess the expected amount of the mineral find in similar formations, and then ascertain the likely occurrence of these formations in unexplored areas. This approach may yield either high or low estimates, depending on whether the fields in the region that forms the basis for extrapolation are a representative sample of all fields.

Uranium resource estimates are relatively poor because so far exploration has been limited. Thus, estimates based on the first approach are likely to severely understate actual uranium resources. The second approach has not been widely used because uranium finds to date have been under such a wide variety of geological settings that the statistical basic for extrapolating to unexplored regions has been questionable.

Table 12-1 reports two attempts to estimate uranium resources. The WEC estimates are based on country surveys combined with data from the International Atomic Energy Agency (IAEA). The IAEA figures are based on a relatively restricted definition of resources. The estimates include reserves plus probable resources. Probable resources are amounts which are thought to be contained in known fields and in extensions of known fields but where there has been insufficient drilling to classify the amounts as reserves. The WEC has attempted to employ a somewhat broader definition for resources but continues to formulate resource estimates largely on the basis of reports from countries with current uranium production. As such, regions of the world with little exploratory

TABLE 12-1. Uranium Resources (10^3 metric tons uranium; 1979 dollars)

Region	*WEC* [a] *$80/kg*	*WEC* [a] *$80–130/kg*	*Cumulative Prod. to Jan. 1, 1979*	*Perry Extension from U.S. Occurrences* [b]
US	1,310	558	260	1,696
CAN & WEUR	695	773	152	4,604
JANZ	337	15	9	
EUSSR	1,855	18	417	4,140
ACENP	166	N.R.	50	1,980
MIDEST	4	N.R.	0	
AFR	810	311	147	7,200
SEASIA	39	8	0	
LA	258	11	0.6	3,600
Total	5,474	1,695	1,036	23,220

N.R. = None reported.
[a] World Energy Conference, *Survey of Energy Resources 1980* (London: WEC, 1980).
[b] A. M. Perry, *World Uranium Resources* (Vienna: International Institute for Applied Systems Analysis, 1978).

activity to date are ignored. Moreover, reports from uranium producing countries are based largely on producing fields and extensions of producing fields and thus ignore the potential for totally new finds within those countries.

In recognition of these limitations, Perry has extrapolated the occurrence of uranium in the United States per square kilometer to the rest of the world. He bases this extrapolation on the fact that there does not appear to be a strong reason to believe that exploitable uranium deposits are any more or less likely to occur across different large land areas. Such an exercise, as one might expect based on the relative exploratory effort in the United States and elsewhere, leads to a much higher estimate of world uranium resources. While high, Perry feels the estimate is believable and in fact may be somewhat conservative. It tends to be conservative since the figure used to estimate occurrence of exploitable uranium in the United States includes only reserves plus probable resources, excluding the possible and speculative categories. Using Perry's method, the inclusion of these categories would increase the U.S. uranium occurrence, and therefore world occurrence, by nearly 40 percent.[8]

The conclusion one must draw from comparing current attempts to estimate uranium resources is that the figures are of a much different character than those for conventional oil or gas. The quality of resource estimates is more in line with those of shale oil, unconventional oil and gas, and coal. However, unlike these resources, the available evidence on uranium resources does not allow one to dismiss out of hand the possibility that dwindling high-grade resources may force a move to lower grade, higher cost deposits, thereby yielding an escalating price for nuclear-based electricity.

Nuclear Technologies

In order to relate uranium resource estimates to potential future energy supplies, it is necessary to specify the uranium-utilizing technologies likely to be in place in the future. Light water reactors (LWRs) are the least efficient nuclear technology in converting uranium to electricity; adding fuel recycling increases the efficiency by about

20 percent. High temperature gas-cooled reactors (HTGRSs) are approximately 38 percent more efficient than LWRs without fuel recycling; the addition of fuel recycling makes the HTGRs 70 percent more efficient than LWRs. The Canadian deuterium-moderated reactor (CANDU) is very similar to the HTGR from the perspective of uranium requirements (see table 12-2). Advanced reactor systems (near breeders) which produce nearly as much fuel as they use (and therefore require very little uranium beyond startup requirements) are conceptually possible, but development effort has not progressed to the stage of the liquid metal fast breeder reactors (LMFBR). The breeder reactor can become an unlimited source of power because it converts ^{238}U to fissile isotopes more rapidly than it consumes these isotopes. ^{238}U is 100 times more plentiful in natural uranium than the main naturally occurring fissile isotope, ^{235}U.

Some studies examining breeder constraints have asked the question: What is the maximum rate of expansion of a breeder program without exogenous fuel inputs? However, the question is of major interest only if the world has a relatively small resource stock of breeder fuel. Perry has argued that the question is of relatively little importance for several reasons. Among them: the uranium resource is considerably larger than the conservative estimates of the International Atomic Energy Agency (IAEA).[9] The growing stock of LWR waste, when reprocessed, offers and will continue to offer a large breeder fuel source. Even extremely rapid nuclear scenarios are not constrained by plutonium availability; the more likely constraint is simply the rate at which breeder technology can penetrate.

The classic nuclear strategy is an initial reliance on LWR, with a transition to breeders, utilizing plutonium from LWR waste reprocessing as the initial fuel endowment for the breeders. LWRs form the initial stage because they were technologically available before breeders and also represent a cheaper source of electric power at presently existing uranium prices. In order to make the breeder competitive with LWRs, uranium prices would have to increase considerably, depending on the assumed capital costs of breeders relative to LWRs.[10] However, from the standpoint of technology, a breeder program could use uranium directly, bypassing LWRs.

TABLE 12-2. Uranium Requirements of Power Reactors

Uranium Requirements per 21 GWyr[a]	*Tons U_3O_8*
LWR, no recycle	6,010
LWR, recycle	4,750
HTGR, no recycle	4,310
HTGR, recycle	1,770
CANDU, no recycle	4,380
CANDU, recycle	1,770
LMFBR, recycle	Negl.

Sources: Wolf Häfele, Project Leader, *Energy in a Finite World: A Global Systems Analysis,* Report by the Energy Systems Program Group of the International Institute for Applied Systems Analysis (Cambridge, Mass.: Ballinger, 1981) and Hans Bethe, "Relative Merits of Alternative Fuel Cycles," in *Nuclear Energy and Alternatives* (Cambridge, Mass.: Ballinger, 1977).

[a]21 GWyr is the electricity produced by one GW reactor running at a 70 percent capacity factor for 30 years.

Nuclear Energy in the 1960s and 1970s

Initially, nuclear power manufacturers used what has become known as a "turnkey" marketing device to induce utilities to construct nuclear plants. Oyster Creek, Browns Ferry, and Indian Point were among the first nuclear plants guaranteed by reactor manufacturers at fixed prices by the time the key was turned for commercial operation. This device was particularly critical for the introduction of nuclear power because of the very high front-end costs, long construction periods, and uncertainty of the new technology. Although a dozen or so turnkey contracts caused reactor manufacturers financial losses, they did usher in the era of commercial nuclear energy both in the United States and throughout the world.

TABLE 12-3 Nuclear Power Plants in Operation and Planned

Region	MW Operational August 1981	Cumulative Planned MW	Cumulative Planned Up to Year
1. United States[a]	56,946	168,100	1994+
2. OECD WEST	57,955	150,536	1990
3. OECD Asia	12,087	22,200	1987
4. Centrally Planned Europe	16,630	38,930	1988
5. Centrally Planned Asia	—	—	—
6. Middle East	—	900	—
7. Africa	—	2,766	1988
8. Latin America	—	2,935	1988+
9. Non-Communist South, East, and Southeast Asia	2,701	15,191	1989
Total	146,319	401,558	1988–94

Source: Based on "The World List of Nuclear Power Plants," *Nuclear News*, vol. 24, no. 10 (August 1981).
[a]Three-Mile Island, unit 2 (880 MWe), is not counted due to the core disruptive accident of 1979.

Installed nuclear capacity grew from 1,400 MW in 1960 to 3,100 MW by 1970 and to 146,000 MW by 1980. The growth rate between 1970 and 1975 was 36 percent/yr; between 1975 and 1980 it was 59 percent/yr.[11] Table 12-3 lists nuclear capacity in operation, under construction, or actively planned in each of the nine designated regions.

Despite the rapid growth of generating capacity, several factors that became important during the 1970s have greatly slowed the number of new orders for commercial reactors. In 1971, the U.S. government ceased providing public financial support. Second, as a result of the Calvert Cliffs case, nuclear plants were subject to the National Environmental Policy Act, requiring a lengthy process of licensing that includes public participation in the licensing process. Third, large nuclear plants that went on-line did not perform exactly as planned as a result both of the debugging process and earlier overoptimism. Fourth, in March 1979, the core disruptive accident at Three Mile Island Reactor No. 2 caused additional public concern about nuclear power, and financial loss for the utility. These factors combined to add greatly to the cost of nuclear power. Finally, increased prices of electricity passed on to consumers as a result of increased costs of generation from all sources, and slower economic growth after 1973 resulted in unexpectedly slow growth in electricity demand. As a result, utilities found themselves with too much capacity planned and under construction. Attempts to roll these costs into the rate base exerted additional pressure on electricity prices. New orders for generating equipment of all types fell off dramatically. No new orders for nuclear plants were placed and several old orders were cancelled between 1975 and 1980. Some manufacturers, including Babcock and Wilcox and General Electric, contemplated withdrawing from the market.

Problems of nuclear power in the United States spread to or repeated themselves in most market economy countries. Sweden placed a moratorium on new nuclear plants. Austria completed a 700 MW plant but did not operate it. The United Kingdom completed all construction by 1981 but had no additional expansion plans on the drawing board. Similarly, Japan will complete most old orders by 1984 without yet having new programs. Only a few countries, notably France, and to a certain degree, South Korea and Taiwan, continued on a program of rapid expansion of nuclear capacity. Expansion in these ocuntries was aided by two underlying conditions. First, the political structures within the countries minimized the regulatory delays in siting and construction that existed in the United States and other countries. Second, national policy was aimed at reducing dependence on foreign sources of energy—reflected through tariffs or import quotas on other fuels and direct support of nuclear power. As a result, nuclear power continued to be a clearly cheaper source of electricity for newly constructed plants. The centrally planned economies of Europe also have political structures which minimize siting and construction delays. However, most have domestic coal resources or assured supplies of fuel from the USSR. Only the USSR had a steady nuclear plant construction program, mostly for large cities far from oil and gas fields. Other Eastern European countries, such as Poland, Czechoslovakia, Hungary, and Rumania purused their plans of building nuclear power at a slow pace.

For a time, nuclear power plants appeared to be a status symbol among developing countries—the steel plants and airlines of the 1970s. More recent experience suggests retrenchment from reactor commitments. Mexico's plans for its first reactor of twenty by 2000 have been deferred as it reconsiders the economics of nuclear power in light of a weak oil market.[12]

As the 1970s ended, nuclear power had been essentially put on hold in most of the world. In 1977, President Carter adopted a policy of limiting nuclear plant sales to the third world as a result of nuclear proliferation concerns, indefinitely delayed LWR spent fuel reprocessing, slowed development of the prototype Clinch River Breeder Reactor (CRBR), characterized nuclear energy in the United States as an energy option of last resort, and launched the International Nuclear Fuel Cycle Evaluation (INFCE). Although the CRBR project went on at the insistence of the U.S. Congress and the results of INFCE supported continued peaceful use of nuclear energy, the policies of the Carter administration probably aided in slowing down nuclear energy penetration in the United States and in the rest of the world. Then in 1979, an operational accident at Three Mile Island (TMI) led to the disruption of TMI-2 core and further created widespread distrust of reactor safety systems. As a result of the accident, additional safety features were required on new reactors and existing reactors were required to be retrofitted.

Nuclear Power Costs

LWR Capital Costs

The rate of increase in the estimated real (inflation adjusted) capital costs of nuclear power averaged 14.0 percent per year for plants begun between 1967 and 1980 (see Figure 12-1.) This compares to an annual real rate of increase in crude oil prices of 20 percent per year[13] and an increase in utility coal prices of 6.8 percent per year over the same period.[14] Several different factors have contributed to nuclear capital cost increases. Among these are unit cost increases in material and labor inputs; increases in the real interest rate; licensing and construction delays and therefore a lengthening of the period over which interest obligations accumulate; and added safety features which require absolutely more material and labor inputs, more costly design features, and a lengthening of the construction period. To begin to understand the extraordinarily rapid increases in estimated capital costs during the 1970s, it is useful to quantify the effects of each of these component cost increases. Note that the dates in figure 12-1 are the date at which the project was begun. Projects begun in 1980 are not expected to go into commercial operation until 1992. Cost estimates are based on current costs of inputs plus expected escalation above expected inflation. Thus, the reported costs should not be viewed as actual costs expended on any project. The estimates are for standardized plants (size, location, fea-

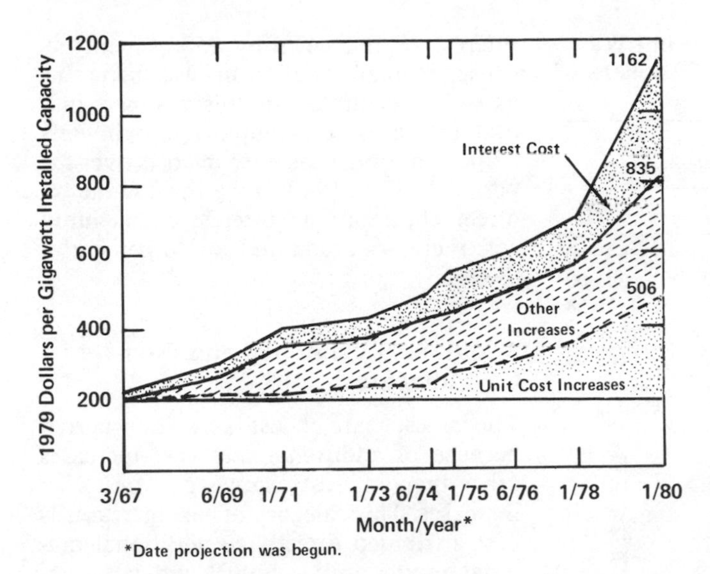

FIGURE 12-1. Real capital cost escalation of nuclear power costs. Based on WASH 1082, 1150, 1230, 1230 revised, 1345, 1345 revised, NUREG, EEDB-1, EEDB-111. See John H. Crowley, "Nuclear Energy—What's Next?" Paper presented at the Workshop on the Electric Imperative, Monterey, California, June 1981. Deflated to 1979 dollars from dollars in year of commercial operation.

tures). (The standardizations are those in NRC-WASH estimates, ERDA-NUREG estimates, and DOE-EEDB estimates as reported in the notes to figure 12-1.)

Interest Charges
The size of the capitalized interest charge is dependent on the underlying capital expense, the interest rate, and the length of time over which charges accrue. During the 1970s all of these factors increased, thus tending to increase the capitalized interest charge (in 1979 dollars) from $13/kW to $328/kW. Table 12-4 disaggregates these various effects. As one can see, if only the construction period had lengthened the absolute increase in the interest charge would have been only $5/kW. While this is small in terms of the actual increase that occurred during the 1970s, it would have represented nearly a 40 percent increase in the interest charge. Still, the total interest charge would have remained a small share of costs. The increases in the rate and noninterest costs separately have about the same effect on interest charges. The increase in the interest rate alone increased the interest charge by $35/kW or 270 percent.

One can turn the problem around and ask how much of a reduction in the interest charge is applicable if various conditions existing in 1967 were to obtain today. The single factor yielding the most dramatic effect would be the interest rate—if 1967 rates applied in 1980, the interest charge would fall from $328 to $73/kW. This would have a fairly significant effect on total capital costs; they would fall from 1,163/kW to 908, or 22 percent. If the construction period itself could be reduced to the 1967 value of 4.1 years, the effect would be smaller but still would result in a capital cost of 1032/kW, 11 percent below the present estimated cost.

Overall increases in interest charges account for 33 percent of the estimated $945/kW cost increase over the period. However, interest charge increases due to a lengthening of the construction period and increases in the noninterest costs should be considered indirectly attributable to safety improvements and other cost-increasing factors. The interest rate effect alone accounted for only 4 percent of the increase.

Unit Cost Increases
Increases in the wage rate and the price of other inputs necessary to construct a nuclear plant generally exceeded the rate of

TABLE 12-4. Sources of Increase in LWR Capitalized Interest Charges (1979 dollars/kW)

	Actual Cost Estimates	
	3/67	*1/80*
Noninterest ($)	205.00	835.00
Interest charge ($)	13.00	328.00
Implicit rate (%/yr)[a]	2.88	11.58
Construction period (yrs)	4.10	6.40
Total cost	218.00	1163.00

1980 Interest Charge at 3/67 Factor Values (Factor Value Substituted)

Rate, noninterest costs, period	13
Rate, noninterest costs	18
Rate, period	47
Period, noninterest costs	48
Rate	73
Principal	80
Period	197
None	328

[a]Calculated by assuming the interest charge given in the source in years of commercial operation dollars is accumulated over the construction period. The implicit rate is the solution to

$$(Y_{max} + I) = Y_0(1 + r)^t + \ldots + Y_i(1 + r)^{t-i}$$

where Y_{max} = Noninterest capital change in year of commercial operation dollars.
I = Interest charge.
t = Construction period in years.
Y_i = Expenditure in years i.

Expenditures in year i are generated using the following formula.

$$\overline{Y}_i = Y_{max}\{1 - [\cos(\pi/2 \cdot i/t)^a]^b\}$$

where $\overline{Y}_i$ = Cumulative expenditure through year i and a and b are constants.
a = 4.0820.
b = 3.2495.

This function has been found to describe the payout schedule of nuclear plants. See Komanoff, *Power Plant Cost Escalation* (New York: Komanoff Energy Associates, 1981), p. 316.

inflation as measured by the GNP deflator during the 1970s. Komanoff has constructed an index of nuclear plant component unit cost increases.[15] Particularly rapid, real increases in the unit cost of inputs occurred during 1974. While the unit cost index of nuclear inputs increased approximately 56 percent over the period, the cumulative effect was to increase noninterest capital costs by 250 percent because the quantity of inputs was increasing as well. Accounted in this manner, unit cost increases contributed approximately 32 percent of the increase in costs over the period 1967 to 1980. If we include the indirect effect on the interest charge unit, cost increases accounted for 46 percent of the total.

Additional Real Actual and Estimated Cost Increases

The largest share of cost increases occurred because of additional real cost increases other than interest charges and unit cost increases. This category of cost increases is best attributed directly to additional material inputs and technical and labor requirements made necessary by increasing safety standards. These types of cost increases directly accounted for 35 percent of the total increase over the period. If one includes the indirect effects of higher interest charges due to construction delays and larger capital costs, the share rises to 50 percent. Various attempts have been made to identify specific cost increases with individual regulatory changes.[16] This has proved difficult because the required changes have tended to have a ripple effect on components throughout the reactor structure rather than being discrete add-on components. Moreover, it has become commonplace for plants to require design changes during construction in response to regulatory changes. It is generally recognized that retrofit costs are higher than if the same features had been in the original design, but it is impossible to judge the magnitude of additional costs imposed by such changes.

Because the costs of plants begun in later years are largely estimated costs, they include expected real escalation. This category includes rather broad expectations about additional design changes as well as unit cost increases in inputs. Thus, some fraction of this cost category should be moved to the unit cost increase category.

TABLE 12-5. Effects of Safety and Regulations on Nuclear Power Plants

Codes and standards	1971[a]	1980[b]
Number of codes and standards nuclear plants must comply with	150	1800
Time requirement		
Time required from commitment to commercial operation	6–7 yr	10–12 yr
Material requirement		
Concrete (1,000 cu yd)	90	153
Rebar steel (1,000 tons)	11	19
Structural steel (1,000 tons)	4.4	10
Cables (1,000 linear feet)	2000	4500
Conduits (1,000 linear feet)	175	425
Labor requirement		
Engineering and services (10^6 hr)	3.4[c]	8.1
Craft labor (10^6 hr)	6.2[c]	16.9

Source: John H. Crowley, "Nuclear Energy—What's Next?" Draft paper presented at Workshop on the Electric Imperative, Atomic Industrial Forum, June 1981, Monterey, Calif.

[a]Typical for 1,000-MW reference design as per WASH-1230.

[b]Material and labor requirements are based on estimates for an 1139 MW reactor. Reported figures are standardized to 1,000 MW assuming straight proportionality.

[c]1972 figures.

Table 12-5 is highly illustrative as an indicator of the magnitude of real increases in basic reactor components. Between 1971 and 1980, the amount of concrete, steel, cables, and conduits required for a 1,000 MW reactor more than doubled; labor requirements increased two and one-half times. Thus, it is not surprising that reactor costs have increased.

The most recent massive reevaluation of nuclear safety requirements occurred as a result of the TMI core accident. D. S. Kettler[17] reports that the Nuclear Regulatory Commission estimated retrofit costs on the order of $25 million (1980 dollars) per unit and the Atomic Industrial Forum (AIF) estimates retrofit costs of $30 to $150 million (1980 dollars) per unit. Kettler argues that an estimate of $15 to $75 million additional for *new* reactors (½ the AIF estimates) may be appropriate since designing in the new requirements is likely to be considerably less costly than retrofitting completed or under-construction reactors. The safety requirement reevaluation after TMI is indicative of the costliness of such actions, particularly in the case of retrofits. However, the range of estimates is indicative of the problems in attempting to estimate or attribute cost increases due to a specific set of design changes.

Komanoff has produced a fairly detailed statistical analysis of the sources of nuclear plant cost increases in plants already completed.[18] As a result, his statistical analysis cannot include information obtained on plants begun but not yet completed. The data set extends into 1980, but all plants were ordered by early 1968 and licensed by 1971. The advantage of this data set is that the costs are much firmer than the estimated costs of plants in various stages of construction. His findings are very interesting, particularly because they challenge engineering estimates of cost savings of multiple unit plants and economies of scale.[19] His statistical results show a 10 percent cost reduction given a doubling of unit size, compared with the industry's thumb rule of 20–30 percent. He finds multiple location of units to result in a modest 10 percent cost reduction per unit and a doubling of architect-engineer experience to result in a 7 percent reduction in costs. While he finds a 20 percent cost penalty for plants with cooling towers, he concludes that the cooling tower variable must be a proxy for some other effect since coal plants do not show a similar penalty. While these results offer an interesting counterpoint to "engineering" estimates, statistical estimates of this type may suffer from equations misspecification or other problems yielding bias in the estimates of undetermined direction or magnitude. More effort needs to be devoted to explaining why the two approaches give different results. As such, the results caution against accepting any single set of estimates as "certain."

The factors Komanoff controls for actually explain very little of the cost increase over the period and, in fact, most work in the opposite direction—i.e., architect-engineering experience tended to increase and later units were more likely to be larger and multiply located. He proposes two alternatives as final variables which explain most of the increase; these are cumulative nuclear capacity and time directly. He correctly identifies these residual costs as primarily induced by safety requirements. However, his use of the model with these variables as a forecast tool presents some conceptual problems. Time itself hardly offers a convincing causal argument for cost increases. Komanoff rejects time as an explanatory variable because of lower explanatory power. His motivation for cumulative nuclear capacity is that as society gains experience with nuclear power, it discovers more of the problems and must legislate safety improvements. This motivation, for purposes of projecting future costs, simply cannot be supported. Future estimates are completely dependent on the functional specification. Komanoff uses a semilog specification which will yield ever-increasing nuclear plant costs as nuclear plant capacity expands. (However, costs increase less rapidly than if a linear specification had been used.) An interpretation consistent with this formulation is that nuclear power cannot be made acceptable at any cost but society only gradually discovers this as capacity expands. (As cumulative capacity approaches infinity, nuclear cost approaches infinity.) While there are those who argue the above case, a statistical approach that assumes the proposition that nuclear power will become ever more costly can hardly be viewed as an approach which either tests the proposition or offers unbiased future estimates.

Komanoff's estimates place nuclear power capital costs at $1,374/kW (1979 constant dollars) in 1988, including interest during construction. While the methodology for arriving at this estimate is clearly faulty, the estimate does not compare unfavorably with our $1,160/kW estimate based on recently reported cost estimates. Komanoff has noted that his recent calculations based on industry estimates put individual cost estimates of plants under construction at 20 to 30 percent above his statistically derived estimates,[20] implying plant capital costs as high as 1800/kW.

Conclusions on LWR Capital Cost Increases

There were tremendous increases in the capital cost of LWR plants during the 1970s. Based on the disaggregation of cost increases, what can be said about the long-term costs of nuclear power? Will costs continue to escalate sharply? A significant contributor to cost increases in plants started later in the 1970s has been the recent steep increases in the real interest rate. The early 1970s saw inflation rates of 6 or 7 percent per year in the United States while the prime rate varied mostly within the 5 to 8 percent range. Using the rates as indicators suggests that real interest rates were on the order of 1 or 2 percent per year. For 1980 and 1981, the inflation rate (GDP deflator increase) averaged 9.3 percent while the prime rate rose to the level of 15 and 20 percent; thus real interest rates were on the order of 5 to 10 percent per year. These represent unprecedented levels. Moreover, the rates appear to be a relatively short term phenomenon resulting from a clash between government fiscal and monetary policy and exacerbated by (perhaps incorrect) expectations of high or higher inflation rates in the future, even though the late 1981-early 1982 period showed a weakening in inflation.

The other element of cost increases that yield to logical argument are unit cost increases in materials and labor. In a mature, developed economy with a near-stationary or stationary population, one might expect an upward pressure on labor costs given that labor is relatively fixed and becoming an ever scarcer input relative to capital. On the other hand, even though the number of man-hours available may be growing scarcer relative to the capital stock, the effective labor supply may be increasing because increased training and other productivity improving forces are likely to be at work. In the long run, such effects would be separate from any productivity increases occurring in the reactor construction industry itself; industries where technological advance is rapid will tend to free up labor in the economy, offering a counterbalancing downward pressure in economy-wide wages. Real wages have certainly risen historically; however, the increases are more likely due to increased skills of workers and innovation in the use of labor (higher productivity) than in any demand pressures. One would not expect sustained increases in produced material inputs unless they are based on a scarce input (labor and energy being potential candidates as scarce resources). Cement, a major input of nuclear plant construction, has high energy requirements. However, typically the indirect energy inputs become a relatively small component of a product as advanced as a nuclear reactor. Energy may account for 40 percent of the price of cement, but cement is a fraction of the cost of mixed concrete, and mixed concrete a fraction of the cost of the concrete structure. A large share of the nuclear plant construction costs (nearly 60 percent) are nonmaterial, nonequipment costs—e.g., labor, engineering services, and so on.[21]

The remaining categories of cost increases are all directly related to safety requirements. It is generally recognized that there is a conceptual basis for shortening the licensing period without compromising reactor safety. However, the political basis for doing so in the United States and the major democracies appears to be absent. Rather, increasingly numerous and stringent safety requirements have steadily lengthened the period. The CONAES study (printed in 1979) reported utility views of further delays and extra safety costs as regulatory tightening that increased the magnitude of nuclear projects and "further escalation" in delays as having "already occurred."[22] The accident at Three Mile Island occurred during the latter part of preparation of the CONAES study and considered assessments of the accident were not available until after the report was published. Thus, the optimistic expectations of utility executives were probably dashed before the book had circulated. The 1980 estimated capital costs per kilowatt electric (in 1979 dollars) had risen to $1,160, compared with capital costs in the range of $680 to $870/kWe (1979 dollars) cited by CONAES,[23] and the construction period had lengthened so that it tended toward the 12-year end of the CONAES range rather than the 9-year end.[24]

It seems fairly unlikely at this time that any major steps will be taken to "rationalize" the regulatory process. Cost escalation estimates for plants in the pipeline, as already noted, include some level of additional safety requirements in the form of retrofits. If now, after TMI, costs of regulation have topped off, the presently estimated future costs of nuclear power may be on the high side. However, given the history of cost escalation resulting from safety requirements, it seems imprudent to believe the estimates are low.

An optimistic view of nuclear power costs might be that noninterest costs will remain stable (after those increases projected to 1992) at 835/kW; the construction period will improve marginally from 6.4 to 6 years, and interest rates fall from 11.6 to 5 percent, in line with rates utilities faced in the early 1970s. On such a basis, total capitalized costs could be as low as 950/kW. A more likely estimate might be no improvement in the construction period and either a further increase in noninterest costs or less of a drop in the real interest rate, or some combination of the two which would generate a total capitalized cost of a LWR of $1100/kW. On the other hand, if cost estimates put forth by Komanoff prove correct, for whatever reason, total capital costs could range to $1,800/kW.

Breeder Capital Costs

Because of the limited experience with breeder technology, breeder costs are highly uncertain. The CONAES study offers two analyses of breeder capital costs and arrives at estimates of 10 to 40+ percent above LWR plants.[25] One approach relied on the fact that 80–90 percent of the breeder plant requires conventional technology more or less identical to a LWR. The remaining 10 to 20 percent, the steam supply system, was expected to be two to three times more expensive than for a LWR. Thus, capital costs should be 10 to 40 percent greater for the LMFBR.

The second approach was based on actual costs or cost estimates for the Clinch River and the French Phenix LMFBRs and expected cost improvements due to economies of scale and increasing construction experience associated with commercialization. Such estimates suggested breeder costs 40 percent above those of LWRs, but the CONAES study added that while the amount of reduction from estimated costs of the prototype Clinch River reactor were not uncommon in analogous industrial development, achievement of such cost goals was uncertain.[26] Using the 10 and 40 percentage figures and LWR costs of $950–1,800/kW suggests absolute cost differences of $95–720/kW between LMFBRs and LWRs.

Chow reports several early (pre-1975) estimates of the breeder cost differential as ranging from $0 to $210/kW (1979 constant dollars).[27] The most recent estimates range from $185 to $1,250/kW (1979 constant dollars). The most recent U.S. DOE estimates place the near-term differential at $670/kW (1979 constant dollars), but the ultimate cost differential is estimated to be just under $500/kW.[28] These figures suggest a capital cost range for the breeder from approximately $1,150/kW up to nearly $3,000/kW if one assumes very high costs for LWRs and a very high differential cost.

Capacity Utilization

Capacity ultilization plays an important role in the cost of generated electricity because the nuclear plant incurs capital charges whether it is running or idle. The capacity utilization factor for nuclear power has also become a fairly controversial number in computing nuclear power economics. Early cost analyses assumed that nuclear power plants would operate at 80 percent of capacity. More recently, the accepted benchmark has been 70 percent. CONAES noted three reasons for downtime, including scheduled maintenance and refueling, unscheduled outages, and lack of demand.[29] Lack of demand is relatively unimportant in explaining downtime of nuclear plants since the capital cost of nuclear makes it a base load source of power. Estimates put downtime due to demand swings at less than 1 percentage point, with some methodologies indicating it may account for up to 2 percent.

Yearly average capacity factors for all reactors in the United States have ranged from 66 percent in 1978 to 51 percent in the first half of 1980.[30] Thus, the average actual rate has never reached the hoped-for 70 percent. Large reactors have tended to have lower utilization factors (a peak of 62 percent was reached in 1978 by reactors over 800 MW). Cumulative experience shows all types of reactors to have averaged a 60 percent capacity factor. Reactors smaller than 800 MW have an averaged 66 percent utilization factor, while larger reactors have averaged 54 percent. The best experience for a group of reactors has been PWR reactors under 800 MW built by Westinghouse. This reactor class has achieved a 70 percent cumulative capacity factor. It has generally been assumed that as more operating experience is gained, the capacity factor would rise. As indicated by

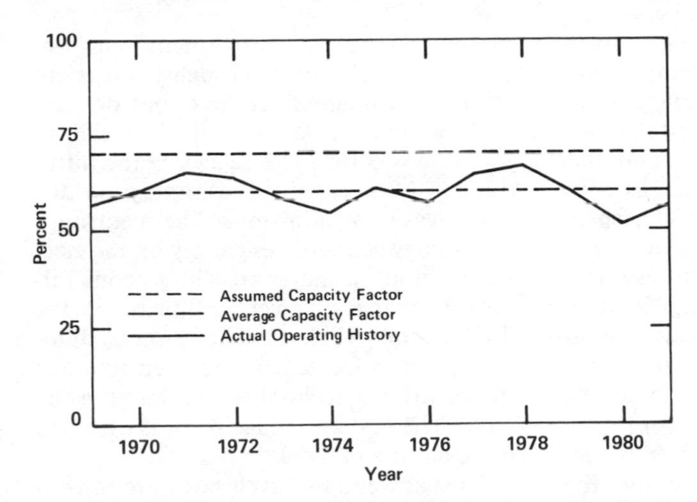

FIGURE 12-2. Nuclear power plant capacity utilization factors. (From Charles Komanoff, *Power Plant Cost Escalation,* Komanoff Energy Associates, 1981 and U.S. DOE *1981 Annual Report to Congress,* Washington, D.C.: U.S. Government Printing Office, 1982.)

figure 12-2, evidence to date has not shown this to be the case. CONAES argues that poor performance of larger reactors may be due to less operating experience with these sizes.[31] This view would lead to the conclusion that the average capacity factor should rise as experience is gained with the new larger plants.

The perplexing evidence, in this regard, has been the falloff in utilization factors since 1978. Komanoff notes the coincidence of this trend with the TMI accident, but argues that problems have intensified, even apart from TMI.[32] However, given the general reassessment of nuclear power safety after the TMI accident, it is likely that a large part, if not all, of the increase in unscheduled outages, even at reactors not directly affected by design changes specific to the TMI reactor, are attributable to the TMI accident. This interpretation does not mean that capacity factors will necessarily return to pre-TMI rates or eventually improve to a 70 percent rate. The reaction to TMI is likely to have been of two types. The first is a once-and-for-all reassessment of safety systems in light of the new information on risk obtained from the TMI accident. The second is an increased willingness on the part of the NRC and utilities to shut down a plant if problems affecting safety are suspected. Even apart from any concern for public safety, TMI demonstrated to utilities the real possibility of reactor problems leading to the loss of a huge financial investment. Reactions of the first type are likely to work themselves through within a few years of the accident. Effects of reactions of the second type may be permanent.

In light of actual historical evidence and the lesson of TMI, it appears overly optimistic to expect that operating factors can reach 70 percent. However, given the still relatively low level of experience with nuclear reactors, and particularly large reactors, some improvement in operating factors might be expected. A reasonable benchmark factor for the future is a 65 percent capacity utilization factor. This is somewhat more optimistic than Komanoff's benchmark of 60 percent.[33] However, the time frame of this analysis and Komanoff's differs substantially; he is looking 10 to 20 years into the future. In looking 70 years out, there is more opportunity for the full learning curve to be realized.

Costs of Nuclear Fuel Cycles

LWR

The light water reactor fuel cycle is commonly broken into front end and back end components. The front end of the fuel cycle consists of uranium milling, conversion, and fabrication of fuel components. The back end of the fuel cycle refers to the handling, processing (if any), and disposal of waste. Several different back-end fuel cycles can be envisioned. The existing back-end fuel cycle is incomplete; spent fuel enters temporary storage (generally at the reactor site) but few final storage (disposal) sites and no reprocessing facilities exist in the United States. France and West Germany have small reprocessing plants but these are insufficient to handle the present global production of spent fuel. The United States has delayed construction of reprocessing facilities on the grounds that spent fuel reprocessing would increase the dangers of nuclear proliferation. The United States has also failed to approve sites for permanent storage or disposal. In addition to domestic spent fuel handling, the United States accepts spent fuel from foreign countries whose reactor technology was supplied by the United States.

Several sets of estimated fuel cycle costs are given in table 12-6. Front end nuclear fuel costs are fairly well established; considerable operating experience exists on which to base cost estimates. Table 12-6 controls for uranium prices at $35/lb. As can be seen from the total front-end cost figures, uranium accounts for about 50 percent of the front-end costs at current uranium prices. The price path of uranium is dependent on resource availability at various grades and the rate of exhaustion. The gradual upward revision of estimates of uranium resources in the United States and the world, combined with much slower expansion of nuclear power than expected, has led most analysts to revise downward future increases in uranium prices through 2000.

Beyond uranium prices, the other major uncertainty in front-end fuel costs is enrichment costs. Komanoff projects enrichment costs to rise from the present $94/SWU (separative work units) to $129/SWU by 2017. The U.S. DOE estimate places the 2000 enrichment cost at $121.70/SWU, which is roughly consistent with the Komanoff estimate, but projects that enrichment costs will then fall to $91.12/SWU by 2020, reflecting the introduction of a centrifuge technology to displace gaseous diffusion. The centrifuge technology economizes greatly on the electricity input, a major cost in gaseous diffusion. Increasing electricity prices in the TVA region (where the U.S. gaseous diffusion plant is located) is expected to make the centrifuge technology competitive by 1990. The first centrifuge plant is expected to be on line in 1989.[34]

The back-end fuel cycle costs are considerably less well known. Permanent storage or disposal costs depend on where and how the spent fuel is disposed. An ideal back-end fuel cycle would consist of spent fuel being temporarily stored at the reactor site to allow the most radioactive and fastest decaying isotopes to decay. From there, the spent fuel would move either to a reprocessing facility or to a permanent storage or disposal site from which it could be removed at some later date for reprocessing if desired.

In table 12-6, costs are given for both a once-through fuel cycle and for uranium recycle with plutonium recovery. The economics of recycling spent fuel depend on the cost differential between the back end of the once-through fuel cycle and of fuel recycling. This differential determines the additional cost imposed by recycling of spent fuel which, together with an estimate of the amount of uranium and plutonium recovered, determines the uranium price necessary to make spent fuel recycling economically attractive. Assuming that a full fuel recycle program results in a life cycle savings of 30 percent of a nuclear reactor's uranium requirements,[35] the various cost estimates of table 12-6 show mixed results concerning the present economies of fuel reprocessing. Of the two studies giving estimates for both back end fuel cycles, Kettler shows approximately a 6 percent cost advantage for fuel recycling whereas Chow

TABLE 12-6. Nuclear Fuel Cycle Costs (1979 dollars)

	Dollar/Unit as Given					mills/kWh				
	CONAES 1979	*Komanoff 1981*	*Chow*	*U.S. DOE*	*Kettler*	*CONAES*	*Komanoff*	*Chow*	*U.S. DOE*	*Kettler*
Front End										
Conversion ($/lb)	2.20	2.25	2.00	2.20	4.86	0.19	0.19	0.17	0.19	0.41
Conversion (UF_6 to UO_2 ($/lb))	a	a	3.45	a	a	a	a	0.29	a	a
Enrichment ($/SWU)	125	94–129[b]	110	91–122[c]	102	2.93	2.21–3.03[b]	2.58	2.14–2.86[c]	2.40
Transport ($/kgu)	a	a	2.16	a	a	a	a	0.01	a	a
Fabrication	125	112.5	120–140	109	127	0.70	0.63	67–78	0.61	0.71
Back End (once through)										
Storage ($/kg/y)	N.G.	11	a	7	a	N.G.	0.19	a	0.11	a
Transport ($/kg)	N.G.	18	16	17	a	N.G.	0.04	a	0.04	a
Fuel disposal ($/kg)	N.G.	652	130	218	260[d]	N.G.	1.41	0.28	0.47	0.56
Back End (recycle)										
Reprocessing ($/kg)	245	N.G.	490–650	N.G.	424	0.91	N.G.	1.81–2.41	N.G.	1.57
Pu credit ($/g)	30	N.G.	29[e]	N.G.	19	(0.53)	N.G.	(0.52)[e]	N.G.	(0.34)
Waste disposal	60[f]	N.G.	60	N.G.	a	0.13[f]	N.G.	0.13	N.G.	a
Total										
Front end (with uranium at $35/lb or $3.54 mills/kWh)						7.36	6.57–7.39	7.26–7.37	6.48–7.20	7.06
Back end (once through)						N.G.	1.64	0.28	0.62	0.56
Back end (recycle) nct						0.51	N.G.	1.42–2.02	N.G.	1.23

Sources: Nuclear fuel costs in dollar/unit as given are from the following sources: Committee on Nuclear and Alternative Energy Systems (CONAES) *Energy in Transition 1985–2010* (San Francisco: W. H. Freeman 1979), p. 274. Charles Komanoff, *Power Plant Cost Escalation* (New York: Komanoff Energy Associates, 1981), p. 266. Brian Chow, "Comparative Economics of the Breeder and Light Water Reactor," *Energy Policy,* vol. 8, no. 4 (December 1980), pp. 293–99. Brian Chow, *Economic Comparison of Breeders and Light Water Reactors* (Los Angeles: Pan Heuristics, 1979). U.S. DOE/EIA, *1980 Annual Report to Congress,* vol. 3 (Washington, D.C.: U.S. Government Printing Office), p. 117. D. J. Kettler, "Nuclear Versus Fossil Cost Post TMI" (New York: Ebasco, 1980). The underlying assumptions for computing costs per kWh are those used in Komanoff. These include inventory cost charges (credits) of 9 percent per year over the average number of years the component is held before actual electricity generation occurs (revenue is received).

N.G. = Not given.

[a]The study does not break out this process as a separate cost item.

[b]Lower figure is the present cost. Higher figure is the estimated cost in 2017.

[c]Higher cost is the estimated year 2000 cost. U.S. DOE projects costs to fall to the lower figure by 2020.

[d]Permanent storage.

[e]Price evaluated within Chow's model, depending on the uranium price. A relative value compared to uranium is set at 0.82 (gPu/lb·u).

[f]Not included in study, assumed equal to Chow's estimate.

shows as much as a 32 percent cost penalty for a spent fuel recycling with uranium prices at $35/lb.

If one takes combinations of once-through and recycle costs from the estimates of table 12-6, the CONAES and Komanoff cost projections make the best case for recycling. The CONAES recycle costs are actually less than the once-through back-end cost estimates of Komanoff. Thus, it would always be economic to recycle wastes. Chow's estimate, noted above, presents the worst case for spent fuel reprocessing. Uranium prices would have to reach $60/lb to make recycling economically competitive with a once-through fuel cycle. Chow presents a fairly convincing case for high costs of reprocessing, citing the West German experience with constructing a large-scale reprocessing plant where cost estimates quadrupled as construction proceeded. Eventually, the chemical companies involved in the project withdrew financial support as it became clear that uranium and plutonium sales would not cover the cost of the operation.[36] However, two factors may make Chow's reprocessing case overly pessimistic. First, as more experience is gained in constructing a large-scale reprocessing plant, costs could fall. Second, Chow's once-through waste disposal costs appear low, thus favoring the once-through cycle. Chow notes that the full cost of a fuel recycle includes any added cost associated with risks of proliferation.[37] He found that fairly small assessments (3 and 7 percent) on busbar costs of electricity increased the breakeven U_3O_8 price to near $200 and $300 per lb, respectively.

The fairly small charge on busbar electricity costs translates to a large increase in uranium prices because of the small share of total electricity cost accounted for by the cost of uranium. The global cost of nuclear proliferation and the contribution of spent fuel reprocessing and proliferation are topics which are unlikely to yield to easy answers. However, it is possible, for our purposes, to avoid this discussion by asking what the likelihood is of ever internalizing proliferation costs in the cost of nuclear power. In fact, it appears highly unlikely. First, attaching such a fee to electricity costs in countries with nuclear weapons would do nothing to decrease the chance of proliferation or compensate the "victims" of proliferation.

One could propose a tax on the reprocessed fuel itself. This would discourage reprocessing (reducing proliferation) in countries with the tax. Unfortunately, such a policy in the United States would be likely to have the same impact as the United States' unilateral actions aimed at delaying reprocessing—i.e., encourage the development of reprocessing facilities abroad. Finally, one might refuse to sell nuclear technology of any kind to countries with a proliferation risk. This last approach has already proved to be unworkable. Thus, there appears to be no real way of internalizing proliferation risk costs in the cost of nuclear power short of an unprecedented international cooperative effort.

Chow also shows that efficiency improvements (lower life cycle uranium requirements) in the LWR tend to increase the uranium price at which reprocessing becomes economic. A 15 percent improvement increased the uranium break even price to over $120/lb.[38]

Breeder Fuel Cycle Costs

Breeder fuel cycle costs tend to be lower than the fuel cycle costs of LWRs because once an initial fuel endowment is made, the breeder generates a plutonium credit rather than requiring a continuous feed of fissile material. Any cost advantage of the breeder fuel cycle depends on uranium prices and the comparative costs of the fuel processing stages of the fuel cycles. The standard economic analysis of the LWR-to-breeder transtion posits the uranium price as the driving variable for breeder introduction. Uranium price increases, gen-

TABLE 12-7. Breeder Fuel Cycle Costs (1979 dollars)

	Dollar/Unit as Given		*mills/kWh*	
	CONAES	*Chow*	*CONAES*	*Chow*
Fabrication ($/kgHM)	985	613–793	2.36	1.47–1.90
Reprocessing ($/kgHM)	430	605–829	0.77	1.09–1.49
Fissile plutonium value ($/g)	30	30[a]	(0.54)	(0.54)[a]
Inventory charge	N.A.	N.A.	1.26	1.03–1.37
Waste disposal ($kgHM)	190[b]	165	0.13[b,c]	0.15[c]
Total			3.98	3.20–4.37

Sources: Committee on Nuclear and Alternative Energy Systems (CONAES), *Energy in Transition 1985–2010* (San Francisco: N.A. Freeman, 1979), p. 274. Brian Chow, "Comparative Economics of the Breeder and Lightwater Reactor," *Energy Policy,* vol. 8, no. 4 (December 1980), pp. 293–299. Mills/kWh calculations for Chow are the authors' based on CONAES assumptions.

N.A. = Not applicable.

[a]Assumed identical to CONAES assumption.

[b]Assumed identical per kWh cost as for LWR reprocessing—see CONAES, p. 274.

[c]Includes an inventory credit. See notes to table 12-6.

erated by exhaustion of high grade resources, eventually bring LWR fuel cycle costs to a level which overcomes the capital cost penalty associated with breeders.

However, intervening transitions may occur. As already discussed, LWR fuel reprocessing may represent an intervening transition. In addition, transitions to higher efficiency LWR technologies with capital cost penalties or the class of near breeder reactors are possible intervening transitions.

Table 12-7 gives some breeder fuel cycle cost estimates. No firm cost data are available.

Operation and Maintenance Costs (O&M)

Operation and maintenance (O&M) costs are a relatively small component of nuclear generating costs. The large majority are independent of the capacity utilization, short of abandonment of the plant. Chow assumes that "fixed" LWR O&M costs are approximately 2.47 mills/kWh (1979 dollars) at 65 percent capacity with variable O&M costs of 0.18 mills/kWh at 100 percent capacity, or 0.11 mills/kWh at 65 percent capacity. Chow assumes breeder O&M costs of 3.22 mills/kWh for fixed costs and 0.11 mills/kWh for variable O&M costs at 65 percent utilization.[39] Komanoff cites O&M costs of 4.11 mills/kWh, but noting recent rapid real increases in O&M costs, assumes future O&M costs of between 5.7 and 6.8 mills/kWh.[40] U.S. DOE assumes LWR O&M costs are 1.8 mills/kWh at a 0.66 utilization rate and breeder O&M costs are 3.3 mills/kWh at a 0.62 utilization rate.

Decommissioning

Nuclear plants must be decommissioned at the end of their useful life. Relatively little is known about the costs since no experience exists on which to base costs. CONAES argues that decommissioning costs are far in the future, thus will be relatively small in present-value terms. An estimate of less than 1 percent of original capital costs in present value terms is cited for LWR.[41] Komanoff estimates decommissioning costs of 2.1 mills/kWh for LWR but his calculations do not discount the decommissioning cost.[42] Doing so at a 5 percent discount rate over 40 years would reduce the decommissioning charge to 0.3 mills/kWh.

Summary of Electricity Generation Costs

Table 12-8 summarizes the component costs of nuclear power. The use of extreme values of cost estimates of each component results in a wide range of generating costs. Table 12-8 does not vary the uranium price component of fuel costs—these are assumed to be $35/lb in all cases. The reported costs are busbar costs. The low and high estimates are quite extreme because each component is an extreme point; typically, the studies reviewed were not necessarily extreme on every cost component.

While we label these extreme values, it must be noted that the range of nuclear electric generating costs at busbar found in the literature do exceed these ranges. In particular, Komanoff has published an estimate of busbar generating costs of nuclear plants coming on line in 1988 ranging from 40.5 mills/kWh (best case, no TMI) to 47.8 mills/kWh (base case, no TMI) to 53.3 mills/kWh (plausible TMI impact), to 72.0/kWh (worst case).[43] We have already noted the bias of the capital cost projection methodology used in the base case and the failure to discount decommissioning costs. These two components account for 6.7 mills/kWh difference in our best estimate and Komanoff's base case, no TMI estimate. Another 3.7 mills is accounted for by Komanoff's lower (0.60) capacity utilization factor. Most of the remaining difference stems from Komanoff's assumption of a 2 percent per year real increase in uranium prices. Despite the methodological problems, his base estimate cost is within our range. However, his TMI cost estimate cases fall outside the range. This is traceable to a compounding of the base case's

TABLE 12-8. Nuclear Electricity Generation Cost Summary (1979 dollars)

	LWR			*LMFBR*		
	Low	*Best Estimate*	*High*	*Low*	*Best Estimate*	*High*
Capital cost (dollars/kW) installed capacity	950	1100	1800	1135	1600	3050
Capacity utilization (percent)	70	65	60	70	65	60
Real fixed change rate (percent)	8	10.3	12	8	10.3	12
Capital cost (mills/kWh)	12.4	19.9	41.1	14.8	28.9	69.6
Fuel cycle cost[a] (mills/kWh), once-through cycle	7.1	7.6	9.0	N.A.	N.A.	N.A.
Fuel cycle cost (mills/kWh) recycle	7.9	9.0	9.4	3.2	4.0	4.4
O&M (mills/kWh)	1.8	5.0	6.8	3.3	9.2	12.5[b]
Decommissioning	0.1	0.2	0.4	0.1[c]	0.3[c]	0.7[c]
Total costs (mills/kWh) once-through	21.4	32.7	57.3	N.A.	N.A.	N.A.
Recycle	22.2	34.1	57.7	21.4	42.2	87.2

[a]Uranium at $35/lb.

[b]No estimates were given at these levels but all studies giving estimates for LMFBR O&M costs indicated them to be at least as high as LWR O&M costs. The best and high estimates are proportional to LWR costs.

[c]Assumed to be 1 percent of capital change—identical to LWR cost.

methodological problems. The rationale of the base case projection methodology is that as society learns about nuclear power (by operating more capacity), it sees the need for more stringent safety requirements. Komanoff then adds additional capital penalties to these base projection costs to account for TMI. The conceptual problem with adding on additional TMI costs, if one accepts the underlying statistical methodolgy and motivation, is why TMI is not simply part of the learning experience. Adding an additional cost penalty for TMI would appear to be double counting the learning experience. For this reason, it appears necessary to dismiss Komanoff's higher cost estimates on methodological grounds. Cost increases above those projected here may occur if additional problems with LWR's cause increased regulation. On the other hand, several forces are at work that tend to reduce costs, including economies of scale, engineering experience, and multiple siting. While any cost reduction due to these factors is likely to be modest, they will offset any increases due to unit cost increases or additional safety requirements. Finally, the escalation in costs imposed by TMI and other safety regulation has caused engineers to reassess the design of the LWR. A drastically redesigned reactor could lead to a lower cost reactor with equal or improved safety features. The research in this area is just beginning.[44]

CONCLUSIONS

The future of nuclear power is extremely uncertain. Costs have escalated tremendously in the past decade. However, costs for competitive power-generating technologies (primarily coal) have escalated as well. How nuclear power fares in the future depends on the relative cost of electric-generating technologies and on the demand for electricity in general. This question can only be answered in the context of a full equilibrium energy model.

This chapter has sought to examine LWR and breeder costs in some detail. We found that unit cost increases (inflation in materials and labor above the general inflation level) were a major contributor to cost increases. These increases were roughly equal to cost increases due to stricter safety standards.

The rapidly escalating costs of nuclear power and the resultant drop in expected future nuclear capacity have made questions concerning the adequacy of uranium resources considerably less pressing than they seemed to be in the mid-1970s. Likely exhaustion of relatively high grade uranium is far enough in the future to give adequate lead time on breeder development if nuclear fission is deemed a necessary energy source into the indefinite future.

While this chapter focused on costs and resources, these issues are only two elements of the nuclear power debate and are not the main subject of discussion. The issues of environmental impacts and public health and safety effects generally take center stage. These issues are unlikely to be resolved as issues of public debate. It is simply a fact that within the population there are individuals whose evaluation of the impact of nuclear power on them personally is negative while others evaluate the personal effects as positive. Marginal individuals may switch as information about nuclear power improves or as technology or resource reconsiderations change the safety characteristics of nuclear power. From a predictive (as opposed to a prescriptive) standpoint, the important aspect is how the institutional setting allows these diverse views to be represented. The 1970s have seen safety concerns reflected as specific construction requirements, protracted hearings on safety and licensing, and demonstrations aimed at blocking construction at individual sites. These avenues are explicitly available or tacitly accepted as mechanisms by which individuals can effect an actual outcome that is near their personally desired outcome.

The ultimate reflection of these concerns has been in higher cost nuclear power, though safety concerns have probably accounted for only half the total increase over the past decade.

A strong current presently exists which holds that a "second nuclear era" requires a drastic redesign of LWR. If so, the second nuclear era awaits the blueprints for new designs. Until these designs are made specific, little can be said about the cost. Extant LWR technology remains an option, though one that is much more expensive than originally envisioned. The existing "moratorium" of nuclear plant orders is a de facto moratorium (rather than a legislative moratorium) which can be traced to weak electricity demand. When (or, for some, if) demand recovers, nuclear power supplied by currently existing reactor technology is an option which can be had at a set of economic, environmental, and safety costs if theoretically better technologies fail to become a reality.

NOTES

1. "The World List of Nuclear Power Plants," *Nuclear News,* vol. 24, no. 10 (August 1981).
2. United Nations, *Monthly Bulletin of Statistics,* ST/ESA/SER.Q/104, vol. XXXV, no. 8 (August 1981); and the National Foreign Assessment Center, *International Energy Statistical Review,* ER IESR 81-007, July 1981.
3. "Nuclear Power Capacity Outside U.S. Soars 25 Percent in 18 Months," *AIF INFO News Release,* February 6, 1980; U.S. Department of Energy, *1980 Annual Report to Congress,* vol. 2 (Washington, D.C.: U.S. Government Printing Office, 1981).
4. World Energy Conference (WEC), *Survey of Energy Resources 1980* (London: WEC, 1980), p. 183.
5. Ibid., pp. 186–190.
6. Committee on Nuclear and Alternative Energy Systems (CONAES) *Energy in Transition 1985–2010* (San Francisco: W. H. Freeman, 1979), p. 231 and Wolf Häfele, *Energy in a Finite World,* Report by the Energy Systems Program Group of the International Institute for Applied Systems Analysis (Cambridge, Mass.: Ballinger, 1981), p. 120 note the problems of mining uranium from low-grade deposits.
7. CONAES, *Energy in Transition,* p. 231.
8. Author's estimate, based on amounts in each resource category reported in U.S. DOE, *1980 Annual Report to Congress,* vol. 2 (Washington, D.C.: U.S. Government Printing Office, 1981), p. 179.
9. A. M. Perry, "The Nuclear Option—Nuclear Resources and Nuclear Development Strategy," Draft Manuscript, Institute for Applied Systems Analysis, Laxenberg, Austria, 1978.
10. IIASA and WEC devoted considerable effort to examining rates of development and timing of introduction of breeders and the implications for the rate of expansion of nuclear power on a global basis. CONAES examined similar issues for the United States where the United States is limited to domestic uranium resources.
11. *Nuclear News,* vol. 24, no. 10 (August, 1981); and Häfele, *Energy in a Finite World.*
12. Douglas Martin, "Mexico's Nuclear Power Plans," *Washington Post,* May 23, 1982.
13. Based on yearly average Saudi crude oil prices as reported in *International Financial Yearbook,* International Monetary Fund, Washington, D.C., 1981 and deflated using the U.S. GNP deflator.
14. Based on c.i.f. costs of coal delivered to utilities reported in U.S. DOE, *1980 Annual Report to Congress,* vol. 2.
15. Charles Komanoff, *Power Plant Cost Escalation* (New York: Komanoff Energy Associates, 1981), p. 315.
16. Ibid., pp. 24–25.
17. D. J. Kettler, "Nuclear versus Fossil Cost Post TMI," EBASCO Services, Inc., September 1980, New York.
18. Komanoff, *Power Plant Escalation,* pp. 197–212.
19. Ibid., p. 20.
20. Charles Komanoff, personal communication, June 1982.
21. CONAES, pp. 265–266.
22. Converted from 1978 constant dollars using the U.S. GDP deflator, ibid., pp. 264–265.
23. Ibid., p. 263.
24. Kettler, "Nuclear versus Fossil Cost Post TMI," and John H. Crowley, "Nuclear Energy—What's Next?" Workshop on the Electric Imperative, Atomic Industrial Forum. Monterey, California, 1981.
25. CONAES, p. 273.
26. Ibid.
27. Brian G. Chow, "Comparative Economics of the Breeder and Light Water Reactor," *Energy Policy,* vol. 8, no. 4 (December 1980), pp. 293–307.
28. U.S. DOE, *1980 Annual Report to Congress,* vol. 3.
29. CONAES, p. 266.
30. Komanoff, *Power Plant Cost Escalation,* p. 250.
31. CONAES, p. 267.
32. Komanoff, *Power Plant Cost Escalation,* p. 252.

33. Ibid., p. 253.
34. U.S. DOE, *1980 Annual Report to Congress,* vol. 2, p. 174.
35. See CONAES, pp. 242–243.
36. Chow, "Comparative Economics of the Breeder and Light Water Reactor."
37. Ibid.
38. Brian Chow, *Economic Comparison of Breeders and Light Water Reactors* (Los Angeles, Calif.: Pan Heuristics, 1979), p. 39.
39. Ibid., assumes a 70 percent capacity factor.
40. Komanoff, *Power Plant Cost Escalation,* p. 268.
41. CONAES, p. 272.
42. Komanoff, *Power Plant Cost Escalation,* p. 272.
43. Ibid., p. 40.
44. A project at the Institute for Energy Analysis is currently exploring this issue. Brian Chow, "Improving Nuclear Economies Without Increasing Proliferation Risks," Pan Heuristics, draft 1982, has suggested possible cost-reducing improvements.

Solar and Wind 13

This chapter reviews the potential contribution of solar thermal electric conversion (STEC), photovoltaic energy conversion, and wind-powered systems. The world resource base of these energy sources is briefly reviewed, and operational limits to this base are developed. These limits include technical factors, cost, land requirements, regional differences in insolation, material needs, and storage. Supply curve considerations are then drawn from forecasts of price-quantity relationships.

SOLAR THERMAL ELECTRIC CONVERSION

There are a number of possible systems for solar thermal electric conversion: parabolic trough, parabolic dish, satellite, total systems (heat and electricity), central receiver, and solar ponds. Based upon research to date, the most viable method for large-scale conversion is the central receiver. In this concept, sunlight falling on flat, two-axis tracking mirrors (heliostats) is focused on a central boiler. The absorbed energy produces superheated steam or hot gases to drive a Rankine steam cycle or Brayton cycle. Projected sizes of STEC systems range to 150 MW. There appear to be no significant technical constraints to the concept; a 10-MW experimental unit is currently under construction.

Resource Availability

Solar energy is an immense resource that is characterized by low energy density and intermittency. Figure 13-1 shows an approximation of world insolation, measured in average annual sunshine hours. The average energy available (in kWh/m^2/yr) to a tracking surface is 0.5 to 0.7 times the number of sunshine hours. As a worldwide average, 1,500 kWh/m^2/yr is available on a tracking surface and 1,150 kWh/m^2/yr on a horizontal surface.[1] However, average radiation varies by a factor of three from arid regions to the less sunny regions of North America and Europe. Furthermore, in northern latitudes, monthly insolation may vary by a factor of ten.

The intermittent nature of the resource requires that either storage be provided or electrical generation be restricted to approximately an intermediate or peak load usage. More will be said about the availability and costs of storage when the supply characterization is developed.

Recovery Constraints

Technical and Geographic Limitations

Although there are no theoretical problems which would preclude STEC implementation, the engineering problems involved in projects of such massive scale are consid-

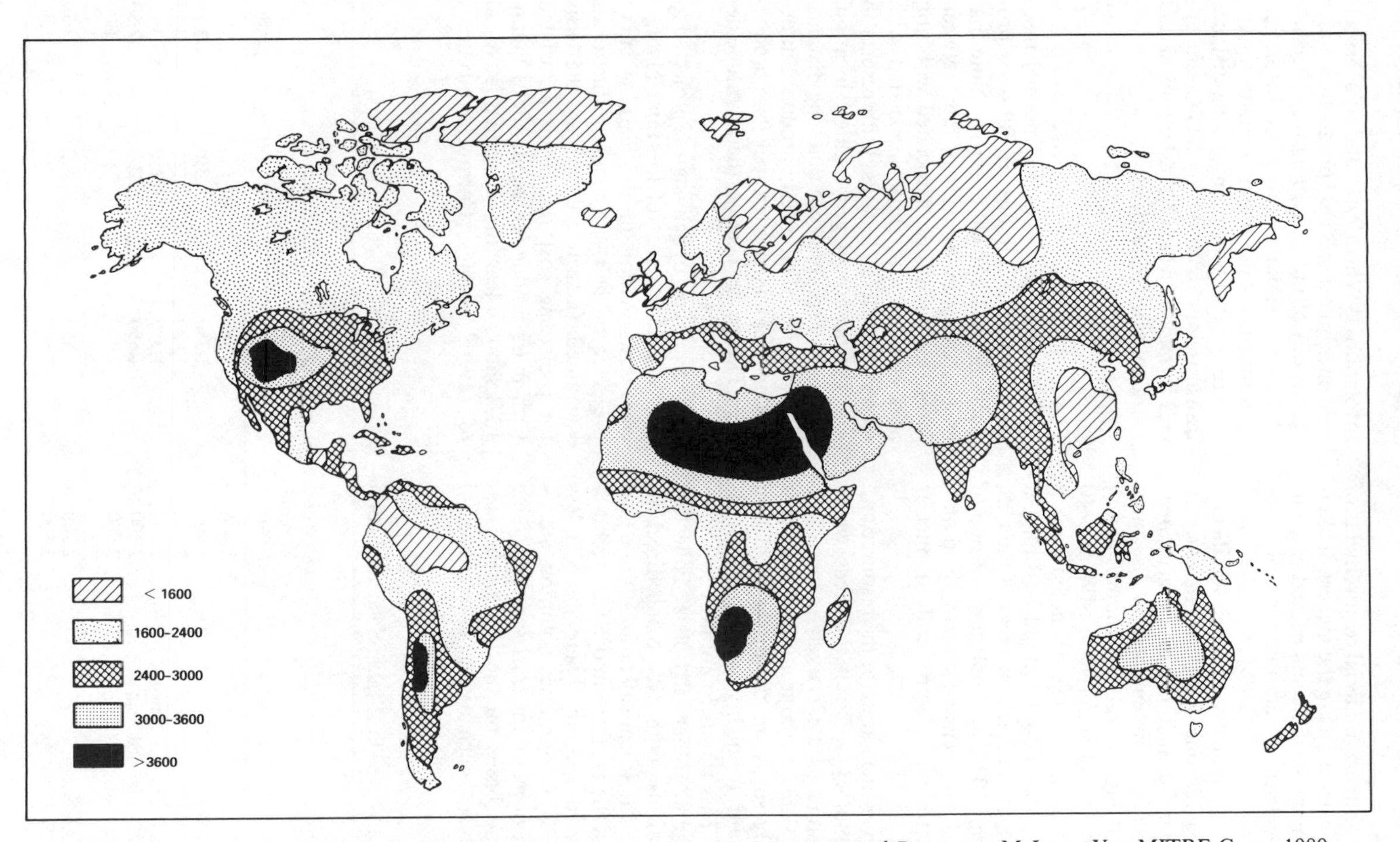

FIGURE 13-1. Annual number of sunshine hours. (Reprinted with permission from Charles A. Zraket and Martin M. Scholl, *Solar Energy Systems and Resources,* McLean, Va.: MITRE Corp., 1980, pp. 6–7. Copyright © MITRE Corporation.)

erable. Labor and material requirements of STEC construction are the highest of alternative energy systems and exceed conventional electrical generation by a factor of seven.[2]

Land requirements, although significant, are not a limiting factor, particularly in light of land availability in hot arid areas where STEC implementation is most likely. Using the rule of thumb that a square mile will supply 100 MW at a 40 percent load factor, it is estimated that all U.S. intermediate load could be supplied using one-thirtieth of 1 percent of the U.S. land area.[3] STEC technologies are one of the least land-intensive sources, third only to oil/petroleum electric and to nuclear systems.

However, the nature of the resource and the technology do limit the areas where STEC can be used effectively.

Since STEC is a concentrating system, it must rely solely on the direct insolation component, the highly collimated solar radiation. This stands in contrast to photovoltaic and other flat-plate collectors which can use both diffuse and direct insolation. These considerations could further limit the application of STEC technology to arid sections where there is less diffusion by clouds, dust, pollution, etc.

The regionality index is shown in table 13-1. It is based on the relationship between a very rough average of total normal insolation (kWh/m^2/yr), and the associated energy cost of STEC production under conditions of industrial maturity. Because of the approximate nature of the insolation conditions, no provision is made for the intraregional substitution of long-distance transmission for declining STEC efficiencies in regions which have both ideal and marginal solar conditions.

Costs

Figure 13-2 shows the relationship between energy cost and total normal insolation under conditions of full-scale production and technical maturity. Note that costs escalate by a factor of four between insolation conditions associated with hot, arid climates and those associated with cloudy, northern climates. With respect to the significance of geographical location, studies have pointed out that long-distance transmission of electricity produced in desert environments is cheaper than producing STEC power near northern latitude demand centers.[4]

Since the cost of heliostats is projected to be one-half the cost of the total plant, a close examination of the learning curve projected for heliostat mass production is warranted. (Learning curves are the basis for projecting cost reductions associated with mass production economies.) Figure 13-3 shows a learning curve that is typical of several studies. Current projections,

TABLE 13-1. STEC Scalar: Solar Insolation—Regional Cost (1979 constant dollars)

Region	*Annual Number of Sunshine Hours*	*Total Normal Insolation (kWh/m^2/yr)*	*Average Energy Costs (mills/kWh) at Maturity*	*Index*
1. US	3,000	2,250	130	1.63
2. OECD West	1,800	1,350	250	3.13
3. JANZ	2,400	1,800	175	2.19
4. Centrally Planned Europe	1,800	1,350	250	3.13
5. Centrally Planned Asia	2,200	1,650	200	2.50
6. Middle East	3,600	2,700	80	1.00
7. Africa	3,300	2,475	90	1.12
8. Latin America	2,200	1,650	200	2.50
9. South and East Asia	2,800	2,100	125	1.56

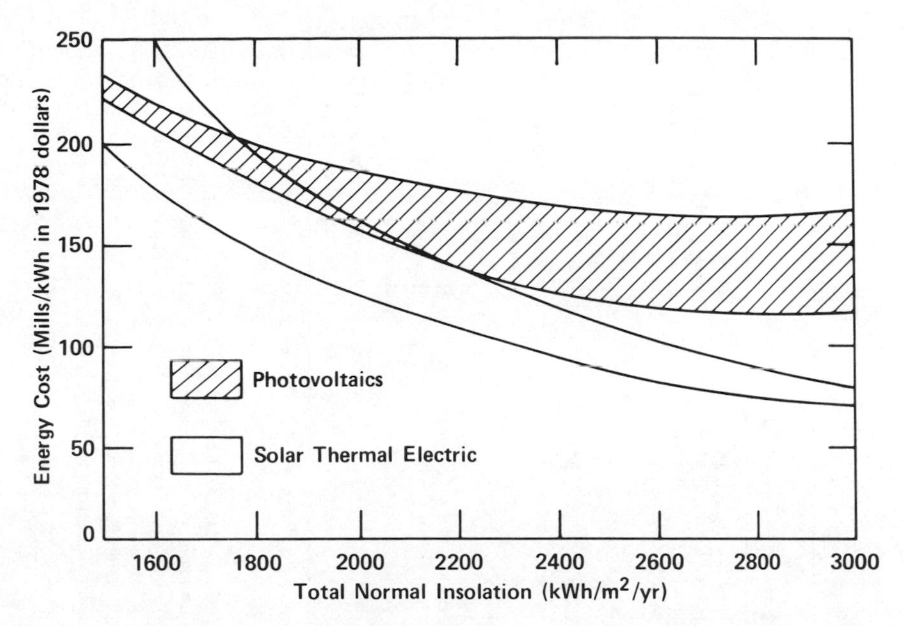

FIGURE 13-2. Solar electric generation delivered energy cost (1978 constant dollars). (Reprinted with permission from Charles A. Zraket and Martin M. Scholl, *Solar Energy Systems and Resources,* McLean, Va.: MITRE Corp., 1980, p. 38. Copyright © MITRE Corporation.)

FIGURE 13-3. Learning curve, heliostats (1980 constant dollars). (From L. L. Vant-Hull, *Second Generation Heliostats,* vol. I, Report No. SAND 81-8177, Huntington Beach, Calif.: McDonnell Douglas Astronautics Co., April 1981, p. 10.)

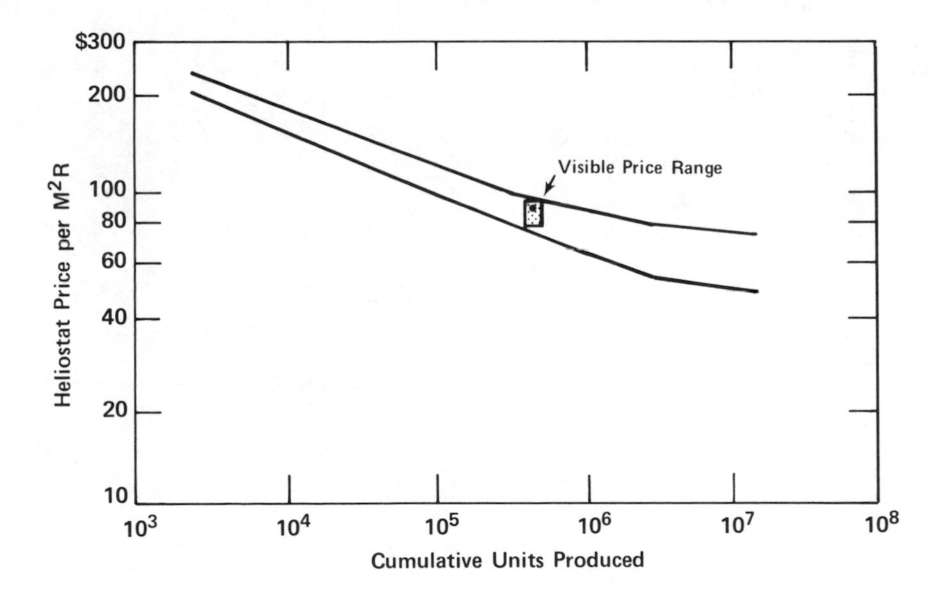

TABLE 13-2. STEC Energy Cost Projections

Study	*Capital Cost $/kW*	*Power Cost mill/kWh*[a]	*Projected Year*	*Size of Unit*	*Notes*[b]
Bennington[c]	4,640	196	1981	10 MW	cf = 0.5
	2,668	157	1990	100 MW	cf = 0.36 fuel saver
	998	59	2025	100 MW	cf = 0.36 fuel saver
	800	34	2025	100 MW	cf = 0.5, intermediate (combined cycle)
Stanford Research Institute[d]	3,770	159	2000	150 MW	Fuel saver
	2,730	115	2020	150 MW	cf = 0.5
National Research Council[e] (CONAES)	9,750	819	1975		15% capital charge/load factor 0.2
	3,250	273	1985		lf = 0.2
	1,820	98	1990		lf = 0.33
	1,560	65	2000		lf = 0.45
	1,560	65	2010		lf = 0.45
A. D. Little, Inc.[f]	2,320	93–174	2000	150 MW	Fuel saver
Stanford Research Institute[g]	2,860	163	1985		20% capital charge, cf = 0.4
	1,430	82	1995	Intermediate load	20% capital charge, cf = 0.4

[a]Calculated from capital cost when only capital costs are given in the study. Assumes fixed charge rate = 18.5%, which includes 3% operation and maintenance costs; assumes cf = 38% unless otherwise noted.

[b]cf = capacity factor; lf = load factor.

[c]G. Bennington et al., "Solar Energy, A Comparative Analysis to the Year 2020" (McLean, Va.: MITRE Corp., August 1978).

[d]U.S. Department of Energy, "Solar Energy, A Status Report," DOE/EIA-0062 (Washington, D.C.: U.S. Government Printing Office, January 1978).

[e]National Research Council, National Academy of Sciences, "Study of Nuclear and Alternative Energy Systems, Supporting Paper 6, Domestic Potential of Solar and Other Renewable Energy Sources," Washington, D.C., 1979.

[f]A. D. Little, Inc., "Distributed Energy Systems: A Review of Related Technologies" (Cambridge, Mass., November 1979).

[g]J. G. Witwer, "Costs of Alternative Sources of Electricity" (Menlo Park, Calif.: Stanford Research Institute, July 1976).

based on a specific production scenario of 50,000 heliostats per year, indicate costs in the range of $85 to $100 per square meter. In order to produce STEC units at a capital cost of $1,700, average heliostat costs would have to drop to about $80 per square meter.[5]

As with all intermittent sources, costs associated with storage present the most serious problems in estimating costs. As long as pumped hydro can be used, the limited capacity of STEC can serve as intermediate load. Without storage, STEC replacement of capacity will be restricted. Storage costs are again a basis for deriving the supply function in a later section. STEC systems have an advantage over wind and photovoltaics in that thermal storage may be used with fewer conversion losses. Although thermal storage is currently expensive, it holds greater possibility of a technological breakthrough than does purely electrical storage.

Energy costs for current and future years are projected in table 13-2; these costs reflect many of the attributes shown in figure 13-2.

Contribution

Table 13-3 summarizes published projections of STEC implementation. Percent of projected supply is used as a parameter in the following analysis, which determines the characteristics of the STEC supply function.

Supply Curves

Table 13-4 presents the costs and output limits of the STEC contribution for the

TABLE 13-3. STEC Implementation Projections

Study	*Quads Displaced*	*Projected Year*	*Percent of Projected U.S. Energy*
CEQ[a]	<28	2000	1.6%
	5–10	2020	7
Bennington[b]	0.3	2000	<1
	2.9	2020	1.5
Bennington[c]	1.0	2000 (Reference case)	<1
	1.6–2.2	2000 (Accelerated)	1–1.6
National Research Council (CONAES)[d]	0	1975	0
	0.001	1985	<1
	0.006	1990	<1
	1.0	2000	<1
	2.5	2010	1.7
Stanford Research Institute[e]	0	1985	0
	0	2000	0
	0	2020	0

[a]Council on Environmental Quality: "Solar Energy Progress and Promise," Washington, D.C., April 1978.

[b]G. Bennington et al., "Solar Energy, A Comparative Analysis to the Year 2020" (McLean, Va.: MITRE Corp., August 1978).

[c]G. Bennington et al., "Toward a National Plan for the Accelerated Commercialization of Solar Energy" McLean, Va.,: (MITRE Corp., January 1980).

[d]National Research Council, National Academy of Science, "Study of Nuclear and Alternative Energy Systems, Supporting Paper 6, Domestic Potential of Solar and Other Renewable Energy Sources," Washington, D.C., 1979.

[e]J. G. Witwer et al. "A Comparative Evaluation of Solar Alternatives; Implications for Federal RD&D" (Menlo Park, Calif: SRI International, January 1978).

TABLE 13-4. STEC Supply Curve Considerations (1979 dollars)

Year	*Step*	*Cost, mills/kWh*	*Cost Components mills/kWh*	*Applicable Quantity Limits*
1975	1	600	600 STEC (Fuel saver)	Up to 5% of electric output of region
	2	628	600 STEC 28 Hydro storage	Add up to 2% of regional hydro capacity to step 1 quantity
	3	650	600 STEC 50 Nonhydro storage	Add up to 15% of electric output of region to step 2 quantity
	4	743	650 STEC 93 Nonhydro storage	From step 3 quantity to electric capacity of region
2000	1	120	120 STEC (Fuel saver)	Up to 5% of electric output of region
	2	148	120 STEC 28 Hydro storage	Add up to 3% of regional hydro capacity to step 1 quantity
	3	160	120 STEC 40 Nonhydro storage	Add up to 25% electric output of region to step 2 quantity
	4	180	140 STEC 40 Nonhydro storage	From step 3 quantity to electric capacity of region
2025	1	70	70 STEC (Fuel saver)	Up to 10% of electric output of region
	2	105	70 STEC 35 Hydro storage	Add up to 3% of regional hydro capacity to step 1 quantity
	3	110	70 STEC 40 Nonhydro storage	Add up to 25% electric output of region to step 1 quantity
	4	130	90 STEC 40 Nonhydro storage	From step 3 quantity to electric capacity of region
2050	1	70	70 STEC (No Storage)	Up to 10% of electric output of region
	2	110	70 STEC 40 Hydro storage	Add up to 3% of regional hydro capacity to step 1 quantity
	3	110	70 STEC 40 Nonhydro storage	Add up to 25% of electric output of region to step 2 quantity
	4	130	90 STEC 40 Nonhydro storage	From step 3 quantity to electric capacity of region

years 1975, 2000, 2025, and 2050. A step function, consisting of four discrete steps, reflects increasing marginal cost in each time period:

Step one projects fuel saver mode costs as well as quantities expressed as a percentage of regional electrical capacity.

Step two incorporates existing pumped storage costs; quantity is limited to a percentage of existing hydropower.

Step three incorporates other storage capability. These are more costly than hydro, but the potential for technological change dictates a rapid decrease in costs. Quantity is limited to use as intermediate load, and is therefore limited to 25 percent of regional output.

Step four represents the stage where further use of STEC requires significant transmission costs from the more ideal STEC sites in each region to the load centers.

Costs shown in tables 13-2 and 13-3 are subject to the caveat that the particular learning curve assumed actually holds. Apart from other problems in this approach, learning curve estimates for new technologies must be based on historical experience in a "similar" industry. In this case the automobile industry was judged to be the most similiar industry for which estimates of the learning curve could be made. There are obvious dissimilarities in the production processes.

The relatively high costs of this technology reflect the opinion of many experts that although the technology is currently feasible, there are other solar approaches which require fewer materials and replacements, and less operation costs and mechanical precision.

PHOTOVOLTAIC ENERGY CONVERSION

Photovoltaic devices convert light to electricity by means of a nonthermal process, using solar cells comprised of semiconductive materials such as silicon. The process works at ambient temperatures and produces no by-products other than heat from the sunlight, which is not converted to electricity.

Photovoltaic conversion has been technically feasible since 1954 and has been employed in limited use in such high-cost applications as remote terrestrial communications and space satellites. The primary R&D effort is directed toward cost reduction because this is the most significant determinant of application. Reputable studies have projected cost reductions of up to two orders of magnitude; under such circumstances, potential application of photovoltaics is substantial in both central generation and distributed energy systems.

Other cost-related issues which await technological resolution involve tradeoffs between: high efficiency/high cost versus low efficiency/low cost materials; concentration devices versus flat plate systems; and tracking versus nontracking systems. Because of the intermittency of solar flux, storage and design load factors become important cost-related issues.

Resource Availability

Unlike solar-thermal systems, photovoltaics need not be concentrating systems. They do not rely solely upon direct insolation, but can use diffuse insolation as well. Therefore, light scattering caused by moderately turbid atmosphere, haze, or pollution is not a severe constraint.

Recovery Constraints

Geographic and Technical

Even at the low efficiencies associated with current technology, photovoltaic devices require little land relative to other sources. Estimates range from 1.25 to 2.25 thousand square kilometers per exajoule per year; these estimates do not account for the modular incorporation of photovoltaic panels into the architecture of structures in the case of dispersed application. Similarly, materials intensity is low compared with other alternative energy sources, although it is higher than that of conventional technology. Glass and steel are the primary requirements among materials which compete for other uses.[6] Studies of material availability of cell inputs have shown that the elements of silicon, zinc, copper, tin, sulfur, and phosphorus impose no limits on photovoltaic cell production. On the other hand, the amounts of cadmium, antimony, and gallium needed for an annual production of a new 10^9W peak capacity could cause supply disruptions.[7]

Technical considerations include development of materials and the material/concentration tradeoff. Table 13-5 summarizes

TABLE 13-5. The Theoretical and Projected Array Efficiencies of Selected Photovoltaic Materials

Material	*Theoretical Efficiency (%)*	*Theoretical Efficiency (%)*
Silicon-crystalline	22	10–14
Silicon-amorphous	15	7–11
Gallium arsenide	24	12–16
Cadmium sulfide	17	7–11

Source: Arthur D. Little, Inc., *Distributed Energy Systems: A Review of Related Technologies,* DOE/PE 03871-01, prepared for U.S. Department of Energy, Assistant Secretary for Policy and Evaluation, Washington, D.C., 1979, p. 2-3.

the efficiencies of materials likely to be used in the years 1985–2000. The primary emphasis of current R&D work with single-crystal silicon is reducing the costs of preparing materials. Thin-film, continuous ribbon technology offers promise of very low-cost cells. Currently, thin film cells have efficiencies below 9 percent. Both cost reductions and efficiency improvements are viewed as equally probable components in the move toward economic viability.

Concentration is an issue, as it is possible to economically use very high efficiency (20–40 percent) cells which are expensive on a per unit basis. Applicable concentrators include flat reflectors, Fresnel lenses, and point focus systems which provide concentration ratios up to 40 times.[8] Studies have shown concentration to be an economic option until the year 2000, when higher cell efficiencies will prove low-concentration systems to be less costly.[9] Similarly, because of the use of diffuse as well as direct insolation, tracking has been shown to be uneconomical.

Costs

Figure 13-2 showed the relationship between energy cost and normal insolation, under conditions of technical maturity and full-scale production. Note that costs increase by a factor of 1.5 between those incurred in optimal photovoltaic locations and those in marginal photovoltaic locations. This cost difference is significantly less than that associated with STEC installations where costs differed by a factor of four. This is because of the relatively higher efficiency of photovoltaics in a marginal climate where insolation is diffuse and because of the lesser efficiency of photovoltaics in the ideal location, relative to the STEC technology.

Cost of the photovoltaic array is normally considered to be 25–35 percent of the system cost, and is the cost component which holds the most promise of reduction through technological breakthrough and mass production. Figure 13-4 shows learning curves for photovoltaic arrays. Solid plots refer to actual changes in price associated with the 85 percent slope. Currently the cost of commercial silicon is about $10/peak watt, a significant reduction from $500/peak watt in 1970.[10] DOE targets call for 56¢/peak watt in 1986 and as little as 11¢/peak watt in 1990. These targets are close to the most optimistic price projections, and would require a breakthrough in thin film cells in order to be realized. Although learning curve analysis can project certain ranges, the rate of further cost changes depends upon technological change, and indirectly, upon extra-market incentives.

Storage technology and associated costs are also subject to breakthrough projections. Energy cost projections in table 13-6 include storage costs based upon sufficient storage to allow for intermediate load use. In most estimates this requires 6–8 hours of storage at prime locations. Such storage provides 60 percent efficiency at storage costs ranging from $40 to $60/kWh. These costs are projected to fall to $30/kWh with technological maturity.

Environmental constraints to photovoltaic systems apply primarily to manufacture. Cadmium, gallium, arsenide, and arsenic are highly toxic and are recognized carcinogens; the nature of required constraints will not be known until mass production is engineered.

Table 13-7 summarizes published projections of photovoltaic implementation.

Supply Curves

Table 13-8 shows costs and output limits for photovoltaic contribution for the base year and for 2000, 2025, and 2050. Increased marginal costs associated with increased contribution are the basis for the step function. Because of the preference for electrical storage, the hydro and nonhydro storage costs parallel those of the WEC supply function. The number of steps vary with the forecast year:

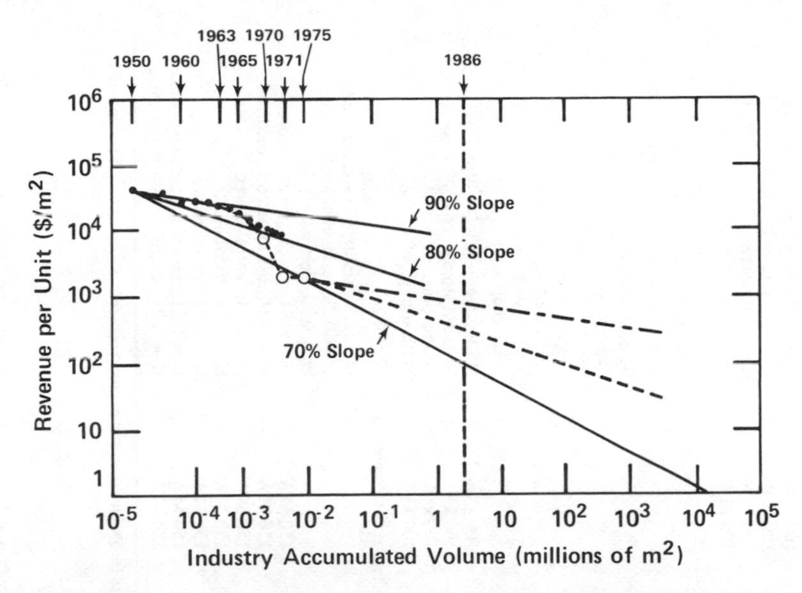

FIGURE 13-4. Learning curves for photovoltaic arrays (1975 constant dollars). (From National Research Council, National Academy of Sciences, "Study of Nuclear and Alternative Energy Systems," Supporting Paper 6, Domestic Potential of Solar and Other Renewable Energy Sources," Washington, D.C., 1979, p. 49.)

Step one projects fuel saver costs and quantities expressed as a percent of regional electrical output.

Step two incorporates the costs of existing pumped storage as the least costly of storage methods. For the year 2050 it is assumed that nonhydro storage is cheaper and hydro storage is therefore deleted.

Step three adds the contribution of nonhydro storage, up to the use as total intermediate load.

Step four exists for the year 2000 as a higher cost function for the cell array, stemming from competition for materials and higher resulting costs.

Regionality Limitations

Although photovoltaic efficiency is not as constrained as STEC by the need for direct insolation, higher costs are associated with areas of lesser insolation. The difference between locations with high levels of insolation and those with marginal conditions is the basis for a cost differential of 1.7. The regionality index for photovoltaic is shown in table 13-9. It is based upon rough averages of regional insolation and assumes that production of solar arrays takes place under conditions of technological and industrial maturity.

WIND

Because of the comparatively long-term experience in the use of wind power plants, future contributions and costs of this source are more certain than those of other renewable sources. Wind systems of various scales have been used over the past 50 years and learning curves indicate that future costs of wind energy conversion systems (WECS) will not differ significantly from costs associated with current technologies.

In projecting the significance of WECS contribution as an energy source, account must be taken of the highly localized qual-

TABLE 13-6. Photovoltaic Cost Projections (1979 dollars)

Study	*Capital Cost $/kW*	*Power Cost mills/ kWh*[a]	*Projected Year*	*Size of Unit*	*Notes*[b]
Bennington[c]	20,880	1,470	1978	100 MW	250 MW storage, cf = 0.3 silicon.
	1,624	114	1990	100 MW	250 MW storage, cf = 0.3, silicon.
	1,160	82	1990	100 MW	Fuel saver, cf = 0.3, thin-film.
	1,368	96	2020	100 MW	250 MW storage, cf = 0.3, silicon.
	672	47	2020	100 MW	Fuel saver, cf = 0.3, thin film.
Stanford Research Institute[d]	6,630	157	1985		cf = 0.20, dispersed, intermediate load.
	3,289	82	1995		cf = 0.20, central, intermediate load.
	1,365	71.5	1995		cf = 0.20, central, intermediate load.
National Research Council[e] (CONAES)	26,000	2,587	1975		Middle estimate
	5,200	520	1990		Highest cost
	1,531	152	1990		Middle estimate
	5,200	520	2000		Highest cost
	1,531	152	2000		Middle estimate
	5,200	520	2010		Highest cost
	723	72	2010		R&D goal
A. D. Little, Inc.[f]	14,112		1977	>100 kW	Without storage, cf = N.A.
	14,868		1977	>100 kW	Without storage, cf = N.A.
	448		2000	>100 kW	With storage, cf = N.A.
	1,123		2000	>100 kW	With storage, cf = N.A.
EIA/DOE[g]	982	80	1990	Central base scale	cf = 0.26
DOE/ET[h]	1,120–1,680	56–78	1986		cf = 0.20, residential application.
	<1,120	<56	1990		cf = 0.20, residential application.
	896–1,120	45–67	1990		cf = 0.26, electricity utility.
Stanford Research Institute[i]	42,895	4,000	1975	150 MW	lf = 0.21, central power station.
	984	99	2000	150 MW	lf = 0.21, distributed intermediate.
	1,054	106	2000	150 MW	lf = 0.21, central power.
	520	52	2020	150 MW	lf = 0.21, distributed.
	639	65	2020	150 MW	lf = 0.21, central power.
	618	62	2000	150 MW	lf = 0.21, central, breakthrough case.
	400	40.22	2020	150 MW	lf = 0.21, central, breakthrough case.

[a]Calculated from capital cost when only capital costs are given in the study. Assumes fixed charge rate = 18.5%, which includes 3% operation and maintenance costs; assumes cf = 38% unless otherwise noted.

[b]cf = capacity factor. lf = load factor.

[c]G. Bennington et al., "Solar Energy, A Comparative Analysis to the Year 2020" (McLean, Va.: MITRE Corp., August 1978).

[d]J. G. Witwer, "Costs of Alternative Sources of Electricity" (Menlo Park, Calif.: Stanford Research Institute, July 1976.)

[e]National Research Council, National Academy of Sciences, "Study of Nuclear and Alternative Energy Systems, Supporting Paper 6, Domestic Potential of Solar and Other Renewable Energy Sources," Washington, D.C., 1979; increased cost in 2000–2010 is due to the use of sites with lower average wind speeds because best sites have been used.

[f]A. D. Little, Inc., "Distributed Energy Systems: A Review of Related Technologies," Cambridge, Mass., November 1979. This represents an average of estimates.

[g]U.S. Department of Energy, *1980 Annual Report to Congress,* vol. 3 (Washington, D.C.: Government Printing Office, March 1981), p. 287.

[h]U.S. Department of Energy, "Solar Energy, A Status Report," DOE/EIA-0062 (Washington, D.C.: U.S. Government Printing Office, January 1978). Converted from peak to rated by factor of 5; based on targets.

[i]J. Witwer et al., "Comparative Evaluation of Solar Alternatives, Implications for Federal RD&D" (Menlo Park: Calif.: Standard Research Institute, January 1978).

TABLE 13-7. Photovoltaic Implementation Projections

Study	*Quads Displaced*	*Projected Year*	*Percent of Projected U.S. Energy*
CEQ[a]	2–8	2000	2.5–6.66
	10–30	2020	14.0–21.0
MITRE Corp.[b]	—	2000	0
	0.2	2020	<1
MITRE Corp.[c]	0.2	2000 Reference	<1
	1.5	2000 High estimate	1
National Research Council (CONAES)[d]	0.004	1985	<1
	0.07	1990	<1
	4.0	2000	2.7
	0.9	2000 Advanced technology	1
	4.9	2010	3.3
	16.0	2010 Advanced technology	11.0
EIA/DOE[e]	—	1978	0
	—	2000	0
	0.2	2010	<1
	0.5	2020	<1
Stanford Research Institute[f]	—	1985	0
	—	2000	0
Behavioral lag	3.4	2020	2.4
No behavioral lag	—	1985	0
	—	2000	0
	3.6	2020	2.6
SERI[g]	0.5–0.8	2000	1

[a]Council on Environmental Quality, "Solar Energy: Progress and Promise" (Washington, D.C., April 1978).

[b]G. Bennington et al., "Solar Energy, A Comparative Analysis to the Year 2020" (McLean, Va.: MITRE Corp., August 1978).

[c]G. Bennington et al., "Toward a National Plan for the Accelerated Commercialization of Solar Energy" (McLean, Va.: MITRE Corp., January 1980).

[d]National Research Council, National Academy of Sciences, "Study of Nuclear and Alternative Energy Systems, Supporting Paper 6, Domestic Potential of Solar and Other Renewable Energy Sources," Washington, D.C., 1979.

[e]U.S. Department of Energy, *1980 Annual Report to Congress,* vol. 3 (Washington, D.C.: U.S. Government Printing Office, March 1981).

[f]J. G. Witwer et al., "A Comparative Evaluation of Solar Alternatives; Implications for Federed RD&D" (Menlo Park, Calif.: Stanford Research Institute, January 1978).

[g]"A New Prosperity," unpublished SERI report, Boulder, Colo., 1981.

ity of the wind resource. Although the resource is large, site selection is a critical and limiting factor in forecasting potential contribution. Specific factors which determine potential are: cost, regional wind speed, intermittency, load schedule, and storage.

The intermittency issue separates WECS application into two modes: a power-displacing nonstorage use in which the benefits of wind power are based upon displacement of fuel (fuel-saving mode), and a capacity-displacing mode which depends upon storage. Major limitations of WECS usage, therefore, stem from the costs and technologies of storage as well as from the demand for a purely power-displacing capacity. The intermittency issue will be considered later in this section.

WECS technologies are initially classified as being of horizontal or vertical axis. As most units currently in use are horizontal axis turbines, the discussion concentrates on this type. Since current designs of wind turbines have coefficients of performance which are 80 percent of the theoretical maximum, design does not appear to be a source of significant economic break-

TABLE 13-8. Photovoltaic Supply Curve Considerations (1979) dollars)

Year	*Step*	*Cost, mills/kWh*	*Cost Components mills/kWh*	*Applicable Quantity Limits*
1975	1	2,500	2500 PV (no storage)	Up to 5% of electric output of region
	2	2,528	2500 PV 28 Hydro storage	From step 1 quantity up to 2% of regional hydro capacity
	3	2,593	2500 PV 93 Nonhydro storage	From step 2 quantity up to 25% of electric output of region
2000	1	102	102 PV (no storage)	Up to 5% of electric capacity of region
	2	130	102 PV 28 Hydro storage	From step 1 quantity up to 3% of regional hydro capacity
	3	171	102 PV 69 Nonhydro storage	From step 2 quantity up to 25% of electric capacity of region
	4	219	150 PV 69 Nonhydro storage	From step 3 quantity up to electric capacity of region
2025	1	50	50 PV (No storage)	Up to 10% of electric capacity of region
	2	85	50 PV 35 Hydro storage	From step 1 quantity up to 3% of regional hydro capacity
	3	88	50 PV 38 Nonhydro storage	From step 2 quantity up to 50% of electric capacity of region
2050	1	45	45 PV (No storage)	Up to 10% of electric capacity of region
	2	73	45 PV 28 Nonhydro storage	From step 1 quantity up to 75% of electric capacity of region

through as it is in other renewable technologies.[11]

WECS systems are also distinguished by size, with small-scale commercial units ranging up to 60 kW, and large-scale units ranging from 0.1 MW to 5 MW. Current applications of both scales of units are centered around the fuel-saving mode, in the absence of a suitably economic storage mechanization.

The following sections deal with resource availability, costs, limits to contribution, and regionality.

The World Resource Base

Figure 13-5 presents a gross estimate of world wind energy resources. It is emphasized, however, that such an estimate cannot reflect local anomalies which create

TABLE 13-9. Photovoltaic: Solar Insolation—Regional Cost (1979 constant dollars)

Region	*Annual Number of Sunshine Hours*	*Total Normal Insolation ($kWh/m^2/yr$)*	*Average Energy Costs (mill/kWh) at Maturity*	*Index*
1. US	3,000	2,250	160	1.19
2. OECD West	1,800	1,350	230	1.70
3. JANZ	2,400	1,800	185	1.37
4. Centrally Planned Europe	1,800	1,350	230	1.70
5. Centrally Planned Asia	2,200	1,650	215	1.59
6. Middle East	3,600	2,700	135	1.00
7. Africa	3,300	2,475	140	1.03
8. Latin America	2,200	1,650	215	1.59
9. South and East Asia	2,800	2,100	165	1.22

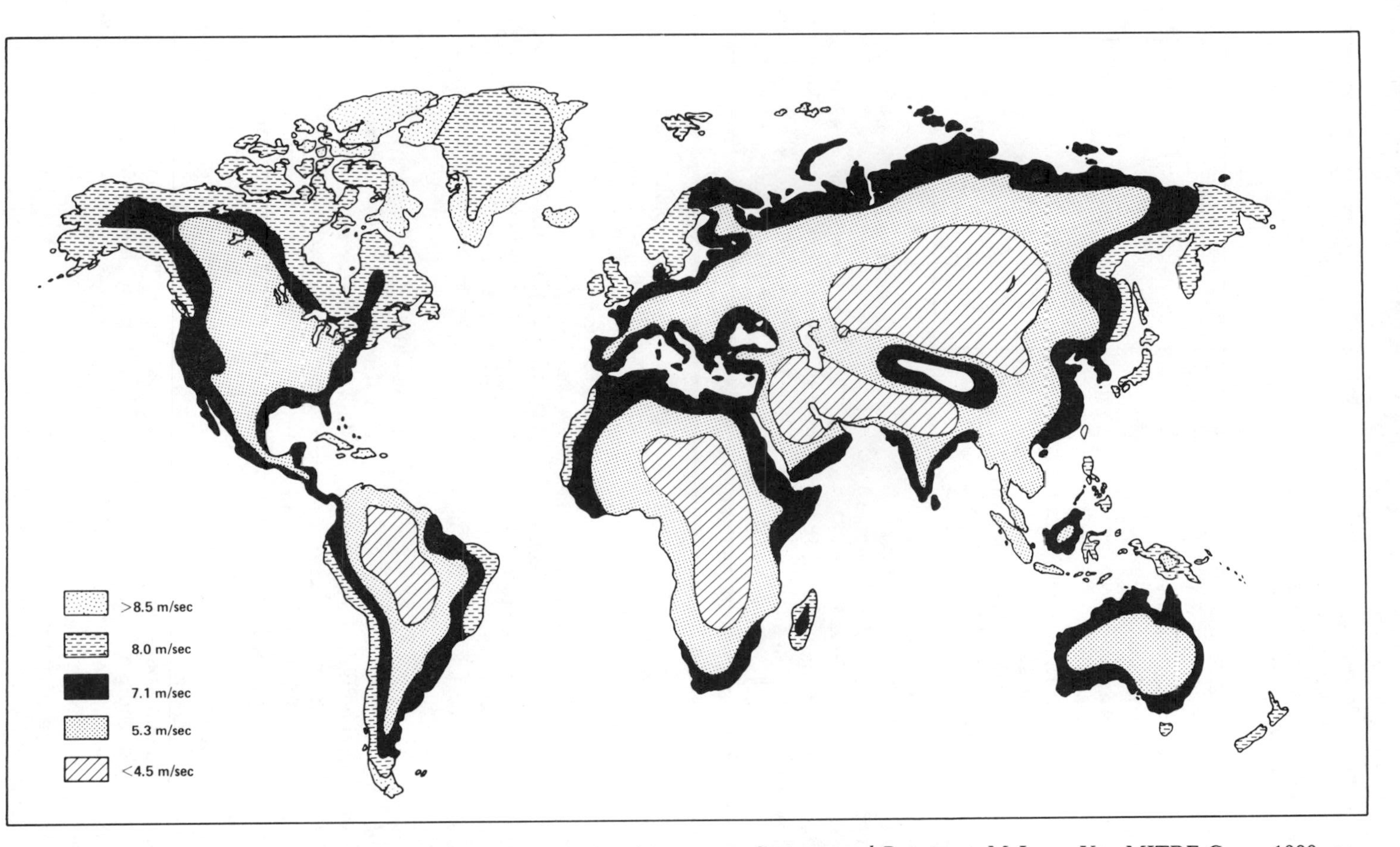

FIGURE 13-5. Annual availability of wind energy. (Reprinted with permission from Charles A. Zraket and Martin M. Scholl, *Solar Energy Systems and Resources,* McLean, Va.: MITRE Corp., 1980, pp. 8–9. Copyright © MITRE Corporation.)

high and steady winds; rather, the map depicts broad trends on each continent. World average wind energy is about 200 $kWh/m^2/yr$ at 10 meters above land masses.[12]

There is considerable sensitivity of power generation associated with smaller variations in wind speed and availability. With respect to wind speed, wind power potential varies as the cube of wind velocity. Figure 13-6 indicates this relationship for specific turbine design. Power availability is also significantly affected by height, as wind power increases with height by a 3/7 exponential factor.[13] Both of these factors pose important limits which are reflected in the supply curves.

As figure 13-5 indicates, there are significant differences from continent to continent in resource availability. This will become a basis for differentiations of regional availability as developed in a later section. Despite the immensity of the resource, technical, economic, and institutional factors significantly constrain the power potential of this resource.

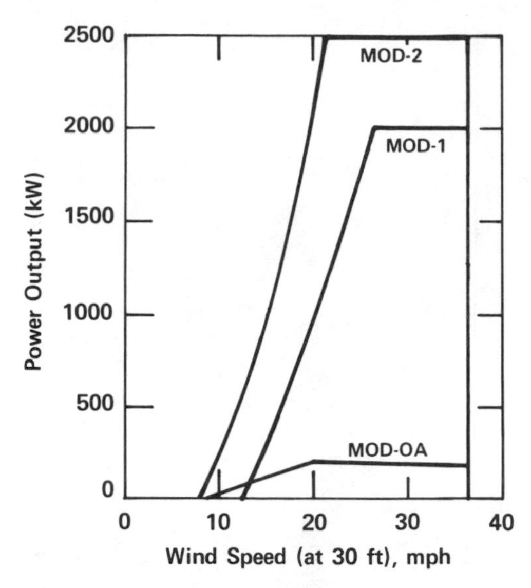

FIGURE 13-6. Power output as a function of wind speed for three wind turbines. MOD-OA has a rated power output of 150 kW at a wind speed of 20 mph. This model is regarded as "first-generation technology." MOD-1 is a larger first-generation technology machine designated to be "cost competitive, safe, reliable . . . for utility applications at moderate wind sites." It has a rated output of 2,500 kW at 18 mph. (From Arthur D. Little, *Distributed Energy Systems: A Review of Related Technologies,* U.S. Department of Energy Report No. DOE/PE-03871-01, Washington, D.C.: U.S. Department of Energy, November 1979, p. 6–21.)

Recovery Constraints

Although the ultimate factor limiting the implementation of WECS is cost, technical and institutional constraints must also be considered. These latter, however, have more impact on costs than in setting ultimate limits on WECS use.

Technical and Geographical

There is no indication in the literature that a shortage of land sites would become an ultimate constraint upon WECS implementation. Studies of U.S. potential indicate that over 157,000 square miles have sufficient wind potential to support WECS; of this area, 14 percent is available for WECS siting and this would support 1 trillion kWh of production, a significant fraction of the approximate 2 trillion kWh used annually. Obviously site availability is not currently a primary constraint.[14]

Yet, because of local variation in wind availability, siting does constrain costs, in that if prime sites are utilized, further installation must be made on sites of lower average wind speed. A Mitre Corp. study indicates that prime sites will be exhausted when approximately 24 percent of the potential WECS market has been saturated.[15] The cost of power may differ by a factor of 8 between a site with a 20 mph average wind and one of 10 mph.[16] Similarly, the CONAES study shows power costs increasing by 50 percent; in that projection additional costs associated with lower wind sites override cost reductions stemming from mass production.

More substantial constraints stem from the storage issue since storage is limited by technological factors. A number of storage systems have been proposed in addition to the lead-acid batteries that are currently used only with small-scale generators. Systems that would be compatible with utility-scale generation include:[17]

- Hydrogen storage systems in which hydrogen is generated by hydrolysis
- Thermal storage in which electrical energy is converted to heat
- Flywheel storage
- Compressed air in underground caverns
- Pumped hydroelectric storage

Of these options, the only one which is currently economic is pumped hydroelectric storage. In constructing the supply curves, this is considered an option for the year 2000, since reversible hydroelectric turbines would have to be in place. Yet this provides one limit to the supply curve of capacity replacement WECS, based upon the projected world hydroelectric capacity. This limit is calculated to be 3 percent of world electrical generation in 2000. Wind generation with storage capacity competes with peak-load generation, currently valued at over 30 mills/kWh. However, if grid-connected wind energy has no storage and is run in the fuel-saving mode, it may have no use during off-peak periods, and accordingly would be valued up to the value of the fuel it displaces.

Two studies give some capacity-replacing credit to nonstorage WECS; these values show a range of 5 to 10 percent of total electrical capacity which can be supplied in this mode.[18] All additional use is valued at fuel-displacement costs. This 5 to 10 percent of system capacity would be supplied in addition to the 3 percent which related to pumped hydrostorage.

Costs

The learning curve slopes in figure 13-7 are 85 percent; this indicates that for each doubling of production, costs will be reduced

FIGURE 13-7. Estimated typical WECS learning curves, in 1978 dollars. (Reprinted with permission from E. L. Eldridge and W. E. Jacobsen, *Distributed Solar Energy Systems,* vol. II, Report No. MTR-79W00021-02, McLean, Va.: MITRE Corp., May 1980, p. 63. Copyright © MITRE Corporation.)

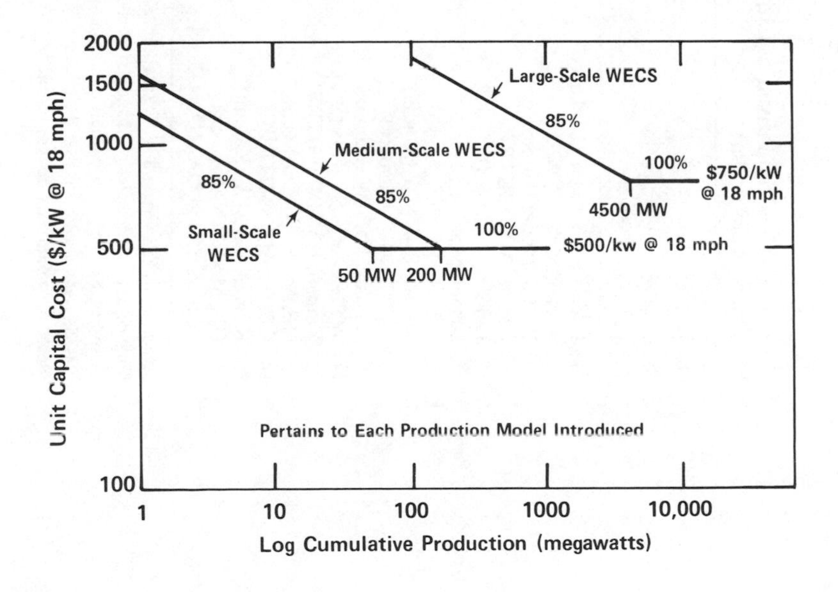

TABLE 13-10. Comparison of Estimated Targeted Installed Costs for Selected Large-Scale and Medium-Scale Wind Machines (1978 dollars)

Type of System[a]	*Type and Specified Rated Wind Speed*	*Power Output at Specified Rated Wind Speed*	*Power Output at 18 mph Wind Speed*[b]	*Price per Kilowatt at 18 mph Wind Speed*[c]
1. Lockheed, 3.9 MW (Lockheed)	HAWT/25.7 mph	3.9 MW	1.87 MW	$ 575/kW
2. Boeing Mod-2 (SWRI)[d]	HAWT/18.0 mph	1.0 MW	1.00 MW	$ 802/kW
3. Boeing Mod-2 (Honeywell)[d]	HAWT/18.8 mph	1.125 MW	1.03 MW	$ 903/kW
4. GE, 1.5 MW (MITRE)	HAWT/22.5 mph	1.5 MW	0.96 MW	$ 947/kW
5. Kaman, 1.5 MW (MITRE)	HAWT/22.5 mph	1.5 MW	0.96 MW	$ 955/kW
6. Schachle (Schachle)	HAWT/37.4 mph	2.7 MW	0.62 MW	$1017/kW
7. GE, Mod-1 (GE)	HAWT/28.0 mph	1.5 MW	0.62 MW	$1155/kW
8. Sandia, 0.48 MW (Sandia)	VAWT/27.5 mph	0.48 MW	0.20 MW	$1318/kW
9. GE Mod-1 (Aerospace)	HAWT/28.0 mph	1.5 MW	0.62 MW	$1348/kW
10. Kaman, 1.5 MW (Kaman)	HAWT/29.2 mph	1.5 MW	0.57 MW	$1636/kW
11. WTG Energy Systems (WTG)	HAWT/25.5 mph	0.2 MW	0.10 MW	$1674/kW
12. Kaman, 0.5 MW (MITRE)	HAWT/22.5 mph	0.5 MW	0.32 MW	$1788/kW
13. GE, Mod 1-A (Bu. Rec.)	HAWT/29.8 mph	2.0 MW	0.73 MW	$1846/kW
14. GE, 0.5 MW (MITRE)	HAWT/22.5 mph	0.5 MW	0.32 MW	$2040/kW
15. Kaman, 0.5 MW (Kaman)	HAWT/22.7 mph	0.5 MW	0.32 MW	$2226/kW
16. GE, Mod-1 (NASA-Lewis)	HAWT/29.8 mph	2.0 MW	0.73 MW	$3715/kW
17. Lockheed Mod OA (NASA-Lewis)	HAWT/19.0 mph	0.2 MW	0.18 MW	$4817/kW

Source: F. R. Eldridge and W. E. Jacobsen, *Distributed Solar Energy Systems,* vol. II (McLean, Va.: MITRE Corp., May 1978), pp. 60–61.

[a]First company listed, designed and developed the system indicated; second participant (in parentheses) analyzed and determined preliminary system costs.

[b]Assumes that the power output of the wind-driven generator varies as the square of the wind velocity.

[c]Based on installed costs after the production of the first 100 to 1000 units in 1978 constant dollars.

[d]Based on preliminary designs for the Mod-2.

by 15 percent. It is therefore projected that current costs exceed ultimate costs by about 50 percent. Most studies indicate that ultimate costs will be reached before the year 2000. Further interpretation of the learning curve implies that large-scale WECS will have capital costs which will not go below $780 kW and small-scale systems will be less than $520 kW (in 1979 dollars and 18 mph wind speed), corresponding to the limits shown in figure 13-7.

Tables 13-10 and 13-11 are derived from studies by the Mitre Corp.; the targeted installed cost refers to costs associated with production of 100 to 1,000 units.[19] These costs were converted to cost per kilowatt and ordered upon an assumption of 18 mph wind speed. Similarly, table 13-12 develops energy costs from capital costs of specific projects. These energy costs are consistent in range with those shown in figure 13-7 where emphasis was given to the relationship between energy cost and wind speed.

Finally, table 13-13 shows a compendium of energy cost projections for specific years as estimated by other studies. Note that cost reductions over time reflect the influence of the learning curve, and that in most studies full production schedules are reached by the year 2000.

Institutional Constraints

Additional physical and institutional constraints to implementation exist, although they are often not readily quantified. These include availability of wind rights; aesthetic considerations; noise; and interference with radio, television, and satellite signals.

Projections of Implementation

Table 13-14 represents a compendium of forecasts of implementation of wind energy in the United States. Because of the diversity of scenarios assumed by each study, the quantity of output is cast as a percent of primary fuel use in electrical generation. This is necessary because all of these studies are of the United States, not of the world.

Regionality

Several factors contribute to interregional differentials of WECS potential. Since nonstorage use of WECS is a function of regional electrical capacity and fuels, this total capacity is the exogeneous variable in implementation of the fuel-saving mode. Similarly, existing and forecast reversible turbine and pumped storage hydroelectric capacity are the basis for the most cost-effective storage.

TABLE 13-11. Comparison of Estimated Targeted Installed Costs for Selected Small-Scale Wind Machines (1978 dollars)

Type of System	*Type and Specified Rated Wind Speed*	*Power Output at Specified Rated Wind Speed*	*Power Output at 18 mph Wind Speed*[a]	*Price per Kilowatt at 18 mph Wind Speed*[b]
1. Wind Engineering Corp.	HAWT/23 mph	23 kW	14.0 kW	$ 420/kW
2. Wind Flower	HAWT/26 mph	18 kW	8.6 kW	$ 474/kW
3. Millville	HAWT/25 mph	10 kW	5.2 kW	$ 490/kW
4. Kaman	HAWT/20 mph	40 kW	32.4 kW	$ 610/kW
5. United Technologies	HAWT/20 mph	8 kW	6.5 kW	$ 925/kW
6. Windworks	HAWT/20 mph	8 kW	6.5 kW	$ 925/kW
7. Dynergy	VAWT/24 mph	5 kW	2.7 kW	$1115/kW
8. Alcoa	VAWT/33 mph	55 kW	15.8 kW	$1150/kW

Source: F. R. Eldridge and W. E. Jacobsen, *Distributed Solar Energy Systems,* vol. II, (McLean, Va.: MITRE Corp. May 1978), pp. 60–61.

[a]Assumes that the power output of the wind-driven generator varies as the square of the wind velocity.

[b]Based on installed costs after the production of the first 100 to 1,000 units.

TABLE 13-12. Installed Capital and Energy Cost Estimates for 1-MW Wind Turbine Systems Purchased in 100-Unit Lots

Study Source[a]	*Capital Cost*[b] *(1977 $/kW)*	*Capacity Factor (%)*	*Energy Cost*[c] *(1977 mills/ kWh)*
Lockheed[d]	1,211	53.7	38.4[e]
G.E.[f]	504	38.0	28.0
Kaman[g]	652	43.0	32.0
Aerospace[h]	588	40.0	31.0
Honeywell[i]	739	43.0	36.3
SWRI[j]	842	59.0	30.1
G.E.[k]	793	38.0	44.0[e]

Source: Arthur D. Little, *Distributed Energy Systems, A Review of Related Technologies* (Cambridge, Mass.: November 1979), p. 6–3.

[a]JBF Scientific Corp., *Summary of Current Cost Estimates of Large Wind Energy Systems,* report prepared for Energy Research and Development Administration, NTIS Report No. DSE/2521-1, February 1977.

[b]Costs taken from reference (a), table B4, and escalated at 8% for two years to represent 1977 costs of all equipment, installation, and site preparation.

[c]Energy cost (*EC*) calculated by reference (a) as follows:

$$EC = \frac{C_c\,(\$/\text{kW}) \times FCR(\%) \times 10^3}{8{,}760\,(\text{hr}) \times cf(\%)}$$

where C_c = capital cost of system.
FCR = fixed charge rate = 18.5%, which includes a 3% component due to operation and maintenance costs.
cf = capacity factor.

[d]Lockheed-California Co., Burbank, Calif., *Wind Energy Mission Analysis,* prepared under Contract No. EY-76-C-03-1075 for Energy Research and Development Administration, April 1976.

[e]These energy costs were calculated using the equation in note (c), but have been normalized (a) to account for variations (the vertical wind speed profile coefficient) assumed in the Lockheed and General Electric studies.

[f]General Electric Corp., *Design Study of Wind Turbines 50 kW to 3000 kW for Electric Utility Applications, vol. II, Analysis and Design,* report prepared for National Aeronautics and Space Administration, NTIS Report No. ERDA/NASA/9403-76/2, December 1976.

[g]Kaman Aerospace Corp., *Design Study of Wind Turbines 50 kW to 3000 kW for Electric Utility Applications, vol. II, Analysis and Design,* report prepared for National Aeronautics and Space Administration, NTIS Report No. DOE/NASA/9403-76/2, December 1976.

[h]Aerospace Corp., *Wind Power for the California Aqueduct,* El Segundo, Calif., March 1976.

[i]Honeywell, Inc., *The Application of Wind Power Systems to the Service Area of the Minnesota Power and Light Company, Executive Summary. October 1976–March 1977. Quarterly Report No. 3,* Minneapolis, Minn., 1977.

[j]C. Ligon et al., *Operational Cost and Technical Study of Large Windpower Systems Integrated with an Existing Electric Utility,* prepared by Southwestern Public Service Co., NTIS Report No. COO-2621-2, April 1976.

[k]General Electric Corp., *Wind Energy Mission Analysis,* prepared under Contract No. EY-76-C-02-2578 for Energy Research and Development Administration, October 1976.

Because of the sensitivity of power costs to wind speed, interregional differentials in this factor are also important. Table 13-15 presents an index based upon regional wind speed. Although it represents an average measure and does not take into account local anomalies, the index does reflect real differences in wind potential and accordingly reflects average cost of implementation.

Supply Curves

Table 13-16 presents the background for supply curve projections for the base year and 2000, 2025, and 2050. A step function

TABLE 13-13. Wind Energy Cost Projections (1979 dollars)

Study	*Capital Cost ($/kW)*	*Power Cost (mills/ kWh)*[a]	*Projected Year*	*Size of Unit*	*Notes*[b]
Bennington[c]					
	2,169	95.4	1979	1.5 MW	Fuel saver mode, cf = 0.48
	1,114	49.0	1981	100 MW	Intermediate load, cf = 0.48
	1,740	76.6	1983	1.5 MW	Fuel saver mode, cf = 0.48
	881	38.8	2020	100 MW	Intermediate load, cf = 0.48
	928	40.8	2020	1.5 MW	Fuel saver mode, cf = 0.48
Stanford Research Institute[d]					
	650	40.3	1985	1.5 MW	Fuel saver mode, cf = 0.43 20% Capital cost factor
	585	37.7	1995	1.5 MW	Fuel saver mode, cf = 0.43 20% Capital cost factor
National Research Council[e] (CONAES)					
	1,138	48.1	1975	0.2 MW	Fuel saver mode/load factor 0.4
	1,138	96.2	1975	0.2 MW	Intermediate/load factor 0.2
	698	29.9	1990	0.2 MW	Fuel saver mode/load factor 0.4
	698	59.8	1990	0.2 MW	Intermediate/load factor 0.2
	668	28.6	2000	0.2 MW	Fuel saver mode/load factor 0.4
	1,040	88.4	2000	0.2 MW	Intermediate/load factor 0.2
	637	27.3	2010	0.2 MW	Fuel saver mode/load factor 0.4
	1,560	178.1	2010	0.2 MW	Intermediate/load factor 0.15
A. D. Little[f]	852	38.4	2000	1 MW	
EIA/DOE[g]					
	1,208	52.0	1985		Base load, cf = 0.35
Eldridge and Jacobsen[h]					
	794	44.1	1980 Technology	Small scale	Calculated at 18 mph wind
	1,759	97.8	1980 Technology	Large and medium scale	Calculated at 18 mph wind

TABLE 13-13. Wind Energy Cost Projections (1979 dollars) (*Continued*)

Study	*Capital Cost ($/kW)*	*Power Cost (mills/kWh)*[a]	*Projected Year*	*Size of Unit*	*Notes*[b]
DOE/ET[i]	1,352	52–104	1975	1 kW	
	780	21–42	1985	8 kW	
	780	31.2	1975	2 MW	
	16–624	10.4–20.8	1985	2–4 MW	
Stanford Research Institute[j]	1,131	47.8	1985	1.5 MW cf = 0.5	
	1,560	65.9	1985	1.5 kW cf = 0.5	
	1,079	45.6	2000	1.5 MW cf = 0.5	
	1,560	65.9	2000	1.5 MW cf = 0.5	
	1,066	45.0	2020	1.5 MW cf = 0.5	
	1,560	65.9	2020	1.5 kW cf = 0.5	

[a]Calculated from capital cost when only capital costs are given in the study. Assumes fixed change rate = 18.5%, which includes 3% operation and maintenance costs; assumes cf = 38% unless otherwise noted.

[b]cf = capacity factor.

[c]G. Bennington et al., "Solar Energy, A Comparative Analysis to the Year 2020" (McLean, Va.: MITRE Corp., August 1978).

[d]J. G. Witwer, "Costs of Alternative Source of Electricity" (Menlo Park, Calif.: Stanford Research Institute, July 1976).

[e]National Research Council, National Academy of Sciences, "Study of Nuclear and Alternative Energy Systems, Supporting Paper 6, Domestic Potential of Solar and Other Renewable Energy Sources," Washington, D.C., 1979; increased cost in 2000–2010 is due to the use of sites with lower average wind speeds because the best sites have been used.

[f]This represents an average of estimates; for individual estimates see table 9-7 in A. D. Little, Inc., "Distributed Energy Systems: A Review of Related Technologies," Cambridge, Mass., November 1979.

[g]U.S. Department of Energy, *1980 Annual Report to Congress,* vol. 3 (Washington, D.C.: U.S. Government Printing Office, March 1981), p. 287.

[h]Averaged from tables XII and XIII, F. Eldridge and W. E. Jacobsen, *Distributed Energy Systems,* vol. 2, McLean, Va., May 1980.

[i]U.S. Department of Energy, "Solar Energy, A Status Report," DOE/EIA-0062 (Washington, D.C.: U.S. Government Printing Office, January 1978).

[j]J. Witwer et al., "Comparative Evaluation of Solar Alternatives, Implications for Federal RD&D" (Menlo Park, Calif.: Stanford Research Institute, January 1978).

TABLE 13-14. Wind-Energy Implementation Projections

Study	*Quads Displaced*	*Projected Year*	*Percent of Projected U.S. Energy*
CEQ[a]	4–8	2000	5–6%
	8–12	2020	9–11%
Bennington[b]	0.0	1985	0.0%
	1.7	2000	1.4%
	6.6	2020	3.4%
Bennington[c]	0.0	1978	0.0
	0.0	1985	0.0
	0.2	1990	<1.0%
	1.7	2000	1.4%
National Research Council (CONAES)[d]			
	0.0	1975	0.0%
	0.5	1990	<1.0%
Fuel saver mode	1.4	2000	<1.0%
	1.8	2010	1.2%
	3.6	2000	2.4%
Storage Mode	19.0	2010	15.0%
EIA/DOE[e]			
	0.00	1978	0.0%
	0.09	2010	<1.0%
	0.21	2000	<1.0%
	0.43	2020	<1.0%
Stanford Research Institute[f]			
Behavioral lag	0.0	1985	0.0%
	2.0	2000	2.0%
	3.8	2020	3.0%
No behavioral lag	0.1	1985	0.0%
	2.3	2000	2.0%
	3.9	2020	3.0%

[a]Council on Environmental Quality, "Solar Energy Progress and Promise," Washington, D.C., April 1978.

[b]G. Bennington et al., "Solar Energy, A Comparative Analysis to the Year 2020" (McLean, Va.: MITRE Corp., August 1978).

[c]G. Bennington et al., "Toward a National Plan for the Accelerated Commercialization of Solar Energy" (McLean, Va.: MITRE Corp., January 1980).

[d]National Research Council, National Academy of Science, "Study of Nuclear and Alternative Energy Systems, Supporting Paper 6, Domestic Potential of Solar and Other Renewable Energy Sources," Washington, D.C., 1979.

[e]U.S. Department of Energy, *1980 Annual Report to Congress,* vol. 3 (Washington, D.C.: U.S. Government Office, March 1981).

[f]J. G. Witwer et al., "A Comparative Evaluation of Solar Alternatives; Implications for Federal RD&D" (Menlo Park, Calif.: Stanford Research Institute, January 1978).

is constructed, representing cost and quantity for four discrete steps:

Step one projects nonstorage-mode costs as well as quantities forthcoming expressed as a percentage of electric capacity of the region. This quantity limitation is based upon the potential of wind power to replace a small percent of capacity, ranging from 5 to 10 percent.

Step two incorporates the use of least costly existing storage technology, specifically pumped storage, and reversible turbine hydroelectric. Quan-

TABLE 13-15. Scalar for Wind Velocity/Cost Relationship

Region	Average Wind Velocity (m/s)	Cost Scalar
1. US	6.1	0.95
2. OECD West	5.8	1.00
3. JANZ	6.5	0.86
4. Centrally Planned Europe	4.8	1.21
5. Centrally Planned Asia	5.3	1.10
6. Middle East	5.4	1.09
7. Africa	5.3	1.10
8. Latin America	5.5	1.06
9. South and East Asia	5.3	1.10

tity is limited to a percentage of hydroelectric capacity in the region.

Step three incorporates other storage technologies. Costs are based upon studies of battery, steam, compressed air, chemical, flywheel, and magnetic storage. Decreases in the cost of the storage component over the projected time period reflect promise of 50 to 80 percent cost reductions associated with advancing technology.

Step four reflects the additional costs associated with the exhaustion of prime WECS sites. Because of the exponential relationship between wind speed and power output, the cost of more or larger turbines outweighs the

TABLE 13-16. Wind Supply Curve Considerations (1979 dollars)

Year	Step	Cost (mills/kWh)	Cost Components (mills/kWh)	Applicable Quantity Limits
1975	1	55	55 WEC (No storage)	Up to 5% of electric capacity of region
	2	83	55 WEC 28 Hydro storage	From step 1 quantity to 2% of regional hydro capacity
	3	148	55 WEC 93 Nonhydro storage	From step 2 quantity to 25% of electric capacity of region
	4	175	82 WEC 93 Nonhydro storage	From step 3 quantity to electric capacity of region
2000	1	39	39 WEC (No storage)	Up to 5% of electric capacity of region
	2	67	39 WEC 28 Hydro storage	From step 1 quantity to 3% of regional hydro capacity
	3	108	39 WEC 69 Nonhydro storage	From Step 2 quantity to 25% of electric capacity of region
	4	128	38 WEC 69 Nonhydro storage	From Step 3 quantity to electric capacity of region
2025	1	39	39 WEC (No storage)	Up to 10% of electric capacity of region
	2	74	39 WEC 35 Hydro storage	From Step 1 quantity to 3% of regional hydro capacity
	3	94	35 WEC 55 Nonhydro storage	From step 2 quantity to 25% of electric capacity region
	4	114	28 WEC 55 Nonhydro storage	From Step 3 quantity to electric capacity of region
2050	1	39	39 WEC (No storage)	Up to 10% of electric capacity of region
	2	79	39 WEC 40 Hydro storage	From step 1 quantity to 3% of regional hydro capacity
	3	94	39 WEC 55 Nonhydro storage	From step 2 quantity to 25% of electric capacity of region
	4	114	59 WEC 55 Nonhydro storage	From step 3 quantity to electric capacity of region

reduction in cost associated with movement along the learning curve. Use of sites with lower average wind speeds is also reflected in a decreased load factor, which further escalates effective costs.

Costs and contributions shown in table 13-16 represent "educated guesses" made within the parameters shown in previous tables. Although there are relatively few technological unknowns in the near term, the combination of nontechnical and storage-related factors make the forecasting of long-term implementation difficult at best.

NOTES

1. C. Zracket and M. Scholl, *Solar Energy Systems and Resources* (McLean, Va.: MITRE Corp., 1980), p. 6.
2. Ibid., p. 25.
3. National Research Council, *Domestic Potentials of Solar and Other Renewable Energy Sources,* Study of Nuclear and Alternative Energy Systems, vol. 6 (Washington, D.C.: National Academy of Sciences, 1979), p. 41.
4. Wolf Häfele, Project Leader, *Energy in a Finite World,* Report by the Energy Systems Program Group of the International Institute for Applied Systems Analysis, draft of chapter 8, pp. 8–20 (Laxenburg, Austria).
5. L.L. Vant-Hull, *Second Generation Heliostats,* vol. I, Report No. SAND 81-8177 (Huntington Beach, Calif.: McDonnell Douglas Astronautics Co., April 1981).
6. Zracket and Scholl, *Solar Energy Systems,* p. 22.
7. Arthur D. Little, Inc., *Distributed Energy Systems: A Review of Related Technologies,* DOE/PE 03871-01, prepared for U.S. Department of Energy, Assistant Secretary for Policy and Evaluation (Washington, D.C., 1979), p. 2–16.
8. Ibid., p. 2–16.
9. J. Witwer et al., "A Comparative Evaluation of Solar Alternatives: Implications for Federal RD&D" (Menlo Park, Calif.: Stanford Research Institute, January 1978), p. 44.
10. J. J. Loferski, "Current from the Sun," *The Sciences* (July/August 1980).
11. Arthur D. Little, Inc., *Distributed Energy Systems,* p. 6–3.
12. Zracket and Scholl, *Solar Energy Systems,* p. 8.
13. Ibid., p. 8.
14. General Electric Space Division, *Wind Energy Mission Analysis,* prepared for the Energy Research and Development Administration, 300/2578-1/1-3, February 1977.
15. G. Bennington et al., *Solar Energy, A Comparative Analysis to the Year 2020* (McLean, Va.: MITRE Corp., August 1978), p. 24.
16. National Research Council, *Domestic Potentials of Solar,* p. 56.
17. MITRE Corp., *Wind Machines,* report prepared for National Science Foundation, Report No. NSF-RA-N-75-051, October 1975, Washington, D.C.
18. J. Witwer et al., *A Comparative Evaluation,* p. 65; and National Research Council, *Domestic Potentials of Solar.*
19. F. R. Eldridge and W. E. Jacobsen, *Distributed Solar Energy Systems,* vol. II (McLean, Va.: MITRE Corp., May 1978), pp. 60–61.

Hydroelectricity and Ocean Thermal Energy Conversion Systems 14

This chapter reviews the prospective contribution of hydropower and ocean thermal energy conversion (OTEC) systems to world electric capacity. Selection of these energy sources was based on two criteria: availability of the resource to a significant number of regions and development of technology to the demonstration or the commercialization stage. The likelihood of a significant contribution by these sources merits a detailed review.

The methodology adopted for this analysis was the following: First, the size and proximate distribution of the world resource was estimated for each source. Then, constraints which affect costs were considered. These include land, material, and technological limitations, as well as institutional and environmental factors.

HYDROELECTRICITY

Hydro power is a relatively benign source of energy that was among the first nonanimal energy sources harnessed by man. Today, nearly all hydropower is converted to electricity. Currently, hydroelectric plants supply 23 percent of the world's electricity. Installed capacity in 1976 represented just over 15 percent of the total hydro potential in the world. Global hydroelectric resources considered to be economically and technologically feasible under current conditions are estimated to be on the order of 2.4 million MW of installable capacity. If the estimated potential capacity was fully utilized, hydropower would contribute about 37.3 exajoules (EJ) of energy annually, assuming a 50 percent capacity utilization rate. The greater portion of the unexploited hydroelectric potential is found in the dveloping world since the developed regions of the world have already exploited a large share of their feasible hydro sites. For example, in OECD West (region 2) 45.9 percent of the estimated installable capacity is presently being utilized, whereas in Africa (region 7) only 2.5 percent of total hydro capacity has been developed.

Several countries obtain most of their electric power from hydro resources. Countries and regions with abundant hydro resources have generally developed them to supply electricity needs, as shown in table 14-1. In South America, 73 percent of the electricity is produced from hydro resources, but this represents only 9 percent of estimated total resources.[1]

This chapter reviews characteristics of the hydropower resource base that are critical to its development through the year 2050. Available data on resources must be termed orders-of-magnitude estimates for much of the world. The only full global accounting of hydropower resources was developed by the World Energy Conference (WEC) and published in 1978. The WEC

TABLE 14-1. Selected Countries Obtaining Most of Their Electricity from Hydropower, 1980

Country	*Share of Electricity from Hydropower (percent)*
Ghana	99
Norway	99
Zambia	99
Mozambique	96
Zaire	95
Brazil	87
Portugal	87
New Zealand	75
Nepal	74
Switzerland	74
Austria	67
Canada	67

Source: Daniel Deudney, "Rivers of Energy: The Hydropower Potential," *Worldwatch Paper 44* (Washington, D.C.: Worldwatch Institute, June 1981), p. 10. Based on UN data reported in *World Energy Supplies.*

assembled estimates of gross theoretical hydro potential; from these they estimated those resources that could be developed under a set of assumptions about the cost of alternative sources of electricity and exclusion of various sites because of national water policies. As such, the reported estimates do not give any indication of price sensitivity of hydroelectric supply. The nature of hydropower resources (uniqueness of major watersheds) does not facilitate generalization of their distribution across grades as, for example, in the case of oil where the conventional assumption is that pool size is distributed lognormally. In addition, the estimates of gross theoretical capacity, as well as exploitable capacity, are subject to revision. The potential error in the original WEC estimates is dramatically illustrated by recent revisions in Brazil's estimated hydro potential. Present estimates put Brazil's exploitable hydro potential at two or more times the WEC estimates.[2] Rather than attempting to update the WEC estimates, which in itself would be a major effort, we have accepted them as published. More extensive efforts to inventory hydro resources on a national level in various countries since the WEC data was compiled are likely to have resulted in considerable revision in some cases.

Hydropower, apart from being an economically competitive, proved technology, is nonpolluting. Often the cost of the dams necessary to exploit a river's hydro power potential can be spread over flood control and recreation uses. These characteristics make hydro power an attractive source of electricity.

The Resource Base and Recoverable Resources

Table 14-2 shows the amounts of the hydro resource base and recoverable resources, by region, for the world. The resource base is a theoretical maximum of the amount of energy obtained from water runoff for the region in a year. As such, it does not distinguish between water collected in large rivers which fall rapidly (high-grade resources) and dispersed, intermittent runoff flows, nor does it account for efficiency of capturing and converting the energy. The estimates are based on land elevation and precipitation runoff.[3] Recoverable resources comprise that part of the resource base considered technologically feasible to develop under a given price and technology regime.

The estimates of recoverable resources presented in table 14-2 reflect technologies and prices of alternative energy types and sources of electricity at the time of the WEC study (1976–78). The application of uniform economic and technological criteria was not possible. The estimates were based on compilation of various national estimates and other sources. Given the availability of information, little more is possible on a global scale at this time.

The estimated total world hydro resource base is etimated at 95.44 EJ. An estimated 2.4 million MW of hydro capacity could be developed globally under the WEC assumptions. Assuming a 50 percent

TABLE 14-2. Resource Base and Recoverable Resources/Annual Production (EJ) and Installed Capacity

		Recoverable Resources		
Region	*Resource Base Annual Generation (exajoules)*	*Installed and Installable Capacity (megawatts)*	*Total Potential Annual Generation (exajoules)*	*Proportion of Resource Base Potentially Exploitable (%)*
US	3.83	170,700	2.30	60.1
OECD West	11.17	312,301	4.67	41.8
JANZ	2.57	84,695	0.99	38.4
EUSSR	15.24	324,207	4.97	32.6
ACENP	9.01	399,388	5.76	63.9
MIDEST	0.64	36,499	0.61	95.4
AFR	15.40	449,424	7.31	47.4
LA	26.60	360,877	6.48	24.3
SEASIA	10.98	291,149	4.17	37.9
Total	95.44	2,429,240	37.26	39.0

Source: Computed from country data presented in WEC *Renewable Energy Resources* (New York: IPC Science and Technology Press, 1978).

capacity utilization puts annual generation potential at 37.26 EJ, or roughly 40 percent of the estimated theoretical hydro capability. Of the remaining resources, some, primarily in the United States, are excluded because of preservation of wild and scenic rivers (allowing development of this energy would increase U.S. recoverable resources by 6 percent), some are located in relatively uninhabitable areas (Alaska accounts for 25 percent of the U.S. resource base and a large but unspecified amount of hydro potential exists in remote areas of the USSR), while some flows are across very flat terrain, making it practically impossible to create sufficient head (a large share of the USSR resource base is excluded on these grounds).[4] Finally, a relatively large share of the resource base must be in the form of very dispersed intermittent runoff and could not plausibly be exploited. These considerations suggest that the 40 percent exploitation of the resource base may be near a practical maximum. Figure 14-1 illustrates the regional distribution of the resource base and recoverable resources.

Table 14-3 presents figures for current operating capacity and planned expansion by 1985. Since lead times for hydro development are likely to be on the order of 10 years, estimates for 1985 are likely to be relatively firm. Shorter lead times are necessary in the case of rewinding, or otherwise upgrading or expanding capacity at existing dam sites. This is likely to represent a relatively small contribution, however. In the United States, 54.7 MW of capacity were estimated to be available from such expansions (including installation of generating capacity at dam sites which previously had none). This would respresent an increase of less than a tenth of 1 percent in existing capacity in the United States.[5] The regional variations in the share of recoverable resources currently being exploited are illustrated in figure 14-2.

Hydroelectric Costs

The costs of developing hydroelectric capacity vary greatly from site to site. A large share of the costs of generating hydroelectricity are attributable to capital costs associated with the initial investment; fuel costs are nonexistent and operating and maintenance costs are relatively low.

Systematic variation in costs from site to

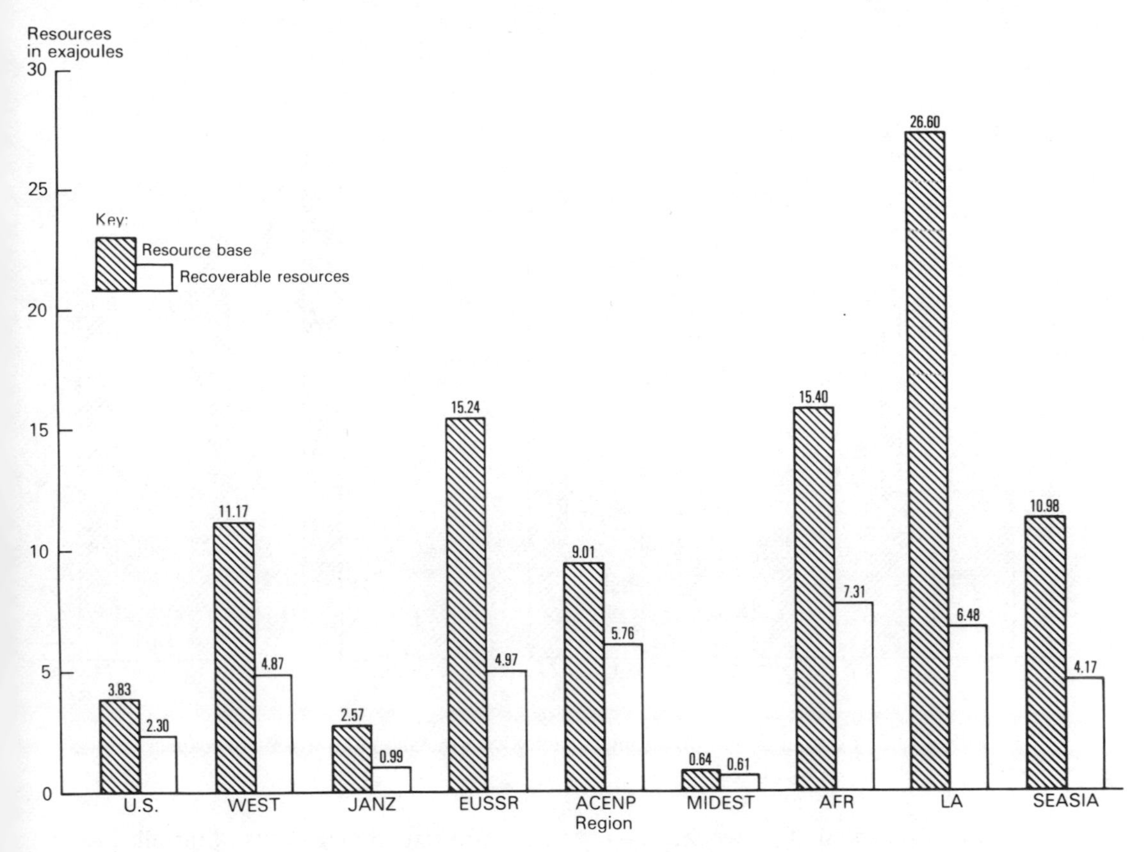

FIGURE 14-1. Comparison of recoverable resources and resource base. (From table 14-2.)

TABLE 14-3. Recoverable Resources

	Recoverable Resources		*Operating Capacity, 1976*		*Amount of Recoverable Resources Presently Developed (%: ratio of installed & installable capabilities)*	*Additional Capacity Operating by 1985*	
Region	*Megawatts*	*Exajoules*	*Megawatts*	*Exajoules*		*Megawatts*	*Exajoules*
US	170,700	2.297	60,000	1.039	35.1	8,200	0.139
OECD West	312,301	4.666	143,409	2.270	45.9	38,428	0.509
JANZ	84,695	0.990	32,023	0.430	37.8	4,701	0.040
EUSSR	324,207	4.977	56,311	1.005	17.4	27,301	0.375
ACENP	399,388	5.760	14,499	0.194	3.6	N.A.	N.A.
MIDEST	36,499	0.614	2,101	0.031	5.8	2,833	0.039
AFR	449,424	7.311	11,058	0.210	2.5	3,770	0.057
LA	360,877	6.483	30,173	0.642	8.4	39,032	0.646
SEASIA	291,149	4.172	16,113	0.243	5.5	13,537	0.161
Total	2,429,240	37.270	365,687	6.064	15.1	137,802	1.966

Source: Same as table 14-2.

N.A. = Not available.

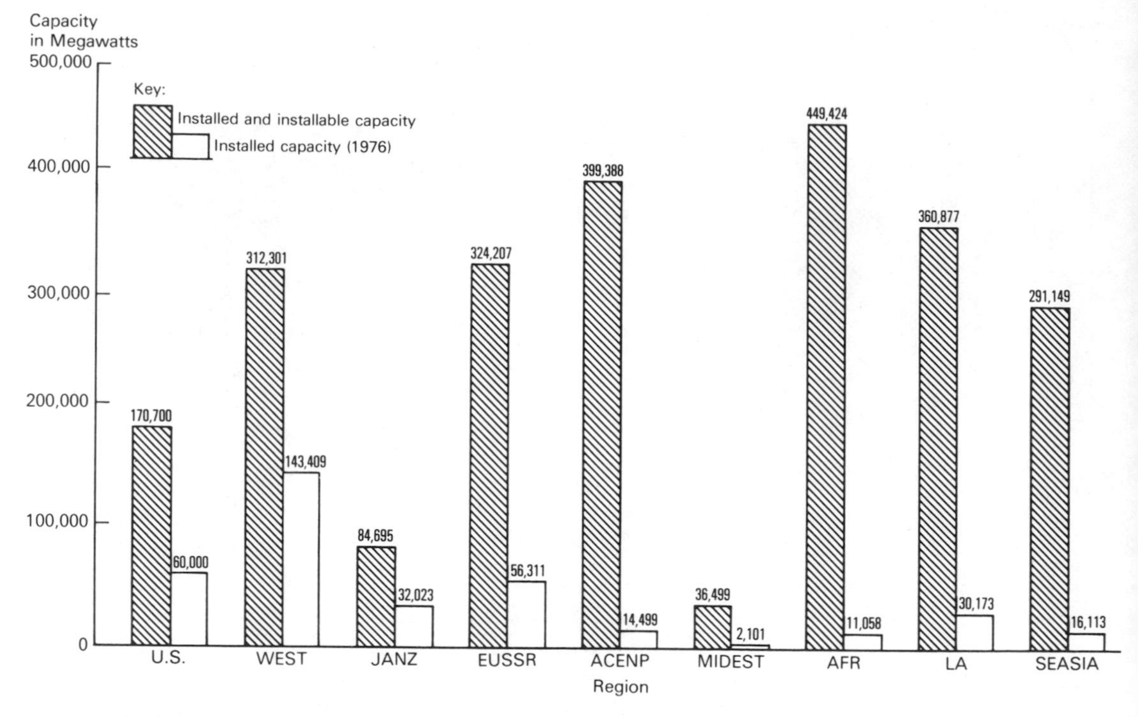

FIGURE 14-2. Comparison of operating capacity with installed and installable capacity.

site is generally attributed to two factors—scale and head. Head is the height of the usefully available column of water above the turbine inlet. The head determines water pressure at the turbine.[6] Higher pressures require less equipment to generate a given amount of electricity and are associated with a lower cost of generation. There are also economies of scale in generation.

The next section examines the relationship between generation costs and head and plant scale. It concludes with a comparison of typical hydro development costs and costs of generating electricity by other means.

Relationship Between Costs, Scale, and Head

Figures 14-3 and 14-4 illustrate the relation between cost and plant scale. The two figures are based on somewhat disparate data. Figure 14-3 indicates average costs for countries or country groups and average plant size in megawatts of installed capacity. The World Bank data are based on two points meant to illustrate the disparity in the costs of generating electricity from large, high-head plants and small, low-head plants. The costs are typical for the developing countries and are based on projects funded by the World Bank. The small, low-head plant cost is typical of 10-MW plant size or less and 5 meter (16.5 ft) or less head. The low cost estimate is for plants from around 1,000 MW and up with 1,000 meter (3,280 ft) head. The U.S. and Japanese points reflect average size of plants (15 MW in Japan and 50 MW in the United States) and average head, though figures for average head were unavailable. The solid lines represent the plant size range over which the costs may be considered applicable.

Japan has achieved very low costs despite its small rivers. Only in the case of very small, low-head installations, as indi-

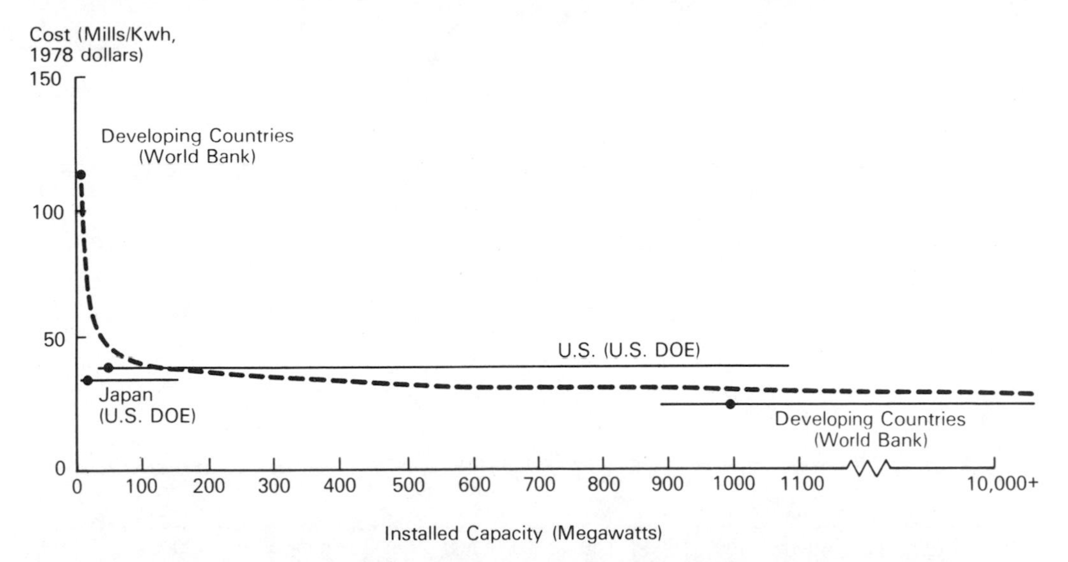

FIGURE 14-3. Cross-country evidence on the relation between plant size and hydroelectric generation costs. *For Japan:* Relative cost factors relative to United States are from *1980 Annual Report to Congress,* U.S. DOE (Washington, D.C.: U.S. Government Printing Office, 1981), pp. 262–263. Weighted average cost factor computed from operation factor and capital factor; weights are based on implied Japanese costs for operations and capital. Relative factor computed to be 0.98 of U.S. cost in mills/kWh. Average size of Japanese hydroelectric installation is from the World Energy conference, *Renewable Energy Resources* (New York: IPC Science and Technology Press, 1978), p. 157. *For developing countries:* Cost per kWh from the World Bank, *Energy in Developing Countries* (Washington, D.C.: World Bank, 1980), p. 43. Approximate size of installation from personal communication with World Bank official, September 1981. Smaller plants are low head (roughly 5 meters) and 10 MW installed capacity or less. Large high head costs are for plant size on the order of 1,000 MW installed capacity with 1,000 meter head. *For United States:* U.S. DOE, *1980 Annual Report,* p. 262.

cated by the World Bank estimate, do costs rise significantly.

Figure 14-4 is based on U.S. data for individual plants. Line segments represent regression estimates for various size categories. The costs are not fully comparable with those of figure 14-3 since they do not include transmission costs, whereas those of 14-3 are full costs through delivery to consumers.

Sixty observations were used to estimate the line segments in figure 14-4. Plant size ranged from 10 MW to nearly 5,000 MW. Head was not explicitly accounted for in the estimates but was not an important variable. (For more detailed presentation and discussion of regression results and method, see the note to table 14-4.)

The estimates confirm the general relationship observed in the cross-country data of figure 14-3. The general conclusion, which appears relatively robust, is that plant scale is important, but only at very small sizes. Most of the economies of scale appear to be exhausted at plant sizes of 50 MW or less. The estimated slope coefficients tend to be small for plants bigger than 50 MW and statistically are insignificantly different than zero at the 5 percent level.

Figure 14-5 is taken from a study by the Electric Power Research Institute and illustrates their findings on the relation between cost and head, controlling for capacity. The approach taken is an engineering study rather than the statistical analysis of

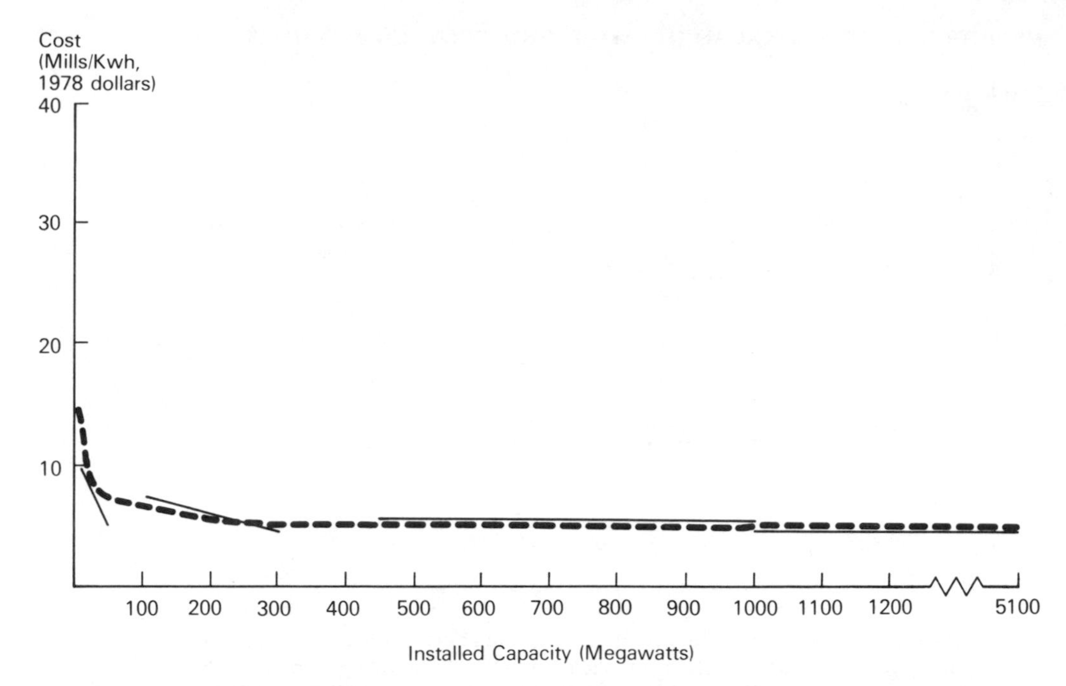

FIGURE 14-4. Cost of hydroelectricity generation versus plant size in the United States. (Data from U.S. DOE, *Hydroelectric Plant Construction Cost and Annual Production Data Expenses—1978,* Washington, D.C.: U.S. Government Printing Office, 1979.) The dashed line is a candidate curve fitting the cost data, sketched from the estimated linear segments. *Estimated Equation* $Y = a + bX$, where Y is cost of generating electric power, including annualized capital cost (dam plus generating equipment), operating costs, and maintenance cost per kilowatt installed capacity and X is size of the power plant in megawatts installed capacity. Capital costs were annualized assuming one 50-year life, straight-line depreciation, and an 8 percent return on investment. Segments estimated for the 1,000+ and 500–1,000 MW categories are based on a sample including the universe of plants in the respective size categories. The data for the 10–50 and 50–500 MW categories consist of a random sample from data reported in the source.

Regression Results

Size category (MW installed)	*Sample size*	*Mean size (MW installed)*	*a*	*B*	*R*	*Head (ft)*
10–50	15	31	44.57 (62.6)	−0.4164 (1.54)	0.16	259
50–100	15	150	39.77 (569.3)	−0.0680 (−1.81)	0.20	464
500–1000	20	938	24.88 (8387.7)	−0.005 (−1.48)	0.005	239
1000*	10	1859	21.72 (90754.7)	−0.0021 (−0.37)	0.21	444

These results were converted to mills/kWh assuming a 50 percent operating capacity.

Interpretation. The intercept is highly significant in all cases, with the 95 percent confidence interval around the estimated coefficient being fairly narrow. (The standard error is less than 1 in all cases.) This indicates that not only are the estimates significantly different from zero but that they are also significantly different from one another, thus supporting the hypothesis of a falling cost curve. On the other hand, the slope estimates are small and, in all cases, insignificantly different from zero. This evidence, taken together, strongly supports the conclusion that economies of scale exist in hydroelectric installations but that nearly all the economies are exhausted at relatively small plant size (less than 50 MW installed). The sample average head for each group (in feet) is also given. Head is often viewed as another variable leading to important systematic cost variance. There is a wide range of head sizes in all the sample categories; the correlation between head and installed capacity was relatively small, thus the bias in b estimates is likely to be small.

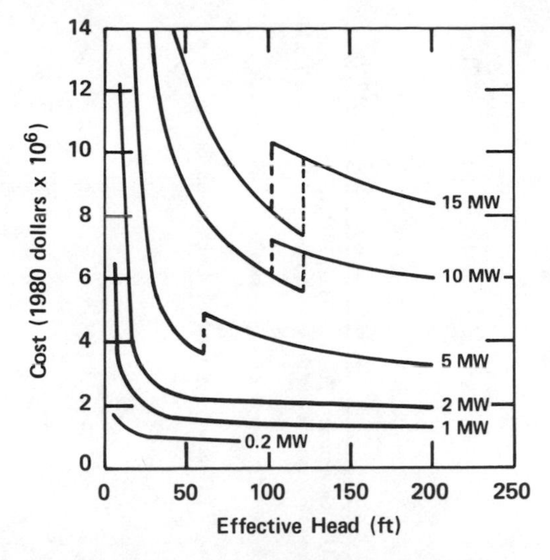

FIGURE 14-5. Head versus cost of hydroelectric plant installation. Assumes use of the least expensive turbine. The dashed lines indicate the need to use a more expensive turbine at a higher head. Three types of small hydro plants were evaluated. These hydro facilities were for plants built at existing dams or a steep drop in a stream. Two of these three types would have these costs and the third type would have costs 7 percent lower than those indicated. (Reprinted with permission from Electric Power Research Institute, *EPRI Journal,* May 1981, p. 51. Copyright © Electric Power Research Institute.)

operating plant data. This allows a study of ranges of the cost curve where few or no plants operate presently. The dashed vertical lines indicate discontinuities resulting from the need for special generators at higher water pressures.

The curves are very similar in shape to the cost versus capacity curves, indicating very high costs at low heads; these fall rapidly once a critical point is reached, and remain roughly constant for higher head levels. The cost plot is a total cost concept rather than cost per unit and should not be confused with figures 14-3 and 14-4, which relate capacity to cost per unit of electricity generated.

The overall conclusion that can be drawn from this examination of head versus cost and scale versus cost is that the long-run supply curve is likely to be relatively unresponsive to price. Both relationships indicate a critical level (a critical head level or capacity level that must be reached before hydroelectric production becomes technologically feasible (low cost). Beyond this point, the cost curve is relatively flat. Moreover, the rising cost range of the curve is limited to relatively low head (less than 15 meters) and small plants (below 10–15 MW capacity). As a result, recoverable resource estimates are unlikely to vary greatly in response to changes in the economic dimension of recoverability. World hydro resources in the mini hydro category are thought to account for only 5–10 percent of total resources.[7]

Cost Comparisons of Electricity Generating Technologies

Table 14-4 compares hydroelectric generating costs with costs of generating electricity from various-sized coal, oil, and nuclear plants. No estimates of operating and maintenance costs were available for mini hydro facilities but these costs are likely to be at least as great and probably considerably more per kilowatt-hour than for conventional hydro. Conventional hydroelectricity tends to be far less costly than other types of electricity, while the mini hydro facilities tend to produce costly electricity—more costly than oil-fired plants. The electricity cost calculations are fairly sensitive to methods used to calculate capital service charges, particularly to the assumed required rate of return. The capital cost calculation could also affect the relative ranking of technologies since hydro costs are primarily attributable to capital costs while costs of electricity generated from oil are largely attributable to fuel costs. A higher required rate of return will tend to reduce the relative cost of fuel and operating and maintenance-intensive technologies, while a low required rate of re-

TABLE 14-4. Cost Comparisons of Electricity Generating Technologies (1975 dollars)

	Mid-1980 Capital Cost ($ per kW installed)	*Capital Cost[a] (mills/kWh)*	*Operation & Maintenance (mills/kWh)*	*Fuel Cost[b] (mills/kWh)*	*Total Cost (mills/kWh)*
Hydro					
Mini hydro, low-head, <10 MW	2,500	57.1	N.A.	—	57.9+
Conventional, various sizes	500–600	11.4–13.7	0.8	—	12.2–14.5
Steam					
Oil					
225 MW	340	7.3			47.0
400 MW	290	6.2	1.0	38.7	45.9
	270	5.8			45.5
1,100 MW	230	4.9			44.6
Coal					
400 MW	490 (+100 for scrubbers)	10.5–12.7			20.4–22
500 MW	450 (+ 90 for scrubbers)	9.7–11.6	4.9	5.0	19.6–21.5
1,000 MW	390 (+80 for scrubbers)	8.4–10.1			18.3–20.0
Nuclear					
600 MW	640	13.7	1.8	11.0	26.5
1,000 MW	550	11.8			24.6

Sources: Varied, including U.S. DOE, *1980 Annual Report to Congress* (Washington, D.C.: U.S. Government Printing Office, 1981), p. 287; J. Dunkerley et al., *Energy Strategies in Developing Nations* (Baltimore, Md.: Johns Hopkins University Press for Resources for the Future, 1981), p. 185; and The World Bank, *Energy in Developing Countries,* August 1980, Washington, D.C., p. 43.

[a]Assuming 8 percent real return, 50-year life for hydro and 30 for all others; 50 percent capacity utilization for hydro and 60 for all others; straight-line depreciation.

[b]Thermal efficiencies of 0.35 for oil and coal and 0.32 for nuclear. Fuel costs per barrel input of $23 for oil (e.g., residual, distillate), $3 for coal ($20 per ton), and $6 for nuclear. See *1980 Annual Report to Congress,* U.S. DOE, 1981, p. 287.

turn will favor capital-intensive technologies.[8]

OCEAN THERMAL ENERGY CONVERSION SYSTEMS

Ocean thermal energy conversion (OTEC) exploits the temperature gradient between warm surface water and cold deep water to operate a Rankine-cycle turbine which drives electric generators. The warm sea water can be used to evaporate a closed system fluid such as ammonia; after powering the turbine, the vaporized fluid is then returned to the liquid state by condensation with the cold sea water.

OTEC units have been conceptualized as large moored floating units. Cold water is pumped through a rigid pipe which is 2,500–3,000 feet in length; the water flow rate is immense: 50 billion gallons per day.[9]

Although the OTEC concept is 100 years old, development is still at the experimental stage. Small-scale units have been tested, but full-scale demonstration plants will not be operational until 1988. Optimistic forecasts expect commercialization by 1995. Conmerical application is envisioned as a baseload generator of up to 400 MW capacity. Discussion here will be limited to OTEC systems and will not consider ocean energy industrial complexes (OEIC), in which the output is used to manufacture on-board electricity-intensive products such as ammonia, inorganic chemicals, fertilizers, and so on.

Although the size of the resource is immense, geographical, technical, and economic factors significantly limit the potential contributions. The following sections consider the world resource base and resource recovery constraints such as technology, material limitation, and costs. Supply curves will be developed.

Resource Availability

Figure 14-6 shows global OTEC thermal resources, measured in terms of temperature gradient. Temperature differentials must be greater than 20°C for OTEC implementation and must be near land masses where demand for output exists. The following section considers the resulting locations for development.

Resource Recovery Constraints

Geography

Given the constraints of a temperature differential and proximity to land, OTEC application has been recommended for islands such as Puerto Rico, Cuba, Hawaii, and the Phillipines, as well as for coastal areas such as the Gulf Coast of the United States, the west coast of Central America, the northeast coast of Brazil, and the equatorial coastlines of Africa and India.[10] A study by the University of Southern California included the above potential as well as locations in Indonesia, Australia, and Malaysia. The analysis considered forty countries lying within 15° of the equator and eliminated countries on a basis of adequate supplies of alternative domestic resources, inadequate demand, and inadequate means of financing full-scale OTEC deployment. A criterion that allowed no more than 15 percent of supply to be accounted for by any single facility was found to be applicable to many of the lesser developed countries.[11]

Other than this study, there appears to be little in the way of site-specific location analysis upon which a regional scalar could be developed. In addition to variation of ocean temperature with depth, site-specific limitations include corrosion and biofouling, seasonal variation of temperature differentials, and local tides and currents.

Technical Limitations

OTEC is still at the developmental stage in which technical limitations are recognized but no attempt has been made to systematically link these limitations to costs of resolving problems. Many of the technological uncertainties stem from the lack of directly comparable systems for many

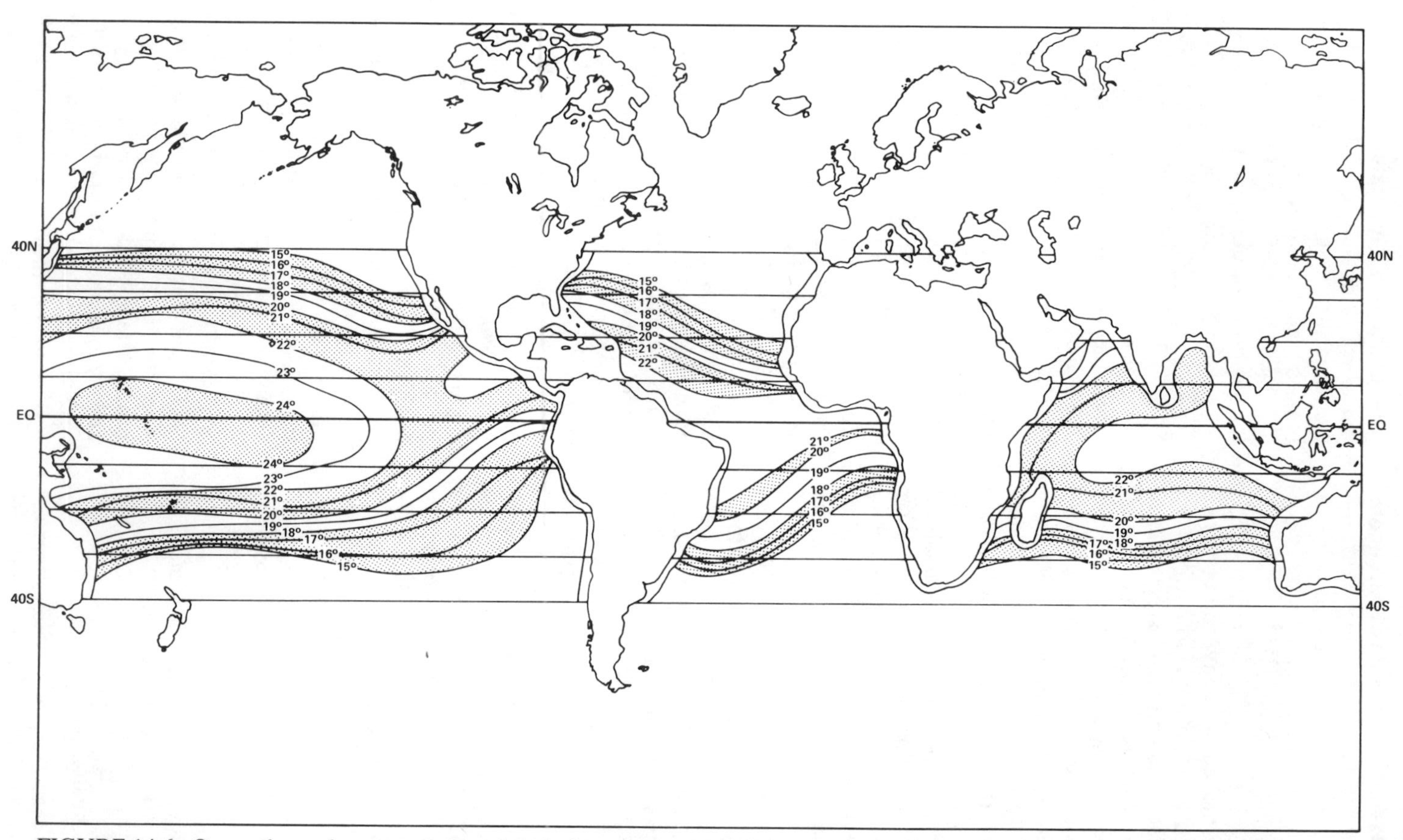

FIGURE 14-6. Ocean thermal resources. (Reprinted with permission from Charles A. Zraket and Martin M. Scholl, *Solar Energy Systems and Resources*, McLean, Va.: MITRE Corp., 1980, pp. 10–11. Copyright © MITRE Corporation.)

of the OTEC components. This category of uncertainty includes size of heat exchangers, the integrity of deep water moorings, and the size of the cold water pipe.[12] The AAAS document, *Solar Energy in America,* is highly critical of OTEC development on these same grounds.[13] The report notes that the low thermal efficiency of the system (about 2.5 percent) necessitates huge heat exchanger surfaces: for a full-scale plant, an area equivalent to 190 football fields is required. There are concerns that leaks, corrosion, and biofouling will in some locations reduce the thin margin of efficiency. The report points out that there are few previous experiences which approach the engineering difficulties expected to be encountered in the stress design of an intake pipe of 50-foot diameter and 3,000-foot length. Similarly, moorings for such a massive structure have not been designed.

Other technological problems, however, do have engineering precedence and have a high likelihood of resolution. These include design of submarine cable energy delivery systems; design of plate-type heat exchangers which use materials that are more readily available than titanium, and studies of estuarian impacts of coastal construction facilities.

Overall, the labor and material intensities of OTEC construction are minimal compared with alternative energy structures. However, as the AAAS authors point out, the use of titanium for the heat exchangers would exhaust the annual U.S. titanium production for the construction of a single commercial OTEC plant. Preventing corrosion by using aluminum heat exchangers is the most likely solution to this resource constraint.

Costs

In view of (or perhaps because of) the immaturity of the technology, recent OTEC capital cost estimates show surprising agreement, ranging from $1,500 to $4,500/kW, not including transmission costs. As is true of similar projects at early stages of development, it is likely that capital costs will

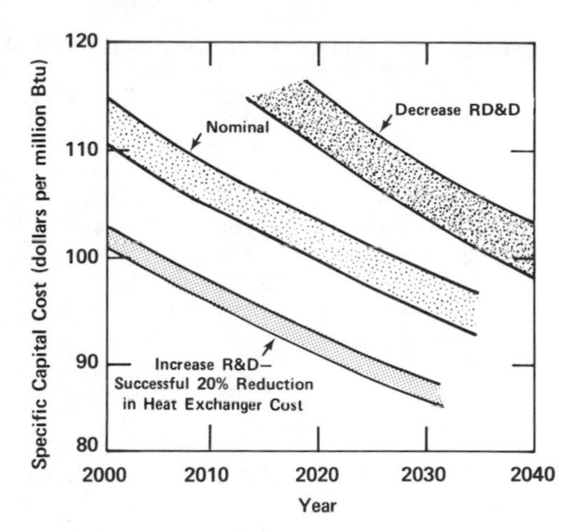

FIGURE 14-7. OTEC specific capital costs (1975 constant dollars). (From J. S. Witwer et al., "Comparative Evaluation of Solar Alternatives: Implications for Federal RD&D," Menlo Park, Calif.: SRI International, formerly, Stanford Research Institute, 1978, p. VIII–20.)

increase as field testing increases knowledge about technological problems.

Studies, however, have implied that the effect of the learning curve will counter this effect. Figure 14-7 shows the results of learning curve effects which allow for cost reductions of 10 to 12 percent over a 20-year period. Presented with the graph is an optimistic case in which a 20 percent reduction in heat exchanger cost is achieved, as well as a pessimistic case where technology development is postponed by 20 years.

Table 14-5 presents a compendium of cost estimates for OTEC. For the reasons expressed above, these estimates are not perceived to be as likely as those of wind, or of solar thermal technology, but more on the order of the photovoltaic estimates.

Other Barriers

Other barriers to commercial development center upon institutional facets of technology deployment. In terms of cost, this is a technology which does not compete favorably with other alternatives; options for fi-

TABLE 14-5. OTEC Cost Projections, 1979 Dollars, Transmission Excluded

Study	*Capital Cost ($/kW)*	*Power Cost (mills/kWh[a])*	*Projected Year*	*Size of Unit*	*Notes[b]*
Bennington[c]	2,552	89.8	1993	250 MW	cf = 0.6, base
	1,496	52.7	2020	250 MW	cf = 0.6, base
Stanford Research Institute[d]	3,608	95.2	2000	Base	cf = 0.8, titanium
	3,478	91.8	2000	Base	cf = 0.8, aluminum
	3,211	84.7	2020	Base	cf = 0.8, titanium
	3,094	81.6	2020	Base	cf = 0.8, aluminum
National Research Council[e] (CONAES)	6,500	178.0	1975	100 MW	lf = 0.9
	3,380	91.0	1990	100 MW	lf = 0.9
	1,625	51.0	1990	100 MW	lf = 0.9 Advanced technology
	3,380	91.0	2000	100 MW	lf = 0.9
	1,430	43.0	2000	100 MW	lf = 0.9 Advanced technology
	3,380	91.0	2010	100 MW	lf = 0.9
	1,430	43.0	2010	100 MW	lf = 0.9 Advanced technology
Cohen & Richards[f]	3,292	105.0	1995	400 MW	High estimate
	2,352	90.0	1995	400 MW	Middle estimate
	1,882	75.0	1995	400 MW	Low estimate
EIA/DOE[g]	2,577	69.7	1995		cf = 0.78, base
Zraket[h]		92.5	First commercial	400 MW	cf = 0.80
		42.6	Ultimate	400 MW	cf = 0.80
DOE/ET[i]	1,560–1,872	29.5–47.3	1990	100 MW	cf = 0.80–0.85

[a]Calculated from capital cost when only capital costs are given in the study. Assumes fixed change rate—18.5%, which includes 3% operation and maintenance costs; assumes cf = 38% unless otherwise noted.

[b]cf = capacity factor; lf = load factor.

[c]G. Bennington et al., "Solar Energy, A Comparative Analysis to the Year 2020" McLean, Va.: (MITRE Corp., August 1978)

[d]J. G. Witwer, "Costs of Alternative Sources of Electricity;" (Menlo Park, Calif.: Stanford Research Institute, January 1978).

[e]National Research Council, National Academy of Sciences, "Study of Nuclear and Alternative Energy Systems, Supporting Paper 6, Domestic Potential of Solar and Other Renewable Energy Sources," Washington, D.C., 1978.

[f]R. Cohen and W. E. Richards, "The Department of Energy Ocean Energy Systems Program," Energy Technology VII, Proceedings of the Seventh Energy Technology Conference, Government Institutes, Inc., June 1980.

[g]U.S. Department of Energy, *1980 Annual Report to Congress,* vol. 3 (Washington, D.C.: U.S. Government Printing Office, March 1981), p. 287.

[h]C. Zraket and M. Scholl, *Solar Energy Systems and Resources* (McLean, Va.: MITRE Corp., 1980).

[i]U.S. Department of Energy, "Solar Energy, A Status Report," DOE/EIA-0062 (Washington, D.C.: U.S. Government Printing Office, January 1978).

nancing will therefore be somewhat restricted. This is particularly true in the case of the lesser developed nations, which by location are prime candidates for this technology. Other problems involve the capital intensiveness of the first few units, as well as uncertainties regarding maritime liability, security, and jurisdiction.[14] Environmental effects of biofouling and local heating are currently speculative.

Potential Contribution and Supply Function

Table 14-6 shows estimates of potential contribution to the United States; these are extrapolated into a rudimentary supply

TABLE 14-6. OTEC Implementation Projections

Study	*Quads Displaced*	*Projected Year*	*Percent of Projected U.S. Energy*
CEQ[a]	1–3	2000	1–2.5
	5–10	2020	7.0
Bennington[b]	—	1985	0.0
	—	2000	0.0
	2.4	2020	1.2
Bennington[c]	0.1–0.4	2000	<1.0
National Research	0.01	1985	<1.0
Council (CONAES)[d]	0.08	1990	<1.0
	1.6	2000	1.0
	3.9	2010	2.6
EIA/DOE[e]	—	2000	—
	0.1	2010	<1.0
	0.3	2020	<1.0
Stanford Research	0.0	1985	0.0
Institute[b]	0.0	2000	0.0
	0.0	2020	0.0

[a]Council on Environmental Quality, "Solar Energy Progress and Promise," Washington, D.C., April 1978.

[b]G. Bennington et al., "Solar Energy, A Comparative Analysis to the year 2020" McLean, Va.: MITRE Corp., August 1978).

[c]G. Bennington, et al., "Toward A National Plan for the Accelerated Commercialization of Solar Energy" McLean, Va.: MITRE Corp., January 1980).

[d]National Research Council, National Academy of Sciences, "Study of Nuclear and Alternative Energy Systems, Supporting Paper 6, Domestic Potential of Solar and Other Renewable Energy Sources," Washington, D.C., 1979.

[e]U. S. Department of Energy, *1980 Annual Report to Congress,* vol. 3 (Washington, D.C.: U.S. Government Printing Office, March 1981).

[f]J. G. Witwer et al. *"A Comparative Evaluation of Solar Alternatives; Implications for Federal RD&D"* (Menlo Park, Calif.: Stanford Research Institute, January 1978).

TABLE 14-7. OTEC Supply Curve Considerations (1979 constant dollars)

Year	*Step*	*Cost, (mills/kWh)*	*Applicable Quantity Limits*[a]
1974	1	178	Zero production
	2	240	Zero production
2000	1	92	Up to 7% of regional electric capacity
	2	120	Up to 15% of regional electric capacity
2025	1	80	Up to 7% of regional electric capacity
	2	90	Up to 20% of regional electric capacity
2050	1	60	Up to 10% of regional electric capacity
	2	68	Up to 30% of regional electric capacity

[a]Applicable only to the following regions:

1. US (and Canada)
3. JANZ
7. Africa
8. Latin America
9. Non-Communist South, East, and Southeast Asia

function in table 14-7. The two steps in the supply function reflect increasing materials costs. The quantity limit in terms of regional electric capacity in step 2 reflects the need for additional baseload in the specified time periods. Because little international site-specific analysis has been performed, no regional scalar can be developed. In place of such a scalar, the note in table 14-7 cites the regions in which OTEC is technically feasible. Other regions have been eliminated upon consideration of the resource.

NOTES

1. World Energy Conference (WEC), *Renewable Energy Resources* (New York: IPC Science and Technology Press, 1978).
2. Communication with Brazilian Embassy, science attache, September 15, 1981.
3. WEC, *Renewable Energy Resources,* p. 154.
4. Discussion based on country notes in WEC, ibid., pp. 157–180.
5. Computed from data in WEC, ibid., p. 161.
6. D. G. Fallen Bailey and T. A. Byer, *Energy Options and Policy Issues in Developing Countries,* World Bank Working Paper No. 350 (Washington, D.C.: 1979).
7. World Bank, *Energy in Developing Countries* (Washington, D.C.: World Bank, 1980).
8. For a full discussion of costing methodologies for energy projects, see Doan Phung, *The Discounted Cash Flow (DCF) and Revenue Requirement (RR) Methodologies in Energy Cost Analysis* (Oak Ridge, Tenn.: Oak Ridge Associated Universities, Institute for Energy Analysis, 1978).
9. J. D. Isaacs and W. R. Schmitt, "Ocean Energy: Forms and Prospects," *Science,* vol. 207, no. 4428, January 18, 1980.
10. C. Zracket and M. Scholl, *Solar Energy Systems and Resources* (McLean, Va.: MITRE Corp., 1980), p. 10.
11. J. Witwer et al., "A Comparative Evaluation of Solar Alternatives" (Menlo Park, Calif.: Stanford Research Institute, January 1978), p. VII–30.
12. Ibid., p. VIII–16.
13. W. E. Metz and A. L. Hammond, *Solar Energy in America* (Washington, D.C.: American Association for the Advancement of Science, 1978).
14. Witwer et al., "A Comparative Evaluation," p. VIII–23.

Biomass 15

Biomass technologies are ultimately solar technologies that utilize the photosynthetic capability of plants to capture and store solar energy. Biomass fuels are distinct from other solar technologies in terms of resource constraints and the final energy carrier product. Direct solar heating and solar electric technologies both suffer from the intermittent nature of the energy source, whereas biomass fuels are easily stored. Like coal, biomass competes as a solid (primarily wood), and as a feedstock which can be converted to liquids and gases; direct solar heating is limited to low-grade heating uses and solar electric competes with liquids and gases only indirectly. The most obvious and most important example is automobile transportation; in the absence of an "acceptable" electric vehicle, solar electricity cannot replace fossil-based transport fuels.

In terms of production considerations, biomass competes with traditional agriculture for prime land. Land availability becomes a serious constraint. In contrast, other solar technologies are less land intensive and optimal areas or locations—deserts or rooftops—do not compete with agriculture.

Despite its exotic sounding name, biomass is not a new source of energy. Wood biomass dominated commercial fuel markets in the developed nations prior to the introduction of coal. Wood, dung, and agricultural wastes remain the second most important energy source in developing nations, accounting for approximately 30 percent of energy consumption.[1] In developed nations, biomass is an important source of energy in the pulp and paper industry. In 1976 it provided 982 petajoules of energy to U.S. paper industry boilers or 1.4 percent of total U.S. commercial energy consumption.[2] Small amounts are also burned in boilers by U.S. electric utilities. Sweden and Finland are even more dependent on wood energy, which supplies 8 and 15 percent of their respective energy needs.[3]

In developing economies, biomass competes directly in a wide variety of energy markets, such as space heating, lighting, cooking, process heat, steam, mechanical drive, and even motive power. In developed economies its markets are narrower, with small amounts used for residential space heating, but virtually none is used for other residential energy services. Its industrial applications are restricted primarily to fueling boilers. There is virtually no transport market for the fuel.

Biomass can be used either directly as a solid, or after conversion, as a liquid or gas. Stobaugh and Yergin are optimistic about the ability of biomass to compete directly with coal as a solid fuel in the industrial and electric utility markets. They argue that with minimal processing (pelletizing) wood can compete both environmentally and economically. The cost of pelletizing wood, however, is not trivial. It adds between $3 and $4 (1979 dollars) per gigajoule.[4]

In addition to the traditional solid fuel markets, biomass can also be converted into liquids and gases. Brazil has undertaken a major program to produce ethanol from sugarcane.[5] While the Brazilian effort is the most widely known, other countries and other processes of ethanol production also compete.[6] Ethanol is not the only end product into which biomass may be transformed. Medium Btu gas and methanol are also available. Specific technologies and costs for biomass conversion are discussed in chapters 8 and 10 of this volume. The focus of this chapter is on likely sustainable global biomass feedstock supplies and associated costs.

RESOURCE BASE

The global biomass resource base consists of all plant matter on land and in the oceans. Most experts peg the energy content of all biomass at between 15 and 20 times the amount human beings currently get from commercial energy sources, although other estimates range from 10 to 40 times.[7] However, in assessing the amount of this energy resource that could potentially be utilized, several factors must be considered. First, biomass production must be viewed in terms of a sustainable flow; the annual addition to the biomass stock represents the biomass resource flow. Fully harvesting this flow without careful management would result in depletion of the soil, thereby reducing annual yields. The potential also exists for carefully managing farms of specialized plant types, thereby greatly increasing the yield. Second, there are energy losses in transforming biomass to useful energy and useful energy carriers. Transformation efficiency may be as low as 10 percent for direct use of firewood in the developing countries.[8] For transformation of biomass to other secondary fuels, the practical efficiencies range from 30 to 40 percent for biogas[9] and around 50 percent for conversion of wood to charcoal.[10] Finally, fully harvesting the world's annual production of biomass could significantly change climate by changing albedo,[11] altering water absorption capabilities,[12] or bringing about significant soil degradation.[13] The annual global production of fuel-equivalent biomass has been estimated as approximately 100 terawatt years per year (TW yr/yr) (roughly $3{,}000 \times 10^{15}$ Btus). Of this, 23 percent is fixed in swamps, grasslands, and tundras; 29 percent in forests; 10 percent in cultivated land; and the remaining 38 percent in oceans.[14] IIASA has argued that no more than 40 percent of energy fixated on land could be "prudently cultivated."[15] This would yield 25 TW yr/yr (747×10^{15} Btus) of energy, but of this, 14 TW yr/yr was estimated to be connected with food or lumber production, of which 2 TW yr/yr could be collected as waste. Thus only 13 TW yr/yr (390×10^{15} Btus) would be available for conversion to useful energy forms. After adjusting for conversion losses, IIASA argues that 6 TW yr/yr (180×10^{15} Btus) of secondary energy fuels (equivalent to high grade coal) would be available. However, this "implies a sophisticated, very careful management of the photosphere."[16]

Table 15-1 contains estimates of biomass availability derived independently of the IIASA estimates and using a somewhat different approach. The figures represent likely upper and lower estimates of the land-based biomass resource base. As reported in Table 15-1, Colin Clark estimated the potential amount of agricultural land in terms of hectare equivalents. (A hectare equivalent is an attempt to adjust land areas to a standardized agricultural potential.) Potential biomass availability is computed on the basis of two separate assumptions. First, it is computed on the basis that all tillable land, as estimated by Clark, is devoted to raising food and fiber and that energy uses are secondary; thus biomass available for conversion to energy fuels is only that collectable as waste from agricultural processes. A second calcula-

TABLE 15-1. Global Energy Potential of Land-Based Biomas: Agricultural Waste and Energy Farms

Region	Land Equivalent[a] (10^6 hectares)	Potential Biomass Availability (10^9 metric tons): Use of Agricultural Waste[b]	Energy Farms[c] (silviculture)	Energy Content[d] (10^{15} Btus)
US[e]	925	1.203	17.113	22.6– 320.9
OECD West[e]	365	0.475	6.753	8.9– 126.6
JANZ	229	0.298	4.237	5.6– 79.4
EUSSR	1,238	1.609	22.903	30.2– 429.4
ACENP	593	0.771	10.970	14.5– 205.7
MIDEST	75	0.098	1.388	1.8– 26.0
LA	2,558	3.325	47.323	62.3– 887.3
AFR	2,292	2.980	42.402	55.9– 795.0
SEASIA	1,452	1.888	26.862	35.4– 503.7
Total	9,727	12.647	179.951	237.2–3374.0

Source: Colin Clark, *Starvation or Plenty* (New York: Taplinger, 1970), pp. 154–60; reprinted in Marylin Chou, David D. Harmon, Jr., Herman Kahn, and Sylvan Wittwer, *World Food Prospects and Agricultural Potential* (New York: Praeger, 1977), pp. 61–62.

[a]*Methodology.* Land classified according to temperature and rainfall. Thornwaite symbols were used:

A–E = abundant rainfall to complete aridity
A'–E' = temperature classification from tropics to perpetual frost (E' = tundra)
r = regular rainfall
s = summer deficient rainfall
w = winter deficient rainfall
d = rainfall deficiency spread over the year

Assumptions. Equivalent in "standard" farmland:

1. "Standard land" = land that is farmed in humid temperate climates, or its equivalent—that is, land regarded as capable of producing one good crop per year or a substantial amount of grazing.
2. Humid tropical land, when well fertilized, is capable of producing grass at a rate five times the best fertilized temperate land (75 tons dryweight per hectare per year versus 15). Two to three crops per year can be grown. A conversion coefficient of 4 is used to convert this land to "standard land."
3. Irrigated land = 1.5 units of standard land.

General use of agricultural methods currently (1970) practiced by the average farmer in the Netherlands, without allowing for any further improvement in agricultural technology, for any food from the sea, or for additional irrigation. 2,763 m^2 required per person for combined agricultural and forestry needs.

[b]Based on estimates for the United States of collectable biomass of 1.30 metric tons of dry matter per hectare, computed from data in U.S. Department of Energy, *Assessment of Industrial Activity in the Utilization of Biomass for Energy* (Springfield, Va.: National Technical Information Service, 1980), pp. 2–25.

[c]Based on estimates for the United States of 18.50 metric tons of dry matter per hectare, computed from data in U.S. DOE, *Industrial Activity,* 1980, pp. 2–25.

[d]Low and high estimates are based on potential biomass available if all land is used purely for traditional food and fiber crops or purely for high-yielding silviculture energy farms. Based on 18.75 × 10^6 Btus/dry metric ton of biomass.

[e]Data available from the source were given in an aggregate for the United States and Canada. On the basis of other data and evidence presented in Clark (pp. 55–63), a value of 81 × 10^6 hectares of "standard land" was estimated to exist in Canada. This amount was subtracted from the United States and Canada aggregate estimate and added to the Western Europe estimate to obtain the figure presented in the table.

tion is made assuming that all land is devoted to energy farms. Both estimates are hypothetical since the assumption that all tillable land could be brought under cultivation is unreasonable—whether as biomass farms or traditional agriculture. Nevertheless, the figures are interesting, the higher estimate because it indicates that intense, specialized cultivation could greatly increase the annual amount of energy fixated as biomass. The lower figure is very similar to the amount of biomass estimated to be available by IIASA after imposing various assumptions of feasibility. For perspective, the estimates can be related to global energy production in 1979.

If an overall 40 percent conversion efficiency is assumed, the lower, waste biomass scenario represents roughly 95 quads of a high-grade secondary energy fuel or approximately one-third of 1979 global energy production. For the energy farm scenario, roughly 1,350 quads of secondary energy would be produced or about 4½ times 1979 energy production.[17] These figures must be taken as *resource base* estimates because no feasibility constraints have been imposed.

The estimates are based primarily on conditions existing in the United States; needless to say, such extrapolation is hazardous at best. On the positive side, the hectare equivalent measure of potential agricultural land standardizes the land to a temperate zone climate like the United States. On the negative side, the agricultural practices of the United States are much more sophisticated than those of most of the developing countries, where land management skills are undeveloped. The waste collection and biomass production estimates, even when applied to the United States, probably impute advances in average management skills. Overall, the collection of agricultural waste normally left as field residues or the intense harvesting of biomass from energy farms removes valuable soil conditioning and fertilizing substances. For this reason, waste conversion using anaerobic digestion processes, which preserve the fertilizer and soil conditioning properties of waste, are likely to provide a more sustainable biomass future even though anaerobic processes may be somewhat more costly than other conversion techniques.

In addition to agricultural residues, energy can be recovered from the wastes of human populations. Table 15-2 presents estimates of the waste produced by urban populations under various assumptions as noted in the table. Only waste produced by urban dwellers is included; it is assumed that the cost of collection effectively bars the use of wastes generated by rural populations. The waste estimate implicitly includes sewage sludge wastes from industry and residential sources and solid waste, again from both residential and industrial sources. The exclusion of human waste in rural areas might be questioned, especially in the case of the developing countries where bogar plants (small scale biogas plants) would easily utilize human waste as well as animal dung. However, bogar plants are unable to recover energy easily from solid wastes, thus reducing practicably useful waste generation per person. Further, industrial wastes are concentrated in cities, thus raising the waste output per capita of city dwellers compared with rural dwellers. Prohibitive costs of rural waste collection have been cited; there are also costs of collecting urban wastes, and for the most dispersed of these, the costs may also become prohibitive. The estimates in table

TABLE 15-2. Global Energy Content of Urban Waste[a] (10^{15} Btus)

Region	*1975*	*2000*	*2025*	*2050*
US	2.6	3.1	3.4	3.5
OECD West	4.9	5.7	6.4	6.7
JANZ	1.5	1.9	2.0	2.0
EUSSR	4.2	5.7	6.2	6.4
ACENP	3.4	6.6	10.1	12.1
MIDEST	0.7	1.5	2.4	2.8
LA	3.3	6.5	8.6	9.9
AFR	1.5	3.7	6.4	8.3
SEASIA	4.3	10.0	17.0	21.7
Total	26.4	44.7	62.5	73.4

[a]Assumes 0.91 dry metric tons of municipal waste produced per year per urban dweller. Municipal waste is assumed to be 75 percent organic matter at 22.046×10^3 Btus/kg of organic matter. These assumptions are based on U.S. estimates reported in U.S. Department of Energy, *Environmental Residuals and Capital Cost of Energy Recovery from Municipal Sludge and Feedlot Manure* (Springfield, Va.: National Technical Information Service, 1980) and U.S. Department of Energy, *The Report of the Alcohol Fuels Policy Review* (Washington, D.C.: U.S. Government Printing Office, 1979).

Regional population projections are those reported in chapter 2 of this study. The following assumptions of percent urban for the various regions are made. Regions US, OECD West, JANZ: 80 percent urban for all years. Regions EUSSR, LA: 70 percent urban in 1975, 80 percent urban in 2000 and thereafter. Region MIDEST: 60 percent in 1975, 70 percent in 2000, 80 percent in 2025, 2050. Regions ACENP, SEASIA, AFR: 25 percent in 1975, 35 percent in 2000, 45 percent in 2025, 50 percent in 2050.

15-2 have not been adjusted in any way for the feasibility of collection. One estimate for the United States is that 60 percent of urban wastes are collectible.[18] In the next section, an attempt is made to assign an economic dimension to waste collection and biomass production.

Before going on to the next section, it should be noted that, like the estimates of biomass availability from agricultural sources, the estimates of urban waste generation are based on very broad generalizations and no fine tuning of estimates has been done for secondary considerations. In the case of agricultural sources of biomass, only land area and a climatic consideration (assumed invariant over time) affect regional availability of biomass; secondary factors such as prevalent plant types, agricultural skills and practices, or potential changes in climate or land availability are not considered. In the case of urban wastes, variance in waste produced per capita, types of waste, and industrial waste generation as they vary by region or income are disregarded. In both cases, data based on U.S. studies and experience are extrapolated to other regions. It is possible to speculate on how regional and time-related variables affect the figures of tables 15-1 and 15-2. However, the more important task is to investigate the likely costs of biomass collection and production.

ECONOMIC DIMENSIONS OF BIOMASS PRODUCTION AND WASTE COLLECTION

Explicit costs are involved in producing biomass specifically for energy conversion. In addition, many biomass "wastes" have alternative uses or, at the very least, costs of collecting and transporting them to a conversion plant. In a few cases, colocation of the waste converter with the waste producer results in a zero cost; in the case of municipal solid wastes that normally must be hauled to landfill areas, the implicit cost may be negative since pyrolytic conversion greatly reduces the volume of the waste that must be transported to distant landfills and buried.

Table 15-3 has been adapted from U.S. Department of Energy figures. The table reports estimates of the total cost of acquiring (producing), collecting, and transporting various biomass sources for conversion to liquid fuels. The table reflects a broad array of costs. Silviculture was determined to be the least expensive land-based energy farming technique.[19] The costs of biomass from silviculture farms compare favorably with many types of biomass waste because the concentrated production on silviculture tracts greatly reduces transportation costs.[20]

TABLE 15-3. Estimated Delivered Costs of Biomass Feedstocks in the United States (1979 dollars per 10^6 Btus except as noted)

	Share of Resource Exploited		
Biomass Source	*0.2*	*0.5*	*0.8*
Wood			
Silviculture	1.90	2.40	4.80
Logging residues[a]	1.70	2.10	4.20
Mill residues[a]	0.45	0.55	1.55
Agricultural residues			
Corn stover	1.40	2.90	4.50
Wheat straw	0.35	1.85	4.10
Grains (1979 dollars/ton)			
Corn	117.00	151.00	175.00
Wheat	146.00	181.00	204.00
Grain sorghum	110.00	140.00	157.00
Sugars (1979 dollars/ton)			
Cane	98.00	105.00	111.00
Sweet sorghum	59.00	62.00	65.00
Municipal waste	−0.10	0.25	0.60
Food processing wastes			
Citrus	2.10	3.60	5.10
Cheese	0.00	0.60	1.30

Source: Except as noted, adapted from U.S. DOE estimates for 2000 in 1977 dollars using the U.S. GNP deflator and from dollars per short ton to dollars per 10^6 Btus assuming 17×10^6 Btus per short ton for wood and agricultural residues, and 20×10^6 Btus per short ton for municipal and food processing wastes.

[a]Adapted from U.S. DOE, *Energy Assessment of Industrial Activity in the Utilization of Biomass for Energy* (Springfield, Va.: National Technical Information Service, 1980) pp. 2-15.

U.S. Department of Energy estimates for the United States for 2000 do not include alcohol production from corn grain because sweet sorghum is substituted for corn grain in this use.[21] In terms of long-run biomass-based liquids, methanol appears to be preferable to ethanol. Ethanol production is based on grains and sugars and thus competes directly with food production. Methanol can be derived from agricultural residues and is therefore complementary with food production in that it can utilize wastes. Methanol can also be produced from wood, making it less competitive in land use because forest crops can be grown on land unsuitable for food crops.

The cost and share estimates reported in table 15-3 should be viewed in the context of the U.S. DOE estimated maximum amounts available for each source reported in table 15-4.

From the data in tables 15-3 and 15-4, it is possible to calculate the amount of waste estimated to be available at zero cost or less; the figure is only 17.4×10^6 dry tons in 1980. This represents only about a third of a quad of raw biomass energy or about one-seventh of a quad of secondary energy equivalent (20×10^6 Btus/dry ton).

For perspective in comparing prices, residential natural gas prices in the United States in 1979 were $3.18 per million Btus and the world price of crude oil was approximately $3.70 per million Btus (both in 1979 dollars).[22] However, the biomass costs in table 15-3 are for the raw biomass feedstock. Conversion losses would be on the order of 60 to 70 percent; thus, the cost of usable biomass Btus is at least 2.5 to 3.5 times the cost of raw biomass Btus. In addition, capital and nonfeedstock operating costs must be added.

ILLUSTRATIVE GLOBAL SUPPLIES OF BIOMASS

It is useful to attempt to assign economic and general feasibility constraints on the resource base estimates presented in tables 15-1 and 15-2. Even very crude estimates can begin to establish whether biomass is a likely major contributor to future energy supplies.

Such estimates require a large number of simplifying assumptions. Begin by assuming that energy crops will not be allowed to displace food and other agricultural crops needed to support a global population of approximately 8.2 billion in the year 2050.[23] Assume that nonagricultural considerations—urban land use, environment, parks, and preserves—result in a constraint of 40 percent of potential hectare equivalent available for traditional agriculture and energy farms.[24] Table 15-5 is based on these assumptions and those detailed in the table notes.

Assuming the energy content of recoverable waste per hectare equivalent of agricultural land is 24.385×10^6 Btus and the energy content of biomass production per hectare equivalent is 346.875×10^6 Btus, the potential energy content of available biomass in 2050 is given in table 15-6 (see notes to table 15-1).

The estimates of shares of particular biomass resources available at different costs, given in table 15-7, have been judgmentally determined based on similar estimates for the United States in table 15-3. Combining feedstock shares available at different costs and estimated total available amounts of biomass feedstocks of various types given in table 15-6 allows one to compute estimates of biomass feedstocks available at different costs.

The figures in table 15-8 have been computed using the derived estimates in tables 15-6 and 15-7. The estimates must be considered as very crude, given the large number of fairly arbitrary assumptions made along the way. Nevertheless, they are of considerable interest. First, as in all tables in this chapter, the costs and energy contents are for raw biomass feedstocks delivered to a plant capable of converting the raw feedstock to a usable energy fuel. Conversion losses are on the order of 60 to 70 percent. The important conclusion is that

TABLE 15-4. Projected Maximum U.S. Biomass Resources Available for Alcohol Production (million dry short tons per year)

	1980		*1985*		*1990*		*2000*	
	Quantity	*Percent*	*Quantity*	*Percent*	*Quantity*	*Percent*	*Quantity*	*Percent*
Wood[a]	499	61	464	56	429	49	549	48
Agricultural residues	193	23	220	26	240	28	278	24
Grains[b]								
Corn	22	—	20	—	8	—	—	—
Wheat	12	—	15	—	17	—	20	—
Grain sorghum	4	—	3	—	3	—	3	—
Total grains	38	5	38	5	28	3	23	2
Sugars[b]								
Cane	—	—	3	—	13	—	13	—
Sweet sorghum	—	—	5	—	56	—	159	—
Total sugars	—	—	8	1	69	8	172	15
Municipal Solid Waste	86	10	92	11	99	11	116	10
Food processing wastes								
Citric	2	—	2	—	3	—	4	—
Cheese	1	—	1	—	1	—	2	—
All other	3	—	4	—	4	—	4	—
Total processing wastes	6	1	7	1	8	1	10	1
Total	822	100	829	100	873	100	1,148	100

Source: Congressional Research Service, Library of Congress, *The Energy Factbook,* 96th Cong., 2d sess., November 1980, p. 750.

[a]Assumes wood from silvicultural energy farms starting in 1995.

[b]Estimates for grains and sugars assume an aggressive development program to establish sweet sorghum as a cash crop. This program would divert land from corn in 1990 and 2000—4.7 and 7 million acres, respectively.

TABLE 15-5. Illustrative Agricultural Land Use in 2050, by Region (hectare equivalents of agricultural land $\times 10^6$)

Region	*Total Productive Land*[a]	*Available Land*	*Used to Produce Food, Fiber, etc.*[b]	*Available for Biomass Crops*[b]
US	925	370	145	225
OECD West	365	146	146	0
JANZ	229	92	56	36
EUSSR	1,238	495	226	269
ACENP	593	237	237	0
MIDEST	75	30	30	0
LA	2,558	1,023	407	616
AFR	2,292	917	443	474
SEASIA	1,452	581	581	0
Total	9,727	3,891	2,271	1,620

[a]From table 15-1.

[b]Assumptions generating these estimates are that 0.277 hectares per capita are required for non-energy agricultural uses, based on Clark's estimated population carrying capacity for the world at "maximum" consumption standards. (Reprinted in Marylin Chou, David D. Harmon, Jr., Herman Kahn, and Sylvan Wittwer, *World Food Prospects and Agricultural Potential* (New York: Praeger, 1977), pp. 61–62; regions use no more than 40 percent of productive land; regions produce food, fiber, and other nonenergy agricultural products sufficient to feed own population up to available land constraints before producing energy crops; food import requirements of land deficit countries are met by land surplus countries and are distributed among them with weights equal to their share of global surplus land available. Population in 2050 is that reported in chapter 2.

TABLE 15-6. Energy Content of Potentially Available Biomass in 2050, by Region (10^{15} Btus)

Region	*Urban Waste*[a]	*Agricultural Waste*[b]	*Land-Based Biomass Farms*[c]
US	3.5	3.5	78.0
OECD West	6.7	3.6	0.0
JANZ	2.0	1.4	12.5
EUSSR	6.4	5.5	93.3
ACENP	12.1	5.8	0.0
MIDEST	2.8	0.7	0.0
LA	9.9	9.9	213.7
AFR	8.3	10.8	164.4
SEASIA	21.7	14.2	0.0
Total	73.4	55.4	561.9

[a]From table 15-2.

[b]Assuming 24.385×10^6 Btus of collectible waste per hectare of agricultural land—see table 15-1 notes and land availability estimates of table 15-5.

[c]Assuming 346.875×10^6 Btus of biomass production per hectare and land availability estimates of table 15-5.

TABLE 15-7. Assumed Shares of Biomass Material Available at Various Costs (1979 dollars per 10^6 Btus)

Biomass Type	*0*	*.50*	*1.00*	*1.50*	*2.00*	*3.00*	*5.00*
Urban waste	0.2	0.5	0.8	0.8	0.8	0.8	0.8
Agricultural residues	0.0	0.1	0.2	0.3	0.4	0.6	0.8
Biomass farms	0.0	0.0	0.0	0.0	0.2	0.5	0.8

a relatively small amount of biomass can be expected to be available at low costs. These estimates are somewhat sensitive to the underlying assumptions but the overall orders of magnitude are unlikely to be changed by varying the assumptions within reasonable bounds.

The biomass available below $1.50 per million Btus is all waste biomass. The critical assumptions for urban waste generation are the population size, percent urban, and waste generation per capita. By 2050 all regions except SEASIA, AFR, and ACENP are assumed to have reached a stationary point of 80 percent urban. Higher levels of urbanization seem unlikely. In the three regions with lower levels of urbanization (50 percent by 2050), it is quite possible that urbanization could have proceeded further by that time. However, even if urbanization reached 80 percent in these regions by 2050, it would increase available urban waste in these regions by only 60 percent or globally by only one-third. Partly counterbalancing this effect is the probably somewhat high estimate of ultimately collectible urban waste. The American Gas Association, which is somewhat optimistic about the prospects for unconventional gas in the United States, estimated that at most 60 percent of urban waste was collectible. Finally, waste production per capita is based on estimates for the United States. If waste generation is related to consumption, both would tend to increase with income level. The poorer regions are unlikely to have reached the present U.S. income levels by 2050; thus waste output could be less per person in these regions. The richer areas will have exceeded U.S. income levels by 2050; thus waste output may be larger than suggested by the

TABLE 15-8. Illustrative Supply Schedule for Biomass Feedstocks in 2050, by Region[a] (10^{15} Btus)[a]

	Delivered Cost (1979 dollars per 10^6 Btus)						
Region	*0*	*0.5*	*1.00*	*1.50*	*2.00*	*3.00*	*5.00*
US	0.7	2.1	3.5	3.9	19.8	43.9	68.0
OECD West	1.3	3.7	6.1	6.4	6.8	7.5	8.2
JANZ	0.4	1.1	1.9	2.0	4.7	8.7	12.7
EUSSR	1.3	3.8	6.2	6.8	26.0	55.1	84.2
ACENP	2.4	6.6	10.8	11.4	12.0	13.2	14.3
MIDEST	0.6	1.5	2.4	2.5	2.5	2.7	2.8
LA	2.0	5.9	9.9	10.9	54.6	120.7	186.8
AFR	1.7	5.2	8.8	9.9	43.8	95.3	146.8
SEASIA	4.3	12.3	20.2	21.6	23.0	25.9	28.7
Total	14.7	42.2	69.8	75.3	193.3	372.9	552.6

Source: Computed from tables 15-6 and 15-7 (see text).

[a]Amounts available at delivered costs equal to or less than those given. Energy content of biomass, unadjusted for conversion efficiency.

Note: Columns may not add due to rounding.

figures in table 15-8. The waste production-income level relation is purely speculative, especially beyond historically experienced income levels. However, across regions, the effects appear to be offsetting. Taking these factors together suggests that the global total for urban waste is unlikely to be much above 1.5 times the estimates of table 15-6.

Similarly, the potential for divergence in actual collectible agricultural residues from estimated values is somewhat limited. The major scope for divergence is in the energy content of collectible wastes per hectare. Collectible waste is largely dependent on the crop type. Table 15-6 is based on an average of about 25×10^6 Btus per hectare of collectible waste. For comparison, other crop residues are estimated to be produced at the following rates: 17.9×10^6 Btus/hectare per year for cotton,[25] 21.0×10^6 Btus per hectare for forest residues,[26] 71.4×10^6 Btus/hectare for corn,[27] 67.1×10^6 hectare for wheat straw,[28] and about 52.9×10^6 Btus/hectare equivalent for sugar cane residue.[29]

While these estimates tend to be higher than the 25×10^6 Btu figure assumed, at least three other factors must be considered. First, the crops listed are the most promising for harvesting residues. Crops with little or no collectible residues will reduce the average. Second, they are estimates for the United States. (The sugar cane estimate is based on data for Hawaii.) Generally, crop production per acre is much higher in the United States than in many other areas of the world. Third, the estimates generally assume full harvesting of residues. Taking these factors into account, it is unlikely that average maximum residue collection could be more than about twice the assumed average.

Finally, the most important category of biomass, in terms of potential availability, is biomass raised specifically for energy production. The major constraints operating to generate the estimate in table 15-8 are that food, fiber, and other traditional agricultural products are produced first and that, of the land potentially suitable for agriculture, 60 percent is reserved for other uses. Such a restriction may be unnecessarily severe. However, various researchers note the undesirability of farming most of the available land surface.[30] The more important dimension of energy farming, however, is the cost. In this regard, silviculture was chosen because it was estimated to be the least expensive source of biomass. The estimates were based on a relatively small silviculture industry projected to possibly exist in the United States in 2000 compared with the large, global industry implied by the estimates given in table 15-6. Pushing a silviculture industry to the limits implied by table 15-6 would almost certainly result in fairly sharp upward pressure on land prices, thereby increasing the cost of producing biomass. For these reasons, the cost estimates are probably somewhat low.

Marine-based biomass harvesting has not been assessed. Data on marine operations are largely unavailable and would be highly speculative.

In assessing these various factors, it is difficult to escape the conclusion that biomass resources are fairly limited in terms of global energy requirements. Energy farms, while potentially a fairly large supplier of biomass, will yield relatively costly end-use fuels; raw biomass delivered to conversion plants is likely to cost at least \$2 per million Btus; after accounting for conversion losses the energy will cost at least \$5 per million Btus plus capital and nonfeedstock operating charges.

NOTES

1. J. Dunkerley, W. Ramsey, L. Gordon, and E. Cecelski, *Energy Strategies for Developing Nations* (Baltimore, Md.: Johns Hopkins University Press for Resources for the Future, 1981), p. 5.
2. D. B. Reister, W. S. Chern, and H. D. Nguyen, *An Economic Engineering Energy Demand Model for the Pulp and Paper Industry,* ORNL/CON-29 (Oak Ridge, Tenn.: Oak Ridge National Laboratory, April 1978), p. 46.
3. R. S. Pindyck, "The Economics of Oil Pricing," *Wall Street Journal,* December 20, 1977.

4. R. Stobaugh and D. Yergin, *Energy Future* (New York: Random House, 1979), pp. 199–200.
5. See for example, "Alcohol: A Brazilian Answer to the Energy Crisis," *Science,* vol. 195, p. 4278.
6. Acid hydrolysis processes have been investigated in the United States, the Soviet Union, and Switzerland. In contrast, sugarcane feedstocks are converted to alcohol via a fermentation-distillation process.
7. Denis Hayes, *Rays of Hope: The Translation to a Post-Petroleum World* (New York: Norton, 1977). p. 187.
8. J. Dunkerley et al., *Energy Strategies,* p. 36.
9. U. S. Department of Energy, *Environmental Residuals and Capital Cost of Energy Recovery from Municipal Sludge and Feedlot Manure* (Springfield, Va.: National Technical Information Service, 1980), p. 7.
10. Hayes, *Rays of Hope,* p. 200.
11. Ibid., p. 202.
12. Ibid.
13. Wolf Häfele, Project Leader, *Energy in a Finite World: A Global Systems Analysis,* Report by the Energy Systems Program Group of the International Institute for Applied Systems Analysis (Cambridge, Mass.: Ballinger, 1981), p. 345.
14. B. Bolin, "Global Ecology and Man," in *Proceedings of the World Climate Conference* (Geneva: World Meteorological Organization, 1979). pp. 24–28.
15. Häfele, *Energy in a Finite World,* p. 345.
16. Ibid.
17. 1979 global energy production as reported in U.S. Department of Energy, *1981 Annual Report to Congress,* vol. 2 (Washington, D.C.: U.S. Government Printing Office, 1981), p. 18.
18. American Gas Association, *The Gas Energy Supply Outlook: 1980–2000* (Arlington, Va., 1980), p. 38.
19. U.S. Department of Energy, *Technical-Economic Assessment of the Production of Methanol from Biomass* (Washington, D.C.: U.S. Government Printing Office, 1979), pp. 2–24.
20. Ibid.
21. Congressional Research Service, *The Energy Fact Book* (Washington, D.C.: U.S. Government Printing Office, 1980), p. 754.
22. U.S. Department of Energy, *Annual Report,* 1981, p. xi.
23. Population estimates by regions are those reported in chapter 2.
24. Häfele, *Energy in a Finite World,* pp. 344, 386, uses the figure of 40 percent, applying it to land-based biomass production. This is a somewhat different assumption from 40 percent of the hectare equivalents of potential agricultural land. Nevertheless, 40 percent represents a reasonable constraint, reflecting the overall problems of harvesting biomass on a large scale without severely disrupting the biosphere. Such concerns are expressed by proponents of so-called "soft energy paths" (see Hayes, *Rays of Hope,* p. 201) as well as others.
25. Based on estimates in U.S. Department of Energy, *Environmental and Economic Evaluation of Energy Recoverable from Agriculture and Forestry Residues* (Springfield, Va.: National Technical Information Service, 1980), p. 2, including field residues and ginning residues converted from pounds per acre to Btus per hectare; 6.985×10^3 Btus/lb assumed, as given in the source.
26. Ibid., p. 32, for a wood pulp plant; includes bark residues and wood residues. Source reported total residues of 14.5×10^3 dry tons of residue from 810 forest acres. For estimate reported in the text it was further assumed that forest growth required 20 years, thus per annum output is 1/20 of the total; the energy content of residues is assumed to be 4.8×10^3 Btus per pound. See Federal Energy Administration, *Energy Interrelationships: A Handbook of Tables and Conversion Factors for Combining and Comparing International Energy Data* (Washington, D.C.: U.S. Government Printing Office, 1977), p. 16.
27. Ibid., p. 58, assuming corn used for grain. Converted from pounds per acre to Btus per hectare, 5.513 Btus/lb assumed, as given in the source.
28. Ibid., p. 80, converted from pounds per acre to Btus per hectare, 5.430 Btus/lb assumed, as given in the source.
29. Based on 12,946 kg/hectare of organic material (see Battelle Columbus Laboratories, *Systems Study of Fuels from Sugarcane, Sweet Sorghum, Sugar Beets, and Corn,* vol. II, *Agricultural Considerations,* Washington, D.C.: U.S. Government Printing Office, 1977, p. 98) at 9×10^3 Btus/lb dry weight (see FEA, *Energy Interrelationships,* p. 16), implying 211.4×10^6 Btus/hectare. However, sugarcane can be grown only in tropical or subtropical climates. To compare this residue output per hectare equivalent, the total residue is divided by 4 (based on Colin Clark's adjustment for land located in tropical areas; see footnotes to table 15-1).
30. AGA, *The Gas Energy Supply Outlook,* p. 38.

Energy Consumption and Carbon Dioxide: A Model 16

The preceding chapters have attempted to give some indication of what can be expected in the way of energy demand and supply in the coming decades. It is apparent that the bulk of energy supplies will continue to be derived from fossil fuels. It has also become clear in the past few years that combustion of these fuels is adding significantly to carbon dioxide in the upper atmosphere. It is not apparent what the long-term effects of this accumulation will be, but indications are that the earth may get warmer and that there will be shifts in global precipitation and temperatures. Consequently, any examination of future energy use should also consider its effects on carbon dioxide production. This section discusses in some detail a model developed to simulate energy-carbon dioxide interactions under various scenarios.

The current state of the modeler's art is such that there is relatively little that one can say about the very long run with much certainty. Elaborate structures cannot be justified over very long periods of analysis, since uncertainty about appropriate values increases with time. This contrasts with short- and midterm models in which fixed factors such as capital stocks and information lags create obvious constraints on energy use and fuel choice.

The long run is very different from either the short- or midterm, and the modeling framework used to assess CO_2-energy-economy interactions must reflect this. In the short- and midterm it may be appropriate to use very detailed process or econometric models. In the very long run, it most certainly is not. In the short and midterm, modelers may feel sufficiently certain of their understanding of first-order effects to grapple with second-order effects. In the very long term, our understanding of even first-order effects is questionable. As a consequence, it is our conclusion that for very long-term modeling, simple models work best.

The term model is used here in its broader sense to imply a consistent approach to developing scenarios. The formal structure developed ensures a basic level of consistency in the energy projections and is designed to facilitate reproducibility and to speed the assessment of alternative scenarios. While the model we have developed is formal in the sense that its equations are explicitly presented and have been translated into computer code, it is neither a particularly large nor a particularly complicated creature. At every decision juncture the simplest possible representation of policy interactions of energy, economics, and demographics have been sought, in the hope that we might create an "open box" rather than the standard "black box" model.

We have also sought to develop this framework with some humility and appreciation for the limitations imposed by the nature of the task at hand and the state of the seer's art. There are uncertainties at every turn. Key exogenous variables, such as population and levels of economic activity, have proved to be extremely illusive forecast targets in the past, and we have no reason to believe that they are now any easier to foretell. Similarly, such model parameters as the price and income elasticities of demand for energy have proved a rich source of economic disagreement.

In short, the future is still impossible to predict. What we hope to do is to construct conditional scenarios that allow the exploration of alternatives in a logical, orderly, consistent, and reproducible manner. This framework is not a crystal ball in which future events unfold with certainty, but rather an assessment tool of specific applicability, which can shed insight into the long-term interactions of the economy, energy use, energy policy, and CO_2 emissions.

MINIMUM MODEL REQUIREMENTS FOR EXAMINING CO_2 EMISSIONS

Despite the desire for simplicity, there are several levels of detail which are required if reasonable energy-CO_2 scenarios are to be constructed: (1) disaggregation by fuel type, (2) very long-term applicability, (3) global scale, (4) regional detail, (5) energy balance, and (6) CO_2-energy flow accounting.

Disaggregation by Fuel Type

All energy is not alike in its emission of CO_2. Nuclear, solar, and hydroelectric power generation contribute no carbon directly to the atmosphere,[1] while coal and western U.S. shale oil are major sources of carbon release. Oil and gas also release carbon in combustion, but are not as important contributors as coal and shale oil (see table 16-1).

In light of the wide disparity among carbon release coefficients of various fuel types, an important element in any carbon release assessment is the composition of fuels consumed over the period.

TABLE 16-1. Carbon Release in the Production and Combustion of Fossil Fuels (grams of carbon per megajoule)

Fuel	*gC/MJ*
Oil	19.2
Gas	13.7
Coal	23.8
Coal gasification	40.7
Coal liquefaction	38.6
Shale oil[a]	41.8
Solar	0.0
Nuclear	0.0
Hydro	0.0

Source: G. Marland, "The Impact of Synthetic Fuels on Global Carbon Dioxide Emissions," in *Carbon Dioxide Review: 1982* (New York: Oxford University Press, 1982).

[a]Western U.S. shale oil from carbonate rock.

Very Long-Term Applicability

The CO_2 problem is long term and requires a long-term analysis. It is unlikely that fossil fuel combustion will culminate in dangerous levels of global warming before the year 2030, even though the policy initiatives necessary to avoid critical accumulations of CO_2 may have to be implemented at much earlier dates. The current terminal analysis date of the model given here is the year 2050.

Global Scale

While some forms of environmental pollution have primarily local effects, the expected climatic changes associated with major CO_2 accumulations do not. The severity and geographical distribution of climatic changes resulting from carbon accumulation depend on the total amount of fossil fuels burned by all global energy users. No major energy consumer can be ignored. Thus, an assessment must include both the centrally planned economies and developing nations.

Regional Detail

Some regions are significant because they are important sources of a major global energy source—for example, the Middle East. Others, such as the European centrally planned economies, are major users of fossil fuels, with uniquely important characteristics. Still other regions, such as North America or the European OECD nations, are important potential sources of energy policy initiatives which would affect carbon emissions. The model uses the breakdown of nine regions that has been followed throughout the book.

Energy Balance

Despite the fact that calculations of carbon release depend directly on the level and composition of global energy combustion, not all energy forecast models can be modified to suit this purpose. For example, much early work in international energy analysis can be categorized as "gap studies." Gap studies forecast supply and demand for energy based on an exogenously specified world oil price path. The general conclusions reached by such studies were that under the price scenarios investigated, there were likely to be deficiencies of global energy supplies. These conclusions were useful in that they provided insights into the energy problem, but neither these studies nor their methodologies were ever intended to address CO_2 issues, and as a consequence are of marginal value in that regard. Such model designs fail to equilibrate global energy supplies and demands across energy use regions. As a consequence, the global oil market may not be in balance and it is impossible to tell how much carbon is released by fossil fuel combustion if production forecasts fall below consumption forecasts. At the very least, an assessment tool must provide a global energy balance to enable consistent CO_2 release scenarios to be generated.

Energy Flow Accounting

Carbon is released at the point of energy combustion, which makes it important to distinguish between primary, secondary, and tertiary forms of energy and to be able to distinguish noncombustion uses (for example, petrochemical feedstocks and asphalt) and flaring. The distinction between primary and secondary energy forms would prove important across scenarios if the role of electricity varied between them and in addition the CO_2 intensity of power plants was either much higher or lower than the economy in general (due, for example, to either heavy dependence on coal—implying high CO_2 intensity in power generation, or heavy dependence on solar and nuclear—implying low CO_2 intensity in power generation). Similarly, it may prove important to be capable of distinguishing coal which is consumed directly from coal used for liquefaction.

The demand for secondary energy types is a derived demand springing from the demand for energy services—tertiary energy. As a consequence, the demand for secondary energy depends, not only on the overall demand for energy and the relative cost of that particular energy modality, but also on the nonenergy costs of transforming that modality into a useful energy form. Thus, coal's low relative cost is frequently more than offset by its associated high capital, labor, and material costs, which often make it a more expensive provider of energy services.

THE MODEL—AN OVERVIEW

Demand

As we have seen, five major exogenous imputs determine energy demand: population, economic activity, technological change, energy prices, and energy taxes and tariffs.

Demographic forecasts are presented and discussed in chapter 2. An estimate of

GNP for each region is used as a proxy for both the overall level of economic activity and as an index of income. While the level of GNP is an input to the system, it is derived from demographic projections of the labor-age population and an assessment of likely labor force participation rates and levels of labor productivity.

Technological change is a parameter which measures the effects changes in the state of knowledge have on the intensity of energy use in the absence of energy price incentives. In the past, technological progress has had an important influence on energy use in the manufacturing sector of advanced economies. The inclusion of a technological change parameter allows scenarios to be developed which incorporate assumptions of either continued improvements or technological stagnation.

The final major energy factor influencing demand is energy prices. Each region has a unique set of energy prices which are derived from world prices and region-specific taxes and tariffs. This is not to imply that all regions are necessarily open to trade in all fuels. The model does have the flexibility to accommodate nontrading regions for any fuel or set of fuels. In fact, it is initially assumed that no trade is carried on between regions in solar, nuclear, or hydroelectric power.

The energy demand module performs two functions. It establishes the demand for energy and its services, and it maintains a set of energy flow accounts for each region (figure 16-1).

The four secondary fuels are consumed to produce energy services. In the three OECD regions (regions 1, 2, and 3), energy is consumed by three end-use sectors: residential/commercial, industrial, and transport. In the remaining regions, energy is consumed by a single aggregate sector.

FIGURE 16-1. Energy flow accounting framework.

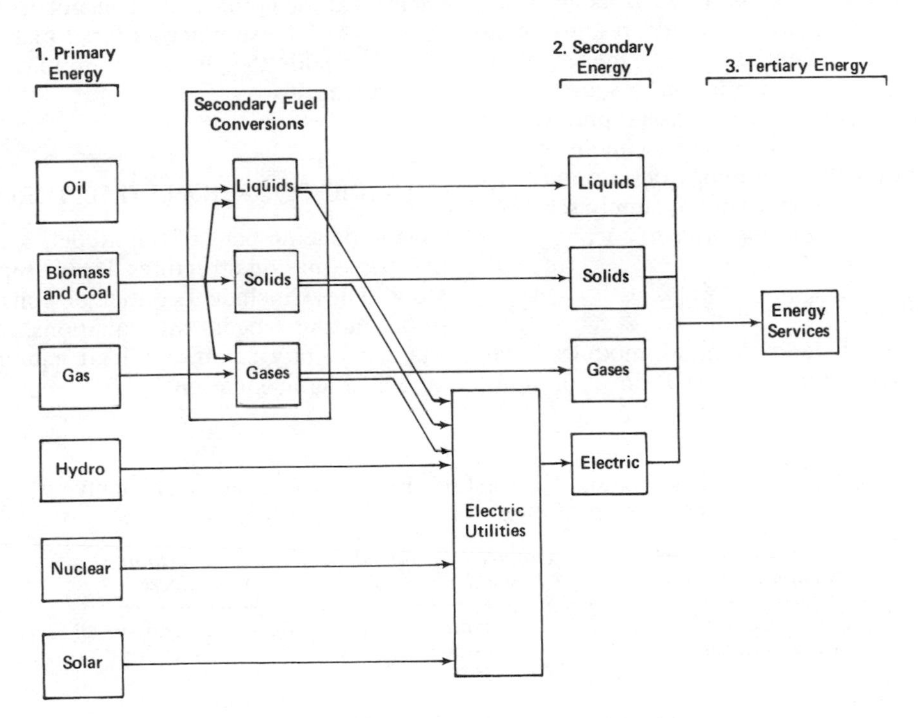

Whereas the flow of energy proceeds from left to right across figure 16-1, the demand for energy is derived from right to left. The demand for energy services in each region's end-use sector(s) is determined by the cost of providing these services, and the levels of income and population. The mix of secondary fuels used to provide these services is determined by the relative costs of providing these services using each alternative fuel. The demand for fuels to provide electric power is then determined by the relative costs of production, as are the shares of oil and gas transformed from coal and biomass. Primary energy demands are then easily calculated from technological parameters. This is shown in figure 16-2.

Supply

The supplies of energy from the six major primary energy types are derived in a less complicated manner than demands. Three generic types of energy supply categories are distinguished, as shown in table 16-2.

The regional supply of resource-constrained conventional and renewable energy modes depends on resource constraints and behavioral assumptions. The supply of unconstrained technologies depends on the technology description (embodied in a conventional supply schedule description) and the price of energy.

ENERGY BALANCE

The supply and demand modules each generate estimates based on exogenous input assumptions and energy prices. If energy supply and demand match when summed across all trading regions in each group for each fuel, then the global energy system balances. But this need not be the case for an arbitrary set of fuel prices. This was one of the chief conclusions of the international energy gap studies. As has already been pointed out, if system-wide supply and demand do not balance, energy prices are adjusted.

First an estimate is made of a set of world energy prices which will clear all the markets. If these prices reduce all energy gaps, then a test is made to see if equilibrium has been achieved. If equilibrium is not yet achieved, then a new set of equilibrium prices are estimated and the process continues. This process is illustrated in figure 16-3, which displays the interactions necessary to achieve a global energy balance.

Once a global energy balance has been achieved, it is necessary to go back through the system and apply carbon emission coefficients at the points of combustion (see figure 16-4). The sum across fossil fuels and regions yields the global carbon release in any period.

GENERIC EQUATION STRUCTURES

In the development of the model, several generic equation structures appear repeatedly. These include Leontief or constant input-output coefficient equations, constant elasticity equations, logit equations and averaging equations.

TABLE 16-2. Distribution of Long-Term Energy CO_2 Assessment Supply Technologies across Supply Categories

Resource-Constrained Conventional Energy	*Resource-Constrained Renewable Energy*	*Unconstrained Energy*
Conventional oil	Hydro	Unconventional oil
Conventional gas	Biomass	Unconventional gas
		Solids
		Nuclear

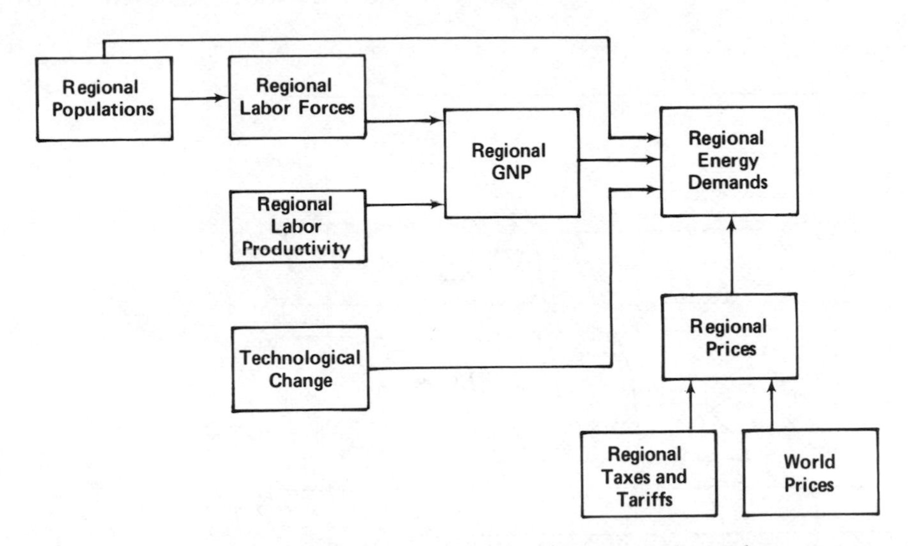

FIGURE 16-2. Key components of long-term, global region, energy demand assessment module.

FIGURE 16-3. The IEA/ORAU CO_2 emissions model.

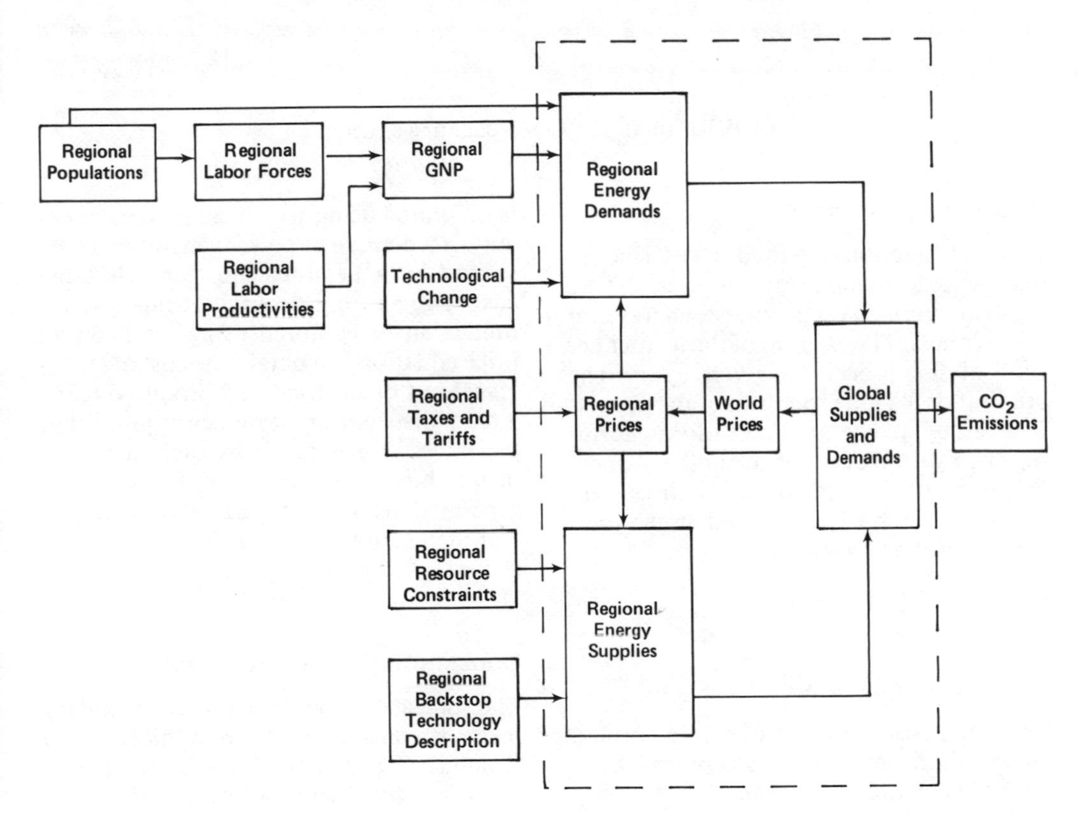

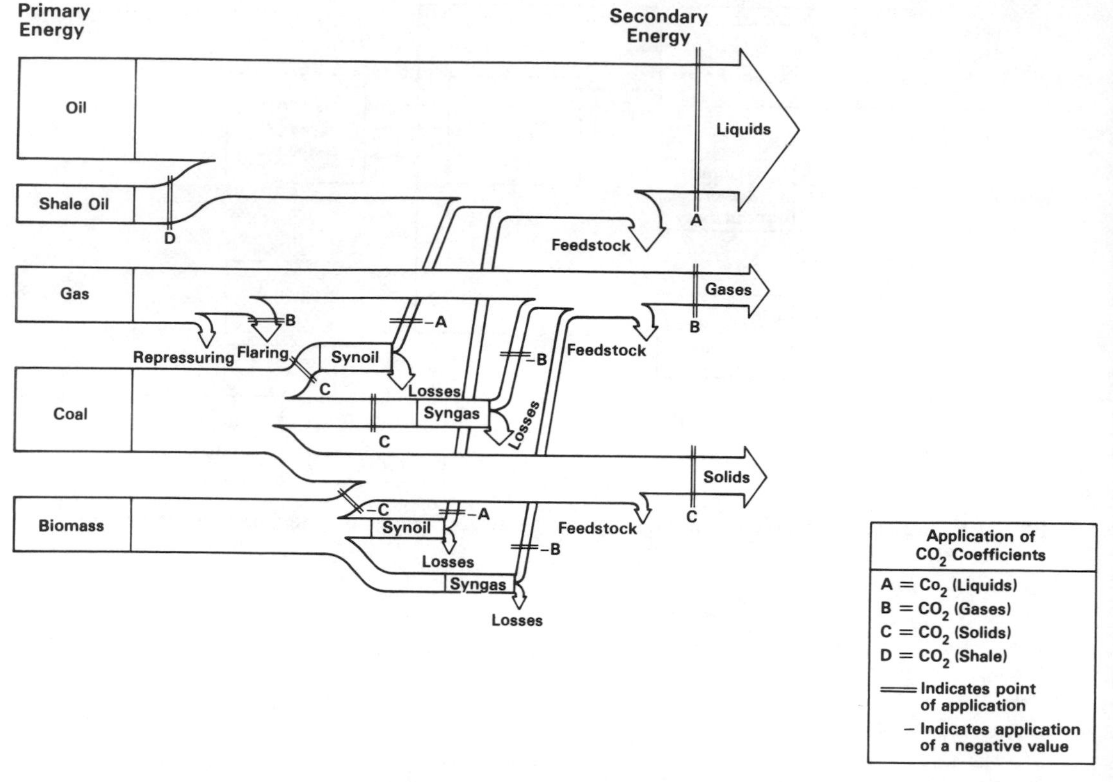

FIGURE 16-4. The CO_2 accounting structure.

Leontief Equations

Leontief equations are used to describe energy transformation technologies. There are two inputs to each process, energy and other inputs. The amount of the former per unit of the process is given by g. The amount of the latter per unit of the process is given by h. The total cost of transforming energy from one type to another, for example coal to gas, P^*, depends directly on the cost of energy inputs P and on the coefficients of the process, i.e.,

$$P^* = gP + h$$

Averaging Equations

In some cases the price of a transformed energy mode may not be captured by a Leontief equation. For example, electricity is generated using six different types of energy. Each mode of power generation is described by a Leontief equation. Thus the cost of generating electricity using a single mode can be captured using the Leontief price equation. To obtain the cost of an average unit of electricity P^* produced using a fixed distribution of modes requires that the prices of generating by each mode, P_i, be weighted by the fraction of total power generated using that particular mode, s_i, and then summed.

$$P^* = \sum s_i P_i$$

Constant Elasticity Equations

The forecast of the demand for secondary fuels is constructed from a hierarchical structure. The demand for total energy services is derived and the share of this market captured by each of the four secondary fuels is determined. The demand for total energy services is derived from a constant elasticity equation, generally written as,

$$Z = AX^B Y^C$$

where the dependent variable Z (for example, aggregate energy service demand) is a function of the variables X (for example, the price of aggregate energy services) and Y (for example, income), and where A is a scale term, and B and C are parameters. The constant elasticity equation gets its name from the fact that the parameters B and C are elasticities. That is

$$B = d \ln Z / d \ln X$$
$$C = d \ln Z / d \ln Y$$

and are thus, respectively, the percentage change in Z (energy service demand) resulting from a 1 percent change in either X (price) or Y (income). Thus B and C would be price and income elasticities respectively.

The constant elasticity equation structure offers a first-order approximation to an arbitrary energy service demand function at a point. That is, it can replicate not only the total demand for service, given an arbitrary set of exogenous variables, but also provides a log-linear estimate of the effect on total demand of a change in these independent variables. It does not provide a second-order approximation of an arbitrary demand function at a point. That is, it does not provide an estimate of the rate at which elasticities change as exogenous variables change for an arbitrary demand function. As we have argued earlier, such an additional level of detail cannot be supported in a very long-term modeling framework.

It is popular to argue that first-order, and in particular constant elasticity, equation structures are inadequate tools for demand analysis. Second-order approximations are preferred. The argument against constant-elasticity equations often is that the elasticities do not change as a function of the independent variables, and that in reality there is no necessity for this particular relationship to be maintained. There are two fundamental problems with this argument. First, there is insufficient information to develop second-order approximations in the context of demand model. Second, even if one could develop a second-order approximation, the current limits of functional forms preclude extending *any* order of approximation to an arbitrary function beyond a single point.

There are no functions that can describe accurately the behavior of elasticities beyond the range of historical experience. And, even within the range of historical experience, different functions impose their own idiosyncracies on measured elasticities. In short, one may safely say that elasticities need not be constant, but little more. One cannot safely predict exactly how or how much elasticities will vary with price and other pertinent variables when these variables stray beyond historical values. In fact, one can challenge *any* demand equation structure on exactly the same grounds.

The constant elasticity equation has the merit of being a first-order approximation to an arbitrary relationship at a point, and furthermore, a well-behaved one. That is, it will never give a priori counterintuitive results. We therefore argue for the reasonableness of such equations in the context of a long-term, global framework.

The Logit Structure

The lower level of the hierarchy, fuel shares, is modeled by a logit structure introduced by McFadden.[2] The key question which the logit equation structure answers is: What share of total energy services are provided by a particular fuel mode? This methodology is used by numerous energy modelers, including Baughman and Jaskow,[3] Anderson,[4] Baughman and Zerhoot[5] Lin, Hirst, and Cohn,[6] and Reister, Barnes, and Edmonds.[7]

The basic idea behind the logit fuel share function is illustrated in figures 16-5, 16-6,

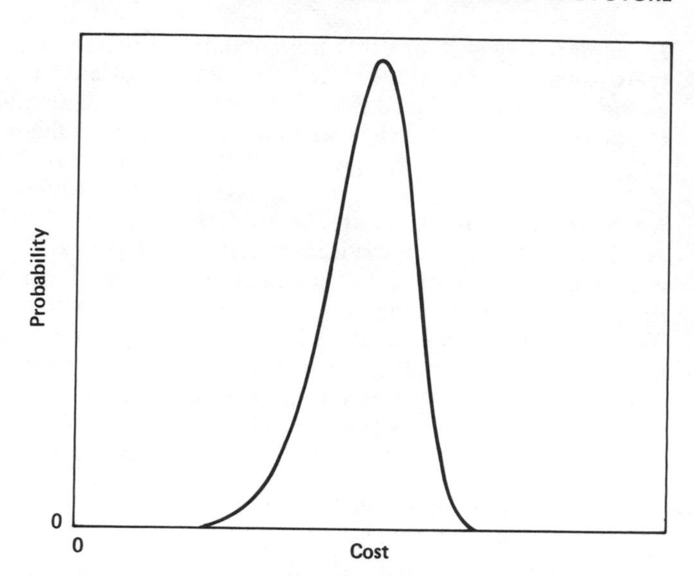

FIGURE 16-5. Cost distribution for a mode of energy service production.

and 16-7. The key concept is that of the cost distribution. That is, in a given region, the cost of a fuel, and hence the cost of providing an energy service using that fuel, varies. This distribution of costs is the result of differing transport costs, tax structures, and associated factor costs. As a consequence, one also observes that the cheapest mode of production does not always capture the entire market. Figure 16-5 shows a cost distribution for a single mode of providing an energy service. The cost distribution shows alternative costs of providing a given energy service and the associated probability that one will observe that cost in a given region. The share of the market captured by that mode depends on the cost of the competing options. For example, natural gas and coal might compete as boiler fuels. Figure 16-6 shows the share that one mode (natural gas) captures as a function of the cost of its competitor (coal). Note that the relevant cost is not simply the fuel cost, but the entire cost of providing a given energy service.

Figure 16-7 displays cost distributions for three options.[8] The cost distribution for

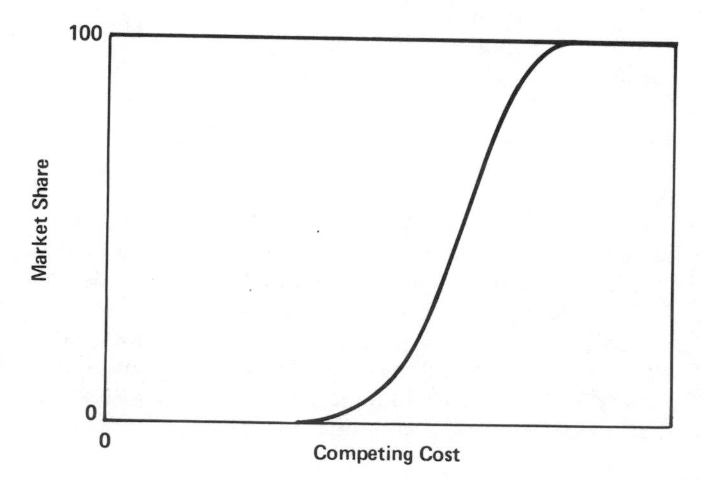

FIGURE 16-6. Market share for a mode of energy service production as a function of the cost of a competing mode.

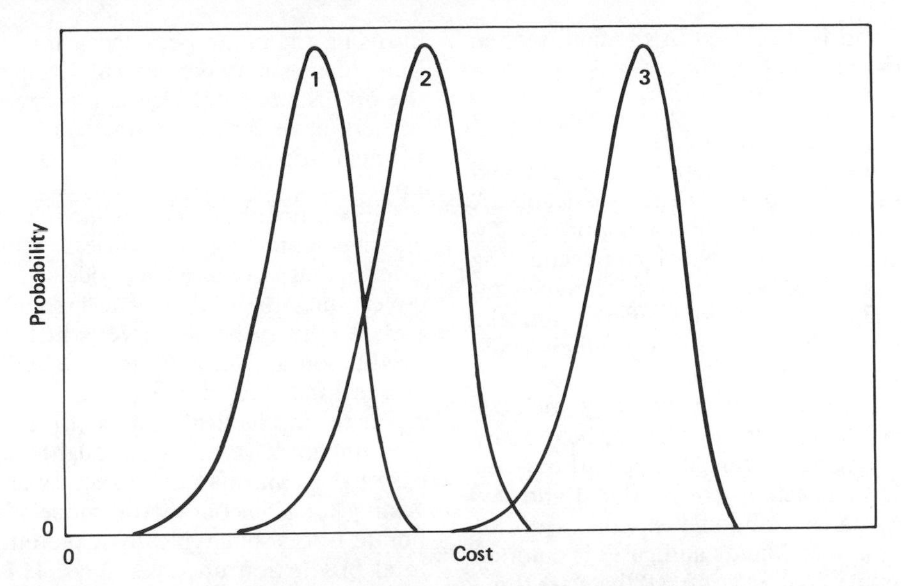

FIGURE 16-7. Cost distributions for three options.

option 3 does not overlap the distribution for option 1; consequently, option 3 cannot capture any of the market from option 1. Options 1 and 2 overlap and will split the market. Option 1 will capture more than half of the market, but option 2 will have a significant share of the market.

The logit share function is attractive because it is a simple, well-behaved function which can replicate real world market behaviors. Market shares for all modes of energy service production must always sum to 100 percent. As the price of a mode of energy service production increases, market share captured by that mode diminishes. In addition, all modes of producing energy services are substitutes for one another. Thus, as the price of a mode increases, it loses some of its market share to each of the other competing modes.

EQUATION SPECIFICATION FOR THE MODEL

The Supply Module

The supply module forecasts supplies and prices of the six major primary energy categories for a given region in a given period. Its inputs are the prices of the three major traded fuels (oil, gas, and coal), and the energy–GNP demand ratio for the two major renewable resources (nuclear and solar). Prices and outputs for the final primary energy category, hydroelectric generation, are introduced exogenously. There are three generically different technologies considered in the module: "Resource-constrained, exhaustible" energy technologies, "resource-constrained, renewable" technologies, and "backstop" technologies.

Four of the six primary energy technologies fall clearly into one of these three categories. Hydroelectric power generation (and including wind and geothermal) is a resource-constrained renewable energy source, while coal, nuclear, and solar are backstop technologies in the time frame of this analysis. The remaining two energy technologies, oil and gas, are hybrids; each has a resource-constrained component and a larger resource category which is unconstrained in the time frame of the analysis and is available only at greater cost. Biomass has characteristics of both resource-constrained renewable and backstop technologies. It is handled somewhat differently and is discussed in the final section of this chapter.

Primary and Unrefined Energy

Each of the three fossil fuels—oil, gas, and coal—forms a primary energy category. A distinction is made between primary energy and a category which, for accounting purposes, has been termed refinable energy. The two categories differ in that the latter includes coal-derived synfuels while the former excludes them.

The primary energy category includes energy at the extraction stage, before any processing has occurred. Thus primary oil and gas consists of conventional and unconventional components. Coal used for conversion to liquids and gases is counted as primary coal. In contrast, there are three related categories: refinable liquids, refinable gases, and refinable solids. Refinable liquids and gases include both primary energy, and the energy content of coal liquids and gases input before the final refining process. Refinable solids include only that coal which is eventually used in the form of solid fuels. It excludes coal production for synfuel conversion. Biomass enters with coal as a solid primary fuel with the potential to be converted to refinable liquids, solids, or gases.

Resource-Constrained Exhaustible Technologies

Oil and gas resources are disaggregated into two grades of resource—conventional and unconventional. The conventional components of oil and gas supply are resource constrained. By definition, all conventional resources are profitable to produce at current price levels. There are, however, real-world constraints which prevent this grade of resource from being consumed immediately. The resource must first be found. This requires search procedures and drilling operations. In the short term, drilling equipment is fixed and the ultimate intensity of its use is limited physically. In the longer term, there are costs of building equipment which are minimized by using the equipment over time rather than in one massive search. Even where the oil resources are well known, resource owners have incentives to disburse their product over time so as to maximize profits.

Conventional oil and gas models can be classified into three categories: extrapolation models, econometric models, and discovery process models. The first of these classes is simplest. A curve is fit between production and time or reserve additions and drilling. Such models mete out a fixed resource mechanically over time. They contain no price effects, though they are easy to use and have had success as forecasting tools.[9] Econometric models incorporate prices, but typically leave ultimate total production unconstrained. This is a distinct disadvantage in a long-term framework. Discovery process models are the most sophisticated representations of conventional oil and gas supply. They individually model the process of exploration, additions to reserves, and production from reserves. While these models have performed well as explanatory and forecast models,[10] they are clearly not simple. Not only must prices, resources, and discovery constraints be incorporated, but so also must expected future prices. Expectations about future prices and reactions of other actors are especially important in the formulation of cartel supply models.

Process discovery models are clearly the most intellectually satisfying of the three modeling categories. They yield insights into the process by which supply is created and are reasonable forecasting tools. They are especially appealing in an economic context since price plays an important endogenous role in the analysis. This is not to say that they are without difficulties. To some extent, the discovery process model pushes all of the interesting questions surrounding production into discovery rate parameters. Thus, "the discovery process model relies on curve fitting just as heavily as any of the curve fitting extrapolation models.[11] A discovery process model will eventually be designed for long-term energy CO_2 assessment.

Initially, however, supply is determined by a simple extrapolation model. The theoretical difficulties of such models are well recognized. They lack any behavioral insights. They are noneconomic in orientation, and the particular function chosen to represent the production time path cannot be justified on physical grounds. Nonetheless, as noted earlier, such models have had good success as forecasting tools. They also are simple and sufficiently flexible to accommodate alternative scenarios for remaining resources. Finally, it is worth noting that the alternative real world supply considerations may be implemented exogenously through the various resource and production rate parameters of the supply schedule.

Production of the constrained resource is handled conventionally via a logistics function, which relates the cumulative fraction of the total resource base which has been exploited, $f(t)$, to time. The relationship is given by

$$\frac{f(t)}{1 - f(t)} = \exp(a + bt) \tag{1}$$

where a and b are parameters, and t denotes time elapsed from an initial period.

This implies that the fraction of the resource exploited by the period t is given by

$$f(t) = \frac{e^{a + bt}}{(1 + e^{a + bt})} \tag{2}$$

The initial resource base to be exploited over all time is denoted by R. The total amount of the resource exploited by time t is given by $R\,f(t)$. This is different from the rate at which the resource is being produced at a given point. The rate of production Q_s is given by,

$$Q_s(t) = f(t)\,[1 - f(t)]\,bR \tag{3}$$

The time path of production is shown in figure 16-8. The initial fraction of resources that were used up in the initial period t = 0 is simply,

$$f(0) = \frac{e^a}{1 + e^a} \tag{4}$$

while initial production is

$$Q_s(0) = \frac{e^a}{(1 + e^a)^2}\,bR \tag{5}$$

The maximum production rate occurs where $f(t) = 1/2$, and $t = -a/b$.

For regions other than the MIDEST, the logistics equation is used to forecast the production of conventional oil and gas.

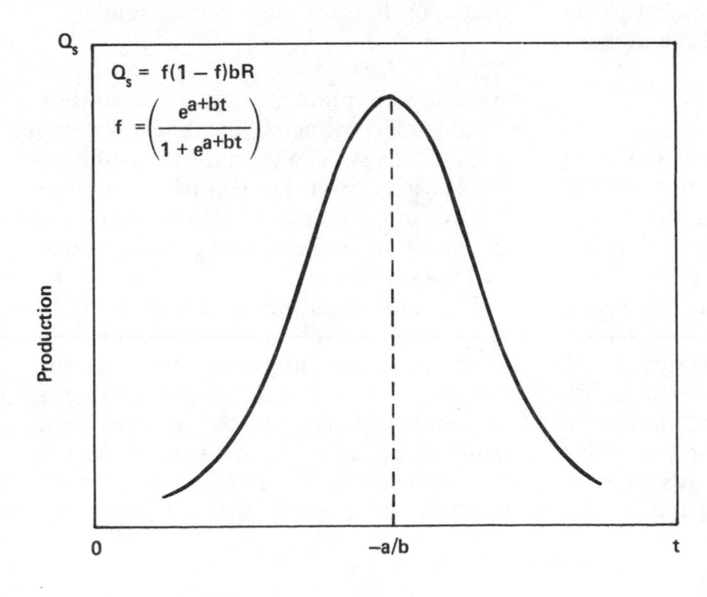

FIGURE 16-8. Production over time of an exhaustible resource using a logistics function.

The level of MIDEST output is assumed to be determined by OPEC policy, and that policy is an exogenous input to the supply module, as is the time profile of the rate of natural gas flaring. For example, a constant level of output is frequently cited as a likely production scenario for this region. Other supply scenarios are also possible.

It is important to point out that while each of the individual regional production time paths may be described by figure 16-8, the global production time paths may not. In fact, the regional pattern of resource distribution is likely to skew the global production time path to the right, with a "fatter" tail and earlier peak than would be obtained from a global logistics representation of production.

For resource-constrained technologies, supply does not respond to price. Production rates are assumed to follow the logistics path, and the total supply is offered without regard to market conditions. The same is not true of backstop oil and gas supplies, which are offered to the market on the basis of market prices and overall rate of economic activity.

The one exceptional resource-constrained technology is hydroelectric generation. Here the level of resource exploitation is given exogenously along with production costs. Both the price and quantity of this resource are passed on to the electric utility components of the demand module.

Natural Gas Flaring

Natural gas is a premium fuel frequently found in association with petroleum. Despite its end-use attractiveness, the market for natural gas was slow in developing, due to problems with transportation and storage of the fuel. As a consequence, associated natural gas was often flared or vented. As the natural gas market has developed, the fraction of gas flared or vented has diminished. In OECD countries, the market has developed to the point where most gas is introduced into a natural gas pipeline system for distribution; flaring and reinjection are not important considerations. Some gas is now liquefied, in a more costly process, and transported by sea. Natural gas markets remain underdeveloped in LDCs, however.

Economic considerations have been important to the development of gas markets. Prices now offer sufficient incentive to market gas which once would have been flared. As a consequence, the fraction of gas flared, capped, and reinjected is expected to continue to diminish. This has been modeled in the framework as follows: The amount of gas that is flared, f, is expected to continue to decline from a present rate of b, to an ultimate rate of a over a period of T years. The transformation is modeled as an exponential interpolation

$$f(t) = a^s b^{(1-s)} \qquad (s = t/T) \qquad (6)$$

where t is the number of years of adjustment already experienced. Note that this equation structure diminished flaring more rapidly in early periods than in later periods. If a and b are equal, as they are in the OECD, the flaring fraction is constant. It is also important to point out that while price is the driving motivation for reduced flaring of gas, it is not included specifically because the magnitude of the gas resource at issue is insufficient to warrant attention to second-order effects which would either hasten or dampen the primary trend.

Backstop Technologies

Backstop technologies are, by definition, capable of producing inexhaustible supplies of energy. The term inexhaustible applies strictly to the context of the analysis. Backstop technologies include unconventional oil, unconventional gas, coal, solar, and nuclear energy.

The traditional use of the term "backstop" implies a resource which can be supplied with an infinitely elastic supply schedule.[12] A backstop technology then is an industrial analog to the perfectly competitive firm in economic theory. That is, the industry is so small relative to the economy as a whole that its production cannot affect the price of its resources or the long-term price of output.

The methodology chosen for use here is somewhat more sophisticated than that used to model the simple backstop concept, but contains that simplification as a special case. The specification used here departs from the simple backstop concept in that it introduces the concept of a normal rate of growth. For a purely produced good with no special input requirements, this norm might be the growth rate of the economy as a whole. For breeder reactors, this norm might be derived from the breeding ratio. For shale oil it might be some other reference rate. If the backstop energy sector attempts to grow more rapidly than its "normal" rate, costs are bid up in the short term. If the sector then returns to its normal or base growth rate of expansion, costs of production fall back toward the long-term backstop price P^*. Backstop supply prices can remain significantly above or below this price only if the sector continues to expand at a rate different from the normal rate.

A relatively simple equation structure is used to relate three parameters: a breakthrough price a below which no outut will be forthcoming; a "normal" backstop price P^* which is determined by the parameter b; and a short-term price elasticity control parameter, c. The supply equation is specified in terms of production costs P and the ratio of output Q_t to a base Q^*, $g = Q_t/Q^*$,

$$P = ae^{(g/b)^c} \tag{7}$$

This equation is depicted graphically in figure 16-9. If the long-term rate of expansion of supply matches the normal rate of expansion, then the backstop technology supply schedule in the long term, $S(Q)$, is infinitely elastic. An infinite amount of supply is available at the price P^*. Over the short run, the industry may expand either more rapidly or more slowly than this base rate. If production exceeds the base level, then short-term costs rise, forcing prices up. If, on the other hand, the industry fails to attain its base production normal rate, prices tend to fall and the industry moves back along its short-term supply schedule. There is a limit, however, to how low prices can fall. There is a shutdown price P_{min} which is given by a.

Note that the short-term supply schedule shifts over time with the base supply level Q^*. Note also that the *marginal* cost P of producing a given supply depends only on the growth rate of output in that period, and that the shutdown price a is independent of output and growth rates.

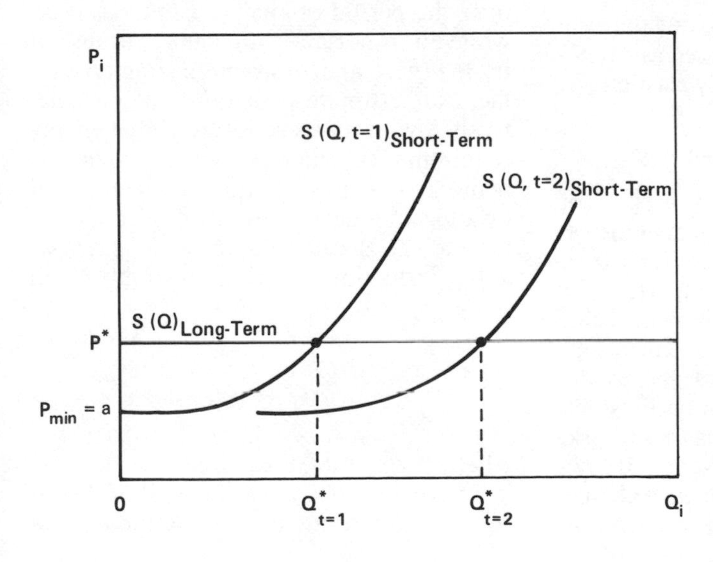

FIGURE 16-9. Short-term and long-term supply schedules for backstop technologies.

The short-term price elasticity of demand E is obtained by logarithmically differentiating equating (7), which yields

$$E = \frac{\alpha \ln Q_i}{\alpha \ln P_i} = \frac{(g/b)^{-c}}{c}$$

Note that the elasticity becomes infinite at the shutdown price ($g = 0$) and completely inelastic as g becomes large.

It is also worth pointing out that in the limiting case, where c approaches infinity, the distinction between the short-run and long-run elasticity vanishes, and the two curves merge.

There are three key parameters to be determined—a, b, and c. The first, a, is simply the short-run, shutdown price of the industry. To determine b and c, first note that at the reference price P^* realized and base production are equal, so that,

$$\ln(P^*/a) = b^{-c}$$

The parameter c may then be expressed in terms of the shutdown price a, the reference backstop price P^*, and the short-term price elasticity of supply E, via

$$c = [E \ln (P^*/a)]^{-1}$$

The parameter b then is found by simple substitution:

$$b = [\ln (P^*/a)]^{(-1/c)}$$

It is finally worth pointing out that the supply schedule can be expressed as a function of price a, Q^* and supply elasticity at the base price E^* by,

$$Q_t = Q^* [\ln (P/a)/\ln (P^*/a)]^{E^* \ln (P^*/a)}$$

Another important characteristic of backstop technologies is technological change. This is less important for a fuel such as coal, and extremely important for a technology such as photovoltaic cells. With technological change, the entire supply schedule shifts downward and to the right. There are a number of ways in which this can be represented. In this framework, technological change is treated as if it lowered the entry price. Technological change is "phased in" over a period of length T. This is described as a decrease in the minimum cost of the technology, a, from an initial value of a_1 to an ultimate minimum of a_2. The transition is carried out in T periods using the formula

$$a = a_1^{(1-t/T)} a_2^{(t/T)}$$

Thus in the initial period ($t = 0$) $a = a_1$ and in the final period ($t = T$) $a = a_2$. (Note, however, that the period of transition can begin and end at any chronological time and that $t = 0$ is used only for expository convenience.)

Resource-Constrained Renewable Technologies

Electricity generated from hydropower, geothermal power, and wind can best be represented in the period of analysis as a category of primary energy characterized by a permanent flow with an ultimately limited contribution to global supplies. Hydroelectricity is, by far, the dominant technology in the category. The resource has an ultimate limit determined by physical constraint. While economic considerations could elicit a marginal supply response, the overwhelming share of the resource could be available at prices below existing electricity prices. Exploitation of the full resource will be gradually phased in over the period of analysis. The resource, while an important contributor because of its low cost and desirable characteristics, faces an ultimate constraint that is relatively low in terms of future global energy requirements. The resource is modeled as being phased in over time as determined by a logistics curve, described in equations (1) and (2). Because the resource is renewable, production in period t is simply given by

$$Q_s(t) = \frac{e^{a+bt}}{\left(1 + e^{a+bt}\right)} R$$

where R is the total resource.

This is not to say that the size of the resource and the rate of exploitation are not

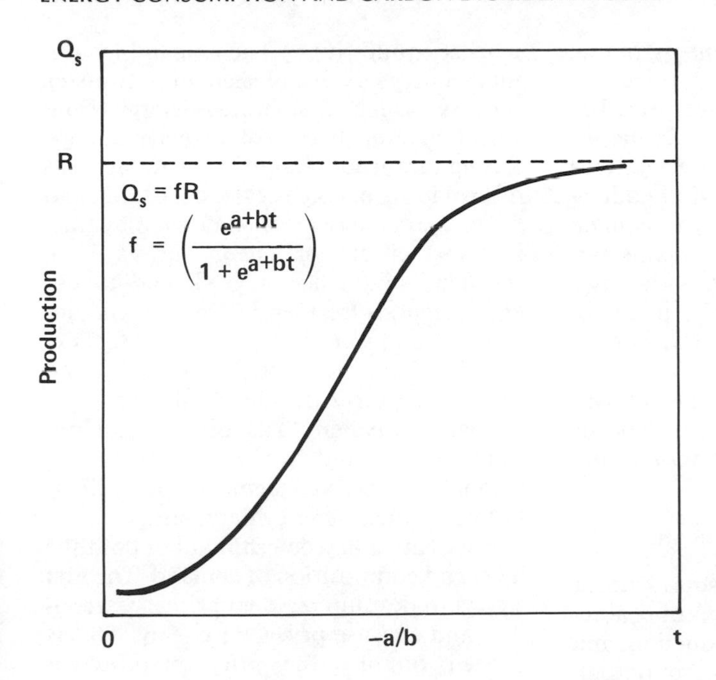

FIGURE 16-10. Production over time of a renewable resource using a logistics function.

dependent on price and profitability. Rather, the size of the resources in question are small by world standards, and the quantitative descriptions of price responsiveness are not well known. Because these economic considerations cannot be expected to add significantly to the degree of accuracy in the estimates of global CO_2 emissions, a first-order approximation of resource-constrained renewable supply was adopted.

The path of exploitation described by the logistics curve is depicted in figure 16-10. The logistics curve describes a path where the share exploited approaches 100 percent of the total resource.

Unrefined Liquids and Gases Production

While coal, nuclear,and solar are treated as pure backstop technologies and hydro is treated as a pure resource-constrained technology, oil and gas are hybrids having elements of both. In addition, unrefined liquids and gases are each made up of three constituents: conventional or resource-constrained supplies, unconventional supplies, and synfuel derivatives of coal. Synfuel derivatives are an intermediate energy good, representing a transformation from one energy type—coal—to another—oil or gas—using energy and resources in the process. Total unrefined supply is the sum of these three elements.

All primary fuels are all eventually refined into secondary fuels. Primary oil is refined into the secondary fuel liquids, while all gas is refined into secondary gases. Coal is unique. Some coal is refined into the secondary product, solids, but some is treated as if it were prerefined into a primary oil or gas equivalent. All three types of oil and gas inputs (conventional, unconventional, and synfuels), are then aggregated and jointly refined into secondary liquids and gases. While a specific coal conversion process may not be decomposable into these two steps, from an accounting standpoint it is treated as if it were.

The conventional component of oil and gas production is obtained from equation (3) from the logistics model or via exogenous assumption. The unconventional supply of the resource is obtained from equation (7). This is accomplished by writing the expression with the energy output ratio as a function of the supply price.

The total supply of coal is determined by equation (7), but as the previous discussion implies, this supply is trisected. One part is converted into primary oil equivalents, one part into primary gas equivalents, and the third part remains as primary coal. The size of the first two components depends upon the prices of oil, gas, and coal, and on the transformation technology.

The cost of producing one unit of synfuel as a substitute for oil or gas depends on the price of coal, the technology, and nonenergy costs. That is,

$$P_{ic} = g_{ic}P_c + h_{ic} \qquad i = \text{oil, gas} \qquad (8)$$

where P_{ic} is the price of a coal substitute for primary fuel i, g_{ic} is the amount of coal required per unit of synfuel production, and h_{ic} is the nonenergy cost per unit of output.

The share of coal allocated to the production of fuel substitute i, S_{ic}, is given by the logit share equation,

$$S_{ic} = \frac{(P_{ic}/P_i)^{ri}}{1 + \sum_{j=1,2} (P_{jc}/P_j)^{rj}} \qquad (9)$$

The supply of refinable fuel i derived from coal, Q_{ci}, thus equals

$$Q_{ci} = \begin{cases} S_{ic}Q_{\text{coal}}/g_{ic} & i = \text{oil, gas} \\ (1 - S_{1c} - S_{2c})\, Q_{\text{coal}} & i = \text{coal} \end{cases} \qquad (10)$$

The Demand Module

The demand module calculates the primary fuel requirements by type for a given region in a given period. The key inputs to the determination of demand are the level of population, level of economic activity (GNP), and prices of primary energy types. (Though prices are exogenous to the demand module, they are endogenous to the analysis framework, a point to be discussed in greater detail in connection with market equilibrium.) The demand for primary energy is established in a two-step process which first traces energy from world market prices for primary energy, through transport and refining, to the costs of providing energy services. The demand for primary energy is then derived by tracing the effects of energy prices back through its influence on GNP, end-use energy demands (residential/commercial, industrial, and transport), secondary fuel demands (liquids, gases, solids, and electricity), through refinery demands for primary equivalent fuels (oil, gas, coal, nuclear, solar, and hydro), and, finally, through the derived demand for synfuels to total primary energy demands.

Prices are a key determinant of both the level and composition of demand. The first step is to develop regional primary, secondry, and end-use prices for energy. This is carried out by the price preprocesser submodule.

The Price Preprocessor

The price preprocessor submodule determines the regional price for each of the three primary fossil fuels (oil, gas, and coal) from the world market prices for each. Next, it calculates regional prices for the four secondary fuels, using the regional prices of all six primary energy types as inputs. Secondary fuel prices are handled in a two-step process. First, the cost of refined fossil fuels (liquids, gases and, solids) is computed. Next the cost of electricity generation is computed using refined (secondary) fossil fuel prices and production costs for nuclear, solar, and hydroelectric facilities as inputs.

The price of traded fuels i in region m, P_{im}, depends on the world market price, P_i, the transport costs associated with that fuel, TR_i, and taxes or subsidies applied to fuels, TX_{im}, in that region. Taxes and subsidies are assumed to be applied proportionately to the landed price of energy so that

$$P_{im} = (P_i + TR_i)\, TX_{im} \qquad (11)$$

In the initial version of this framework, only fossil fuels are assumed traded across the regional boundries. Interregional trade in secondary electricity is almost nonexistant. The cost of producing secondary fossil fuel type j using the associated refinable energy input is given by

$$P_{rj} = P_j g_j + h_j, \quad j = \text{liquids, gases, solids} \qquad (12)$$

where g_j is the unrefined input necessary to produce one unit of secondary fuel j, and h_j are the nonenergy refining costs associated with a unit of secondary fuel j.

Electricity is handled separately. Electricity is generated using the three refined fossil fuels and the three primary electricity sources—nuclear, solar, and hydro—as energy inputs. The cost of producing a unit of electricity using one of these six fuels is denoted P_{ej} and is calculated in a manner analogous to that used in equation (12),

$$P_{ej} = BP_j P_j g_{ej} + h_{ej} \qquad j = 1, \ldots, 6 \qquad (13)$$

where j is either refined fossil fuels or primary electricity, g_{ej} is the fuel requirement for a unit of electricity, h_{ej} is the associated nonenergy cost of generation, and BP_j is a scale factor relating the average cost of refined fuel j to the price paid by electric utilities.

Five fuels compete via cost for a share of the electricity generating market—secondary liquids, gases, and solids, nuclear, and solar. Hydro electricity supply is determined exogenously, as is its cost. The market share of the five fuel types is determined by a logit framework. The share of the market captured by the jth supply technology is determined by expected cost and associated probability density function, via

$$S^*_{ej} = \frac{b_{ej} P_{ej}^{r_e}}{Z} \qquad i = 1, \ldots, 5 \qquad (14)$$

where

$$Z = \sum_j b_{ej} P_{ej}^{r_e}$$

and the parameter b_{ej} is the base market share of fuel j, and r_e is a measure of the variance of the cost function distribution. The market share for fuel j is related to the market share associated with hydro, S_{eh}, via

$$S_{ej} = (1 - S_{eh}) S^*_{ej} \qquad i = 1, \ldots, 5 \qquad (15)$$

The cost of generating a unit of electricity depends on the mix of modes used. Denote the electricity cost as P_e; then the relationship between P_e and modal prices is

$$P_e = \sum_{i=1} S_{ej} P_{ej} \qquad (16)$$

Equations (12) and (13) define the derivation of prices, while equations (14) and (15) determine the market shares.

The final function performed by the price preprocessor is the computation of energy service prices. There is one energy service price for each end-use sector, and an overall aggregate energy service price. There are two classes of detail considered in the model. OECD regions distinguish three separate end-use sectors—residential/commercial, industrial, and transport. For these regions there are three separate energy service prices. Non-OECD regions are not disaggregated and contain only one end-use sector. The cost of energy services in a sector is given by

$$P_k = \sum_j S_{jk} P_{jk} PK_{jk} \qquad (17)$$

where P_{jk} is the cost of providing energy services to end-use sector k using secondary fuel type j (the secondary fuel types are liquids, solids, gases, and electric), P_{jk} is the sum of direct energy and nonenergy costs divided by a base price, BP_{jk}:

$$P_{jk} = (P_j g_{jk} + h_{jk}) \,/\, BP_{jk} \qquad (18)$$

and PK_{jk} is the relative cost of services provided by fuel j to the overall service price. Again, the use of price indices necessitates common units of measure.

The overall aggregate energy service cost PS is a weighted sum of individual energy service prices, or

$$PS = \sum_k S_k P_k / BPS \qquad (19)$$

where *BPS* is the base price of aggregate energy services.

Having completed the computation of all secondary and tertiary (energy service) prices, the demand module uses these in determining secondary and tertiary energy demands.

Determining Secondary Energy Demands The GNP is one of the principal determinants of energy demands, but energy can conversely have an effect on the GNP. To reflect this, the base case GNP is adjusted for the overall level of energy service price to allow for this two-way interaction. This is accomplished through a simple elasticity relationship,

$$Y = GNP \cdot PS^{r_y} \tag{20}$$

where r_y is the percentage change in the GNP resulting from a 1 percent increase in the cost of providing energy services, and Y is the adjusted GNP level.

This formulation yields a first-order approximation to the impact of energy on GNP. These effects would likely change with both the levels of energy prices and GNP, and these changes in turn would likely go on at a nonconstant rate. Unfortunately, there are no clear empirical or theoretical grounds upon which to determine either the direction or magnitude of second- and third-order GNP feedback effects. Even the magnitude of first-order effects is in question, though for most values of expected energy prices, magnitudes of r_y are expected to be relatively small.[13] As a consequence, the representation of the energy price feedback given in equation (20) was deemed to be as accurate as current empirical research can reasonably support.

The total demand for energy services is determined using income and energy service price elasticities, r_{pk} and r_{yk}. The demand for energy services for the residential/commerical and transport sectors in OECD economies is given by

$$E_{sk} = P_k^{r_{pk}} \cdot X^{r_{yk}} \cdot POP \tag{21}$$

where E_{sk} is the total demand for energy services, X is a per capita GNP index, and POP is the population size index. Non-OECD regions and the OECD industrial sector are indexed to the total level of economic activity, Y, so the computation becomes

$$E_{sk} = P_k^{r_{pk}} \cdot X^{r_{yk}} \cdot Y \tag{22}$$

The fraction of energy services provided by each fuel type depends on the relative cost of providing those services, and the level of income. Both income and price effects are considered. These are derived from the logit share structure so that the service share for fuel j in sector S_{jk} is given by

$$S_{jk} = b_{jk} \cdot P_{jk}^{r_{pk}} \cdot X^{r_{yjk}} / Z_k \tag{23}$$

where

$$Z_k = \sum_j b_{jk} \cdot P_{jk}^{r_{pk}} \cdot X^{r_{yjb}}$$

and where price and income elasticities are determined by the power terms r_{pk}, and r_{yjk}, and b_{jk} is the base service share captured by the fuel j.

The demand for each secondary fuel in a sector is identically equal to the product of total service demand with the fuel service share and this value multiplied by the fuel requirement per unit service, divided by the level of technological improvement or

$$F_{jk} = g_{jk} \cdot S_{jk} \cdot E_{sk} / TECH_{jk} \tag{24}$$

The level of technological progress is added to account for the fact that technological progress has acted to conserve energy even when energy prices fell. Needless to say, this factor may or may not be an important source of energy conservation in the future. This specification allows for both continued progress and stagnation to be explicitly considered.

The region's total demand for the secondary fuel type j is found by simply summing over the sectors:

$$F_j = \sum_k F_{jk} \tag{25}$$

The share of total energy services produced in the region by the kth sector is found by computing

$$S_k = b_{sk} \cdot E_{sk} / Z \quad (26)$$

where

$$Z = \sum_k b_{sk} \cdot E_{sk}$$

and b_{sk} is the base case energy service weight.

The Determination of Primary Energy Demands

Primary energy demands may be inferred from the demands for secondary energy sources, and the information on fuel transformations. Fossil fuel demands for primary energy depend on the demands from end-use sectors, the electric utilities, and synfuel conversions from coal. The demand for fossil fuels for electric utility generation, in turn, depends on the demand for electricity.

Primary energy demands are calculated in three steps. First the electric utility demand for refinable fossil fuels is calculated by

$$E_{uj} = g_{je} S_{je} F_e g_j \qquad j = 1, \ldots, 6 \quad (27)$$

where E_{uj} is the electric utility demand for secondary fuel j, e denotes the secondary fuel, electricity, g, S, and F having their conventional meanings. This demand in turn must be adjusted for synfuel conversions. There are several alternative accounting procedures that could be adopted to distribute the demand for refinable oil between conventional oil (or gas) and synfuels from coal alternatives. The convention adopted here assumes that all synfuels from coal are consumed domestically and that all imports of oil (or gas) are of conventional oil (or gas). Refinable energy demand must be adjusted by the share of total demand that cannot be met by coal for domestic synfuel production.

$$\left(1 - \frac{ES_{\text{coal}} \cdot S_{i,c}/g_{ic.}}{F_i + E_{ui}}\right) = 1 - S_i$$

Thus, sectoral primary energy demands are given by

$$E_i = (F_{ik} + E_{uik}) \cdot (1 - S_i) \quad (28)$$
$$i = \text{oil, gas}$$

where k is the sectoral index.

Coal demand, of course, is the sum of direct plus indirect demands, the indirect demands coming from both the demand for electricity and the demand for synthetic liquids and gases from coal.

$$E_{\text{coal}} = F_{\text{coal},k} + E_{u,\text{ coal},\ k} \quad \text{(direct)}$$
$$+ \sum_{j=\text{oil, gas}} S_j \cdot (F_{jk} + E_{ujk}) \cdot g_{ic} \quad \text{(indirect)} \quad (29)$$

The total demand for primary fossil fuels is found by aggregating the primary energy demands for end-use sectors and electric utilities. That is,

$$E_i = E_{ui} + \sum_k E_{ki} \qquad i = 1, 2, 3 \quad (30)$$

It remains only to compute the fossil fuel equivalent value for primary electricity. This is done by multiplying each of the three benign forms of electric power generation—nuclear, solar, and hydro—by the average primary energy used by fossil fuels to produce energy, C. That is,

$$E_i = C \cdot F_{ei} \qquad i = 4, 5, 6 \quad (31)$$

The average primary fossil fuel requirement is computed by first summing all fossil fuel inputs to electric power generation, and then dividing it by the total power generated:

$$C = \left(\sum_{i=1}^{3} E_{ui}\right) \Big/ \left(\sum_{i=1} S_{ie} \cdot F_e\right) \quad (32)$$

Energy Balance

It is an identity that the quantity consumed must equal production. The framework must accommodate this reality. So-called "gap studies" take a price as exogenously given and then calculate the resulting supply and demand. These are generally not equal, but there is no mechanism by which equality can be achieved other than by allowing a residual fuel to provide a backstop.

Such is not the case in the context of the assessment tool framework. Markets for oil, gas, and coal are international and as a consequence, there is an interdependence between price and the resulting supply and demand in that market. By assumption, nuclear, solar, and hydro do not trade and are available as specified in the supply and demand modules. There is no problem with markets clearing for these three fuels. They clear as identities.

The oil, gas, and coal markets are different. They must be cleared by a more complicated mechanism. A set of market prices for these three fuels must be found which bring production and disbursements into agreement. The methodology employed to derive these prices is relatively simple, and involves a search procedure begun at an arbitrary set of prices.

The Market Equilibrium Search Procedure

At the initial prices, world supplies and demands for the three fossil fuels are calculated. A measure of the disparity between supply and demand is calculated by the difference between the natural logs of both sides,

$$X_i = \ln Q_i^D - \ln Q_i^S \qquad (33)$$
$$i = \text{oil, gas, coal}$$

If the gap between supply Q^S and demand Q^D is sufficiently small, then the market is assumed to clear. If the initial prices are not sufficiently close to equilibrium, then an estimate of the new equilibrium prices is made, based on price elasticities of demand.

Denote the price elasticity of demand by U_{ij}, and the price elasticity of supply by V_{ij}, where

$$U_{ij} = \frac{d \ln Q_i^D}{d \ln P_j} \qquad (34)$$
$$V_{ij} = \frac{d \ln Q_i^S}{d \ln P_j}$$

Now for each fuel

$$dX_i = \sum_j (U_{ij} - V_{ij})\, d \ln P_j \qquad (35)$$
$$j = \text{oil, gas, coal}$$

An estimate of equilibrium prices can be obtained by calculating exactly how much prices need to change to reduce excess demand to zero. The necessary change in prices is given by setting dx_i equal to $(-x_i)$, and calculating

$$\begin{bmatrix} d \ln P_{\text{oil}} \\ d \ln P_{\text{gas}} \\ d \ln P_{\text{coal}} \end{bmatrix} =$$
$$\begin{bmatrix} W_{\text{oil, oil}} & W_{\text{oil, gas}} & W_{\text{oil, coal}} \\ W_{\text{gas, oil}} & W_{\text{gas, gas}} & W_{\text{gas, coal}} \\ W_{\text{coal, oil}} & W_{\text{coal, gas}} & W_{\text{coal, coal}} \end{bmatrix}^{-1} \begin{bmatrix} -X_{\text{oil}} \\ -X_{\text{gas}} \\ -X_{\text{coal}} \end{bmatrix} \qquad (36)$$

where $W_{ij} = U_{ij} - V_{ij}$.

New prices are calculated by $P_{\text{new},i} = P_{\text{old},i}$ $(1 + d \ln P_i)$. These new prices are in turn used to compute a new gap measure, which is tested for closeness to equilibrium.

Calculation of Elasticities

Elasticities can be calculated either on the basis of numerical or analytical procedures. The latter have the advantage of being faster to calculate once derivatives have been obtained, but possess the disadvantage that calculating the derivative may be a lengthy, intricate, and tedious procedure. Furthermore, derivative procedures make model modifications more difficult since not only must supply or demand changes be instituted in the model core, but in addition the effects of these changes in model structure must also be traced through the model derivatives. This makes model transport more difficult as well. As a consequence, model run time has been sacrificed in order to obtain malleability.

The procedure used is a simple one. The computer model has been encoded in double precision Fortran. Derivatives are obtained by sequentially varying each price by a small amount, and noting the resulting difference from the unperturbed run.

Emissions and Biomass Supply

Calculation of regional energy demands by fuel type is only an intermediate step in the process of computing carbon dioxide emissions. The two are, however, closely linked. Carbon emissions are calculated by first adjusting energy consumption to reflect energy combustion, and then applying the CO_2 conversion coefficients displayed in table 16-1. These emissions coefficients are applied at the point of oxidation, thus appropriately measuring carbon emissions.

Two adjustments are necessary to convert energy consumption by fuel to energy combustion by fuel. First, consumption must be adjusted for nonfuel uses. This adjustment primarily affects liquids which are used as petrochemical feedstocks, asphalt, road tar, waxes, and lubricants. To a lesser extent it affects gas, which is also used as a chemical feedstock and as carbon black. Small amounts of coal are also diverted. Second, an adjustment for the solids, liquids, and gases derived from biomass must be made.

Biomass must be treated separately from other fuel forms because of its unique position in the carbon system. Each ton of carbon released by biomass into the system originally absorbed that ton of carbon from the system. In fact, there is no net gain or loss on balance. While there may be a zero net gain in airborne CO_2 over the biomass cycle, there may, of course, be dynamic effects. Over the short term, biomass stocks may be drawn down and CO_2 stored in these stocks may be released into the atmopshere. Without replacement of these stocks, biomass would have the same impact as a fossil fuel. Similarly, a constantly growing biomass system without biostock drawdowns would effect a net reduction in airborne CO_2 due to lagged plant growth. For simplicity, the impact is assumed to be neutral, that is, there is assumed to be a zero net CO_2 emission by biomass.

Nonfuel uses of energy carriers are assumed to grow proportionally with the level of economic activity in each region. The mix of solids, liquids, and gases is also assumed constant.

The adjustment for biomass requires a somewhat richer treatment. Biomass and coal have much in common. Both can be consumed as solids or converted to either liquids or gases. The demand for coal and biomass is derived from the same sources, and the price of coal is assumed to govern the price of biomass feedstocks. The supply of biomass is developed using the methodology described in chapter 15, in which the availability of feedstocks depends on the price of biomass and a resource base. Agricultural residue and urban waste are aggregated into a single base which in turn is assumed proportional to the level of economic activity. The share of that base which is exploited is a function of price. The share, S, is given by

$$S = \begin{cases} 0.1 + 0.2 \cdot P & 0 \leqq P \leqq 1 \\ 0.2 + 0.1 \cdot P & 1 < P \leqq 4 \\ 0.8 & P > 4 \end{cases}$$

where P is the price of biomass in 1979 constant dollars per million Btu. Note that while the biomass waste resource base varies by region and GNP, the rate of exploitation varies only with the price.

In contrast, the biomass resource base from biomass farms is invariant with respect to GNP and varies only by region. Again, the exploitation rate is assumed to depend only on the price of biomass feedstocks, via

$$S = \begin{cases} 0.0 & 0 \leqq P \leqq 1.5 \\ -0.6 + 0.4 \cdot P & 1.5 < P \leqq 2 \\ -0.4 + 0.3 \cdot P & 2 < P \leqq 4 \\ 0.8 & P > 4 \end{cases}$$

where S is the share of the resource base exploited, and P is the price of biomass feedstocks in 1979 constant dollars per million Btu.

The total amount of biomass supplied is the sum of biomass from waste plus biomass from biomass farms. Since biomass is assumed to compete directly with coal in the solids market, biomass is subtracted from total solids to obtain total coal production. Coal contributions to secondary solids, liquids, and gases are all adjusted proportionately before CO_2 coefficients are applied.

The CO_2 coefficients in table 16-1 apply to conventional fuel combustion. Backstop technologies and lower grade conventional resources are known to have somewhat different coefficients. In some cases these differences are significant. Western shale oil in the United States is not technically shale, and its recovery and use are associated with approximately 150 percent more CO_2 production than conventional petroleum. The methodology developed has the flexibility to apply different CO_2 coefficients to backstop and conventional technologies for each of the nine world regions.

The Regionalization of CO_2 Emissions

Carbon dioxide is released when fossil fuels are combusted. It is therefore possible to trace the contribution of each region and fuel to total CO_2 emissions. This is a somewhat more involved calculation than it might first appear. Direct consumption of fossil fuels and indirect consumption via electric utilities can be easily measured. In addition, however, the CO_2 released from carbonate rock when western U.S. shale oil is mined must also be added. Furthermore, CO_2 released from coal in conversion to oil and gas must be added in the producing region. Carbon dioxide released from shale oil production is assumed to be the difference between conventional oil and the total shale oil release quoted in table 16-1. Similarly, for coal conversion, CO_2 release is assumed to be equal to the difference between conventional oil and gas and the total CO_2 released from the coal used in the process.

NOTES

1. Though even these technologies contribute some carbon indirectly via their use of equipment that requires fossil energy in its manufacture.
2. See D. McFadden, "Conditional Logit Analysis of Qualitative Choice Behavior," in P. Zarambka, ed., *Frontiers of Econometrics* (New York: Academic Press, 1974), or R. W. Barnes et al., *The Oak Ridge Industrial Model Volume II—Model Description* (Oak Ridge, Tenn.: Oak Ridge National Laboratory, July 1981), pp. A19–A25 for the rigorous development of the theory of logit share equations.
3. M. L. Baughman and P. L. Joskow, "The Effects of Fuel Prices on Residential Appliance Choice in the United States," 50(1) *Land Economics* (February 1974), pp. 41–49.
4. K. P. Anderson, "Residential Energy Use: An Econometric Analysis," R-1297-NSF (Santa Monica, Calif.: Rand Corp., October 1973).
5. M. L. Baughman and F. S. Zerhoot, "Interfuel Substitution in the Consumption of Energy in the United States; Part II: Industrial Sector, "Energy Laboratory Report 75-007 (Cambridge, Mass.: Massachusetts Institute of Technology, 1975).
6. W. Lin, E. Hirst, and S. Cohn, "Fuel Choices in the Household Sector," ORNL/CON-3 (Oak Ridge, Tenn.: Oak Ridge National Laboratory, October 1976).
7. D. B. Reister, R. W. Barnes, and J. A. Edmonds, "The Oak Ridge Industrial Model," in *End Use Modeling and Conservation Analysis,* proceedings from the Electric Power Research Institute-sponsored meetings, Atlanta, Georgia, November 1980.
8. Figure 16-7 illustrates one of the key assumptions required to derive the logit fuel share function—the assumption that all of the options have the same cost distribution. If each option could have a different cost distribution, calculation of the market shares would require a numerical integration.
9. See, for example, K. S. Deffeyes and I. D. MacGregor, "World Uranium Resources," *Scientific American,* vol. 242 (1), pp. 66–76; and E. A. Cherniavskey, *Long-Range Oil and Gas Forecasting Methodologies Literature Survey,* BNL 51216, Brookhaven National Laboratory (August 1980).
10. See Cherniavskey, *Long-Range Oil and Gas Forecasting.*
11. Ibid., p. 11.
12. See, for example, W. D. Nordhaus, *The Efficient Use of Energy Resources* (New Haven, Conn.: Yale University Press, 1979), pp. 10–14.
13. These issues are discussed from several perspectives in C. J. Hitch, *Modeling Energy-Economy Interactions: Five Approaches* (Washington, D.C.: Resources for the Future, 1977). All of these studies focus on the United States. The relationship between energy and the economy is less well understood for many of the world's other regions.

Global Energy and CO_2 to the Year 2050 17

Energy consumption and the environment have been closely linked in the past and are likely to be even more closely associated in the future. Environmental concerns range over the entire energy system from problems associated with mining and extraction to those associated with transformation and consumption. The problems of acid rain, land reclamation, water pollution, and nuclear waste disposal have become all too familiar.[1] Carbon dioxide (CO_2) is a less familiar, but no less ominous problem looming on the long-term horizon.

Carbon dioxide is an important by-product of the combustion of fossil fuels. It is a nontoxic, colorless gas with a faintly pungent odor and acid taste. Since carbon dioxide is nontoxic and does not appear to be associated with the formation of other undesirable compounds, it is not commonly thought of as a pollutant. But CO_2 plays an important role in the determination of the global climate. The presence of CO_2 in the atmosphere produces a "greenhouse effect," allowing incoming sunlight to penetrate, but trapping heat radiated back from earth. Man's ability to significantly affect CO_2 levels through use of fossil fuel gives rise to the possibiliby of an unprecedented global problem.

The CO_2 question is a unique environmental question in a number of ways. Unlike other environmental issues in which the problem is restricted to a region, CO_2 mixes rapidly with the earth's atmosphere, yielding a common concentration. Thus, all regions of the world could experience global warming, though it is expected to be greatest in the earth's polar latitudes and least in the equatorial regions. In addition, changes in the earth's temperature are expected to trigger changes in climatological patterns. There will be both winners and losers in this shift, but there is little certainty as to how the gains and losses will be distributed. Finally, there may be discontinuous changes brought about by global warming. Because the polar regions will likely experience the greatest warming, temperature increases may result in a warming sufficient to melt the Antarctic ice shelf, which, sitting atop the south polar land mass, could add enough water to the earth's oceans to raise sea levels an average of 25 feet. Such an event would alter the world's coastlines, leaving most of the world's great port cities below sea level.

Preindustrial concentrations of carbon dioxide are estimated to have been in the range of 270 to 295 ppm. By 1980 the concentration had increased to 339 ppm. It is widely argued that at atmospheric concentrations of 600 ppm or higher, global temperature and climate changes would almost certainly become significant and be cause for concern.[2] However, climatic change in response to CO_2 concentrations significantly less than 600 ppm may present problems of adaptation for human societies. Human activities which produce CO_2 have been increasing. These include forest clearing, cement manufacture, and the combustion of fossil fuels. It is the latter of these activities which has received the most attention.

If CO_2 emissions from fossil fuels were to continue to expand at a rate of growth similar to the postwar (1950–73) average of 4.5 percent per year, then CO_2 concentrations of 600 ppm would almost certainly be encountered within 50 years.[3]. This raises two key questions. First, will CO_2 emissions continue to grow at rates sufficient to cause airborne CO_2 concentrations to reach "critical" levels within a reasonably short time frame? And, second, assuming an affirmative answer to the first question, what will be the likely consequence for global temperature and climate? To date, most CO_2 research has focused on the second question.[4] In this chapter we focus on the former question.

We begin by assessing the CO_2 emissions implied by the base-case scenario described in the following chapter. Energy production and consumption numbers can be translated into a regional and fuel-specific set of carbon emissions by applying the methodology described in Chapter 16. In this chapter we describe the historical and base-case relationships between energy and CO_2 emissions, report sensitivity results using alternative carbon release coefficients, and perform some preliminary investigations of energy policies aimed at reducing carbon release and delaying the date at which critical atmospheric carbon dioxide levels are reached.

ENERGY AND CO_2

Carbon emissions are strongly linked to energy, but not all energy consumption results in CO_2 release and not all activities that release CO_2 release carbon at the same rates. Nonfossil fuels—nuclear power, solar power, and hydroelectricity—release no carbon, while the combustion of liquids, gases, and solids release CO_2 at varying rates (see table 17-1). In addition, some mining activities, particularly the extraction and retorting of oil shale in carbonate rock formations, release significant amounts of CO_2. Similarly, the conversion of coal into liquids and gases release CO_2 in the transformation process as well as when the resulting synfuel is burned.

TABLE 17-1. CO_2 Coefficients in the IEA/ORAU, Long-Term, Energy-CO_2 Model (teragrams of carbon per exajoule)

Fuel	*Carbon*
Liquids	19.2
Gases	13.7
Solids	23.8
Carbonate rock mining	27.9

A considerable literature exists concerning appropriate values for CO_2 coefficients. Those in table 17-1 were calculated at the Institute for Energy Analysis, Oak Ridge Associated Universities (IEA/ORAU) by Gregg Marland and Ralph Rotty. They are representative of an average global fuel for each given type.

As a consequence, the relationship between carbon release and energy consumption depends heavily on the mix of fuels. Over the post-World War II period (1950–73), energy consumption grew more rapidly than the rate of CO_2 production. For every percent increase in energy consumption, CO_2 production rose on average 0.92 percent (i.e., the energy elasticity of carbon was 0.92). This less than proportional growth in CO_2 can be attributed to the global fuel switching that occurred during this period. Coal use was gradually replaced by oil and then finally by gas as well. Each of these switches represents a substitution toward a fuel with lower CO_2 emission per gigajoule of energy.

In the base case, this trend continues through the year 2025. During this time frame, both energy consumption and CO_2 production growth rates slow dramatically from those achieved in the pre-oil embargo period (1945–73). Growth in energy consumption drops to 2.5 percent per year during the remainder of this century, increases slightly to 2.6 percent between 2000 and 2025, and finally diminishes to 2.4 percent by 2050. This represents a halving of the pre-embargo energy consumption growth rate. Carbon emission growth rates show an even more dramatic change. During the 1975 to 2000 period they fall to a mere third of their earlier level, averaging an annual growth rate of only 1.5 percent. The rate increases in the next quarter century to 2.3 percent, and jumps again between 2025 and 2050 to 3.1 percent per year. The result is a CO_2 elasticity that falls from a historical 0.92 to 0.61, rises to 0.88 for the period 2000–2025, and jumps to 1.3 for the next quarter century.

The volatility in the CO_2 elasticity can be directly attributed to the shifting energy mix. During the 1975 through 2000 period, direct electric energy sources gain market shares, so that by 2000 they are responsible for 20 percent of all primary energy production. This corresponds almost exactly to the loss in market share for oil. Since there is no CO_2 released from these direct electricity technologies, carbon release grows far less rapidly than energy consumption.

The 2000 to 2025 period shows new trends. While direct electricity production continues to gain in market share, so too does coal, while oil and gas lose shares. Coal's share grows less rapidly than direct electricity so that overall carbon emissions grow less rapidly than does energy consumption.

Between 2025 and 2050, things change dramatically. Coal's share, which has grown throughout the post-1975 period, reaches 45 percent. Its increased share is not counterbalanced by direct electricity gains. As cheap hydroelectric potential is fully exploited, all direct electric growth is accounted for by solar and nuclear. Thus the declining shares of oil and gas all go to coal. Moreover, a large share of primary oil is obtained from carbonate rock, adding a major new source of CO_2 emissions. The result, as indicated by the CO_2 elasticity, is that emissions grow considerably faster than energy use. The projected case represents a reversal, after 2000, of the trend toward slower growth in emissions compared with growth in energy consumption.

The switch among various fossil fuel sources translates into considerable variance in the average carbon release per exajoule of fossil fuel burned. We calculate the average CO_2 release for all fossil fuels to be 19.85 TGC/EJ in 1975. This is about the same release rate as for conventional oil. In 2000 the carbon emissions rate declines by about 4 percent to 19.10 TGC/EJ. By 2025, it is 8 percent higher than in 2000 and 4 percent higher than in 1975. In 2050 the carbon release rates rises to 25.3 TGC/EJ, 28 percent higher than in 1975 and 33 percent higher than in the year 2000. The 2050 average for all fossil fuels exceeds the carbon release rate for coal, which is frequently thought of as a ceiling emissions rate for fossil fuels. This finding points to the evident need to disaggregate by fossil fuel.

Overall, the average fossil fuel becomes much dirtier over time. By 2050, coal provides 62 percent of all fossil energy, gas provides a mere 9 percent, and oil provides 29 percent. Moreover, shale oil from carbonate rock accounts for over half of the primary oil and as a consequence the average carbon emission from oil exceeds that for coal by 42 percent, making oil the highest CO_2 polluter per unit of energy.

Carbon Emissions by Source

The sources of CO_2 emissions change almost as dramatically between 1975 and 2050 as does the output of CO_2 (figure 17-1). If carbon emissions are metered by geographical source, thus attributing carbon emitted in the production of shale oil to the region in which it was mined, while attributing the CO_2 released from combus-

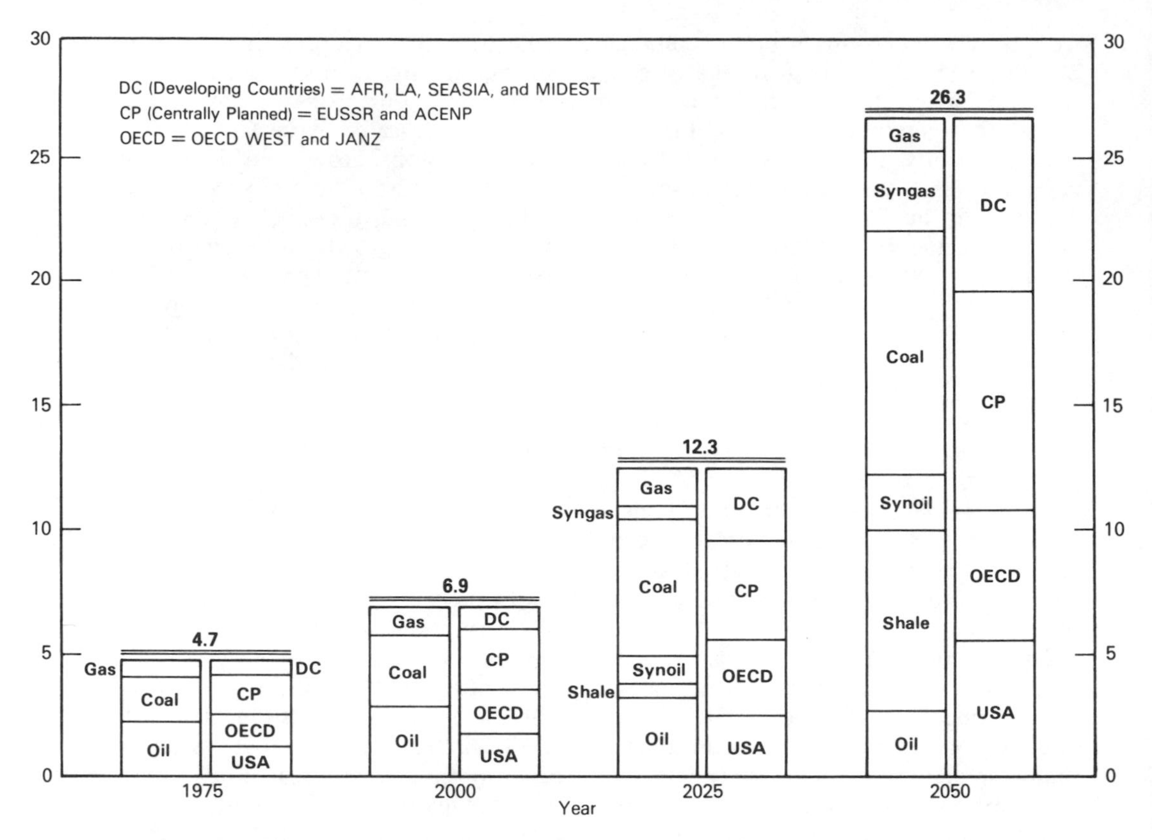

FIGURE 17-1. Global CO_2 emissions from fossil fuel combustion by supply mode and point of emission: 1975–2050 (billions of tonnes of carbon).

tion to the consuming region, the U.S. share of carbon emissions falls from over a quarter to one-fifth of total CO_2 between 1975 and 2050. The same general trend holds true for the other OECD countries. The geographic center of carbon emissions moves steadily south over this period, with developing regions (including the Asian centrally planned region) increasing their share of carbon pollution from less than a quarter to just under a half. The non-Communist developing regions move from less than one-seventh of total carbon release to almost one-quarter. Thus the developed world (including the European centrally planned economies) reduces its contribution of carbon from over two-thirds to slightly over one-half.

As impressive as the shift in geographical source of CO_2 pollution is the shift in supply source. Almost half of all CO_2 released in 2050 comes from the production and use of supply technologies which were not operating on a global scale in 1975—shale oil, synoil, and syngas production. Almost 30 percent of all carbon is released in mining and conversion processes associated with these three technologies. The production of biomass reduces carbon emissions by 7 percent in 2050 over what it would otherwise have been, but it is the rapid mushrooming of the three new polluting technologies between 2025 and 2050 which, together with the increased use of coal for electric power generation, are responsible for the virtual explosion of CO_2 over that period.

The world suffers both from an increasing demand for energy stemming from economic and population growth, and from a shift toward fuels with higher carbon emissions.

Comparison with Other Studies

Three other major energy studies have focused specifically on the energy-CO_2 interaction.[5] Nordhaus has addressed the issue using a linear programming technique to assess optimal strategies for the control of CO_2 emissions. Häfele's group at the International Institute for Applied Systems Analysis (IIASA) worked from a system analysis perspective to construct two plausible energy-CO_2 scenarios.[6] Rotty and Marland used a modified logistics curve approach. The relationship between Hafele, Rotty–Marland, Nordhaus, and this study is depicted in figure 17-2.

Several things are quite clear from an examination of figure 17-2. First this study, Rotty–Marland, and the Häfele (or IIASA) studies are in substantial agreement about rates of CO_2 emissions over the next 50 years. Our base case scenario for CO_2 emissions is near the IIASA low scenario in 2000. By 2025 it lies midway between the IIASA high and low scenarios, and very near the Rotty–Marland result. After 2025 our emissions scenario sharply diverges from the Rotty–Marland projections.

In comparison with the Nordhaus study, this study is topologically very similar, but much lower. Both studies see an accelera-

FIGURE 17-2. Comparison of annual global carbon emission rates (billions of tonnes of carbon).

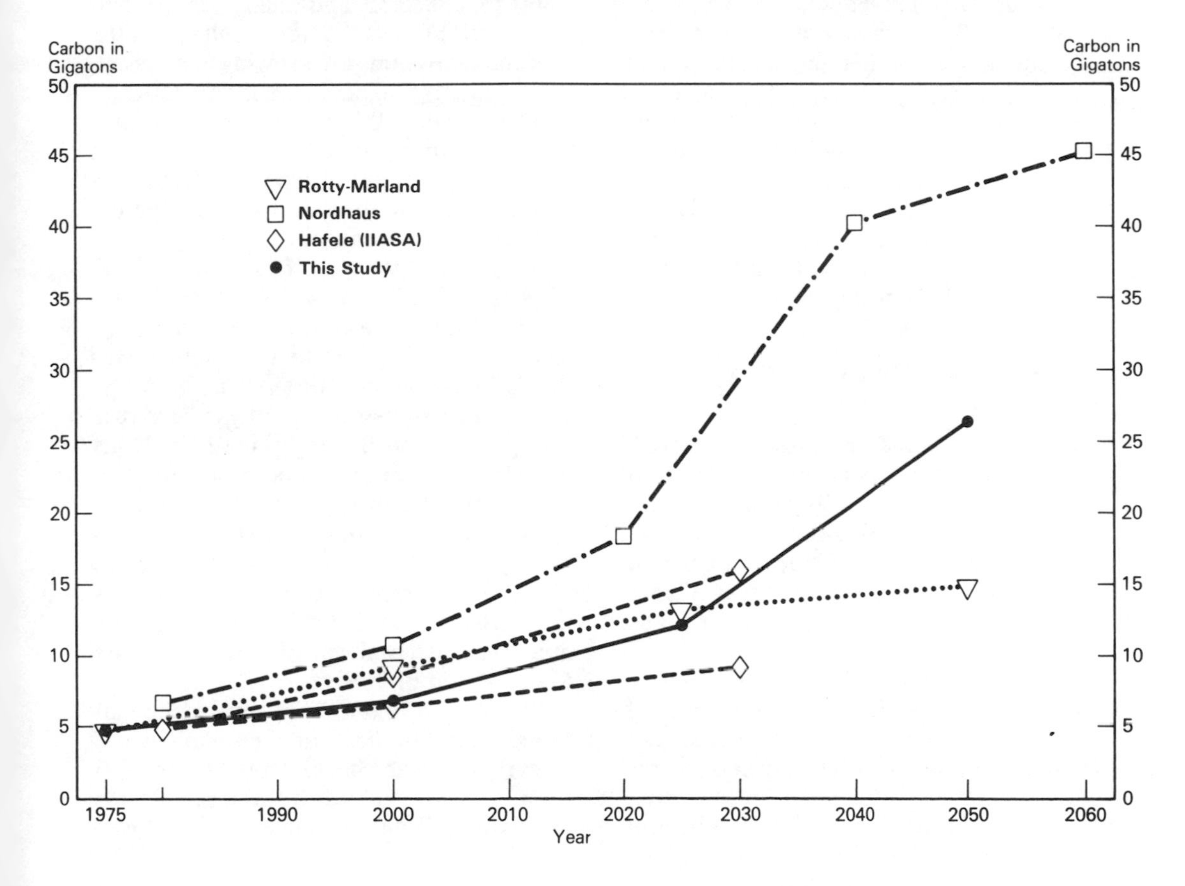

tion in the rate of emissions growth over time, with a very rapid growth after the year 2025. However, the Nordhaus scenario has far greater carbon emissions than this study. In fact, Nordhaus' carbon emissions in 1980 are at a level our findings indicate for the year 2000. Part of the difference between the two scenarios can be explained by the fact that Nordhaus uses higher carbon release coefficients,[7] but much of the difference seems traceable to real differences in energy-use scenarios.[8]

All three scenarios lie well below the 4.5 percent per year historical growth rate for carbon. In fact this study, and much of the CO_2 band between the IIASA high and low scenarios, lies below the 2 percent per year growth rate that Hansen et al.[9] use as a low growth benchmark. The very slow growth in carbon emissions through the remainder of this century (1.5 percent per year) will delay the date of first detection, which may prove vital if carbon emissions escalate sharply soon thereafter, causing rapid temperature and climate change. Thus, the tendency to view the future as one of slowing population and GNP growth, slowing energy demand, and an early transition to renewable fuels provides a seemingly misleading view of carbon emissions. The time path of CO_2 release from fossil fuels seems a much more complex matter.[10]

Emissions and Atmospheric Concentrations of CO_2

Annual CO_2 emission rates can be converted to a corresponding set of atmospheric concentrations by simple integration if an airborne fraction (share of CO_2 sequestered in the atmosphere rather than in biomass, oceans, or other sinks) is known.[11] These are calculated in table 17-2.

Four representative airborne fractions have been selected for integration since the coefficient is not a constant and in fact varies with emissions growth. Moreover, scientists have, as yet, failed to reach agreement on a correct value for the airborne fraction for any given emissions growth rate.

TABLE 17-2. Base Case Atmospheric CO_2 Concentrations for Alternative Airborne Fractions—1975 to 2050 (parts per million)

	Airborne Fraction			
Year	*0.4*	*0.5*	*0.6*	*0.7*
1975	339	339	339	339
2000	365	373	379	386
2025	410	427	445	463
2025	496	536	575	614

The first thing to note is that in all cases the CO_2 concentration is increasing at an accelerating pace. Where the growth in CO_2 concentration between 1900 and 1975 averaged in the neighborhood of 0.15 to 0.25 percent per year, this growth rate escalates each quarter century in the model, going to 0.38, 0.55, and finally 0.91 percent per year. This acceleration results from the fact that emissions are growing at an accelerated rate between 1975 and 2050.[12]

The so-called "doubling date" (the date at which atmospheric CO_2 concentrations reach 600 ppm) can also be calculated. It is most interesting to note that changing the airborne fraction causes little change in the doubling date, which varies between 2067 (f = 0.4) to 2049 (f = 0.7). (We will use the 0.4 and 0.7 airborne fraction as extreme values to generate a "doubling window" for any given emissions path.) Varying the emissions rate about the base case changes the results only moderately as well. If 25 percent is added to incremental emissions rates after 1975, the "CO_2 doubling window" shifts forward approximately 5 years and becomes 2045 to 2062.[13] A 25 percent decrease in incremental emissions has the effect of shifting the doubling date backward by about 5 years to the period of 2053 to 2072.

The critical period is similarly unaffected by the application of extreme values of carbon release coefficients. Table 17-3 shows extreme values of these coefficients for fossil fuels and oil shale mining in car-

TABLE 17-3. Alternative Carbon Release Coefficients for Fossil Fuel Combustion and Oil Shale Mining (teragrams of carbon per exajoule)

Source	*Low Coefficient*	*High Coefficient*	*This Study*
Liquids	18.2	20.6	19.2
Gas	13.5	14.2	13.7
Coal	23.7	23.9	23.8
Oil shale[a]	10.2	83.7	28.4

Source: G. Marland, "The Impact of Synthetic Fuels on Global Carbon Dioxide Emissions," in W. Clark, ed., *Carbon Dioxide Review: 1982* (New York: Oxford University Press, 1982).

[a] Carbon release from oil shale mining is computed as the difference between the extreme shale oil coefficient and corresponding extreme liquids coefficient.

bonate rock compared with the coefficients adopted in this paper. The disparity is rather small for fossil fuels, but large for oil shale mining.

Using all extreme low coefficients yields a critical doubling window of 2050–2070; substituting all high coefficients moves the period forward to 2045–2060. We intepret this result as indicating relative insensitivity of the doubling date to variation of CO_2 release coefficients within plausible ranges.

POLICY RESPONSES TO CO_2 LOADING

The CO_2 issue is a global "commons" problem.[14] The global costs of a single actor's decision to dispose of carbon in the atmosphere is not fully borne by the actor. As a result, there is insufficient incentive to curb emissions. The global nature of the problem means that even when the actors are viewed as nations the problem remains. Viewing the actors as nations does transform the problem from one of actors with truly neglible contributions to CO_2 emissions to one where single nations or nation blocks contribute a significant amount of CO_2 to the atmosphere. As a result, the private (individual country's) incentive to reduce emissions unilaterally may be nonneglible. In the context of the global disaggregation into the nine regions of this study, the United States is presently the largest contributor of CO_2 to the atmosphere, accounting for 27 percent of global emissions in 1975.

Once the problem is defined as a commons problem, the natural route might be to ask what private incentive (in terms of dollars per unit carbon emissions) exists for the United States (or other region) to control emissions. This path presents considerable problems since it requires some assessment of the damage function associated with CO_2 emission. It is a severe understatement to say that the damage is uncertain. We opt for a second approach, asking what effect fairly significant unilateral policy actions by the United States would have on the level of atmospheric carbon dioxide. The nature of the problem suggests two generic policy goals. The first is holding CO_2 concentrations below a given level. The second is extending the date at which a given level is achieved, thereby increasing the time available for adaptation to climate changes. We adopt the value of 600 ppm atmospheric concentration as the critical level. One can take the results as indicative of like experiments for other critical carbon levels; however, we caution the reader that the necessary transform is not one of linear proportionality.

Table 17-4 gives basic results for the base case (labeled 1) and three CO_2 policy cases (labeled 2, 3, 4).[15] Run 2 represents a "naive" political response to CO_2 emissions. In this run, fuel taxes, proportional to current average carbon release, are placed on U.S. oil, gas, and coal consumption at the point of end use. Thus, U.S. residential, commercial, industrial, transport and electric utility customers pay higher prices. The tax rates used in the experiment were: coal, 50 percent; oil, 39 percent; and gas, 28 percent.[16]

Run 2 can be considered a "naive" policy response because it implicitly taxes high CO_2 emitting liquids and gases from shale oil and coal at the low conventional

TABLE 17-4. Effects of CO_2 Taxes and U.S. Coal Export Curtailments on Carbon Emissions: 1975–2050

Model Run	*Year* 1975	2000	2025	2050
1. *Base case*				
Global CO_2 emissions[a]	4.7	6.9	12.3	26.3
U.S. CO_2 emissions[a]	1.2	1.7	2.5	5.4
U.S. GNP[b]	1.52	2.84	4.57	6.80
2. *Low end-use U.S. CO_2 tax*				
Global CO_2 emissions[a]	4.7	6.9	12.0	25.7
U.S. CO_2 emissions[a]	1.2	1.5	1.9	4.4
U.S. GNP[b]	1.52	2.84	4.55	6.78
3. *High U.S. CO_2 tax w. coal export curtailment*				
Global CO_2 emissions[a]	4.7	6.6	11.0	22.3
U.S. CO_2 emissions[a]	1.2	1.3	1.3	2.0
U.S. GNP[b]	1.52	2.83	4.53	6.75
4. *High global CO_2 tax w. U.S. coal export curtailment*				
Global CO_2 emissions[a]	4.7	6.6	9.6	15.7
U.S. CO_2 emissions[a]	1.2	1.6	1.8	2.2
U.S. GNP[b]	1.52	2.84	4.56	6.77

[a]Billions of tonnes of carbon.
[b]Trillions of constant 1975 U.S. dollars.

oil and gas rates and it does not tax or restrict U.S. exports of fossil fuels.

There are two remarkable results. First, the CO_2 taxes reduce U.S. carbon emissions by significant amounts—23 percent in the year 2050. In terms of U.S. gross national product, the costs are minimal. But despite the fact that the absolute reduction in carbon emissions grows steadily, the percentage reduction in U.S. emissions declines over the period from 2000 to 2050. The reason for this reduced percentage impact can be traced to effects of the nonoptimal tax policy which failed to account for changing carbon release of fossil fuels. In addition, the United States, whose domestic consumption of oil is greatly reduced by the tax, continues to produce liquids and solids for export. Since much of the carbon is released in the production process, the taxes on oil consumption take a reduced bite by 2050.

It is conceptually possible under some parameter assumptions for such non-optimal taxes to have the perverse effect of increasing CO_2 emissions. Parameters would have to be such that a heavy switch from solids to shale oil and synfuel occurred in response to the taxes (i.e., high substitution elasticities between fossil fuels but low substitution elasticity toward electricity and higher CO_2 release from unconventional fuels). We have not explored the possibility of achieving such perverse results under favorable (but realistic) parameter assumptions. The other remarkable result is that global carbon emissions are reduced by much less than the U.S. reduction, both in relative and absolute terms. This is because the United States is only one of several major CO_2 polluting regions, but also because the decreased U.S. energy demand resulting from the CO_2 tax lowers world energy prices, which in turn spur energy consumption in other regions. Without this latter effect, one would have expected that the 14 percent reduction in U.S. carbon emissions in 2000 would reduce global CO_2 by about 3 percent, yet the actual impact was a mere 0.1 percent. The decline in U.S. demand for fossil fuels drives the world oil price down by about 6 percent in

2000, which in turn increases global demand for fossil fuel and hence increases CO_2 emissions outside the United States. Over time, however, the United States becomes a smaller total energy consumer and its decreased demand has a smaller impact on price. By 2050, the U.S. CO_2 tax reduces world oil prices by only 2.7 percent from the base case level. As a consequence, the global impact of the tax increases from a 0.1 percent reduction in global CO_2 release per year in 2000 to a 2.3 percent reduction in 2050. In terms of policy goals (avoiding 600 ppm or delaying the arrival date), the action is highly ineffective. It has a net result of reducing atmospheric CO_2 concentrations by only 2 ppm in 2050; the effect on delay in the doubling date is measured in weeks.[17]

The results of run 2 demonstrate the need for much stronger measures. Runs 3 and 4 report experiments representing such measures. They differ from run 2 in that tax policies are designed as optimal tax strategies in the sense that the taxes are effectively placed on carbon release rather than final consumption.[18] As a result, shale oil and synfuel production are taxed at proportionally appropriate levels.[19] They also differ in that U.S. coal exports are prohibited. Finally, all tax rates were doubled.[20] In run 4 the taxes were applied in all regions. The results are presented in table 17-4.

Results of run 3 indicate a somewhat greater impact on global carbon emissions than in run 2, as expected; global carbon emissions fall by 18 percent in 2050. This level easily falls within the 25 percent bands which we constructed around our base case. U.S. carbon emissions are dramatically affected, however, and fall 63 percent from the base case rate in 2050. In contrast to the first energy tax case, in which the suppression of U.S. carbon emissions resulted in a smaller reduction in global CO_2 release, the embargo of coal exports reduces global coal supplies, and hence the global release of carbon as the world shifts toward cleaner energy sources. It is interesting to note that as a result of the tax almost no increase in carbon emissions occurs in the United States between 1975 and 2025.

Still, the doubling window is hardly moved; the tax delays the doubling window from 2050–2065 in the base case to 2052–2071. We note that the reduction in CO_2 emissions is most pronounced at low airborne fractions. These results, stemming from rather severe policy actions, strongly suggest that there is little that the United States can do alone to dramatically change the rate of global carbon buildup in the atmosphere.

Run 4 was designed to examine the potential for international cooperation to yield a substantial delay in CO_2 doubling.

Extending the tax to all consuming regions reduces emissions by 40 percent from base-case levels in 2050. As should be expected, U.S. carbon emissions fall, but less than in the case of unilaterally applied taxes. The result follows from the additional drop in global fossil fuel prices brought on by the slack worldwide energy demand. In fact, oil prices are lowered to the point where shale oil is only a minor global contributor to carbon emissions. Similarly, the U.S. loss in GNP is still small, amounting to only 30 billion dollars per year in 2050.

In terms of the doubling window, the global tax pushes the earliest doubling date out somewhat less than a decade, to 2058–2083. The final effect in terms of a policy goal of significantly delaying the doubling date is rather disappointing given the severity of the tax. Moreover, the global tax case, while an interesting model experiment, should probably be viewed as an academic exercise given the likelihood of achieving the global cooperation needed to implement such a policy.

CONCLUSIONS

An examination of our base-case energy-CO_2 scenario makes several conclusions quite clear. First the "doubling date" is known with some certainty, and it is much later than most early studies found (which assumed a continued 4.5 percent per year increase in the rate of CO_2 emissions). Our base case pushes the doubling window back from the period 2021–2035 to 2049–2067.[21] (See figure 17-3.)

Second, having pushed the doubling date back three decades, CO_2 tax policies are unlikely to have much effect on further delaying the CO_2 doubling period. A massive U.S. policy push with a 100 percent tax on coal consumption and a curtailment of U.S. coal exports moves the window less than 5 years. Even under an admittedly unrealistic scenario of global cooperation in taxing CO_2 emissions, the doubling window shifts less than a decade. This is despite the fact that CO_2 emissions are reduced by 40 percent in 2050. The costs of such a program in the year 2050 are estimated at 380 billion constant 1975 U.S. dollars per year in lost global GNP, though this represents just 1 percent of the total global GNP in 2050.

Third, any real shift away from the base-case scenario will require either reduced global GNP growth or an energy evolution from oil and gas toward CO_2-benign technologies such as solar and nuclear rather than toward synfuels from coal and shale oil. The economics of energy seem to favor a mixed path, with an increasing reliance on coal and shale oil as conventional oil and gas run out. Slower GNP growth is certainly at odds with the development programs of most of the world. As a consequence, the period 2025 to 2050 is likely to see an explosive growth in CO_2 emissions following a period of very slow emissions growth. This combination of lull and storm could delay the date of first detection of CO_2 effects but at the same time present a perilous period after 2025 in

FIGURE 17-3. Comparison of CO_2 "doubling windows."

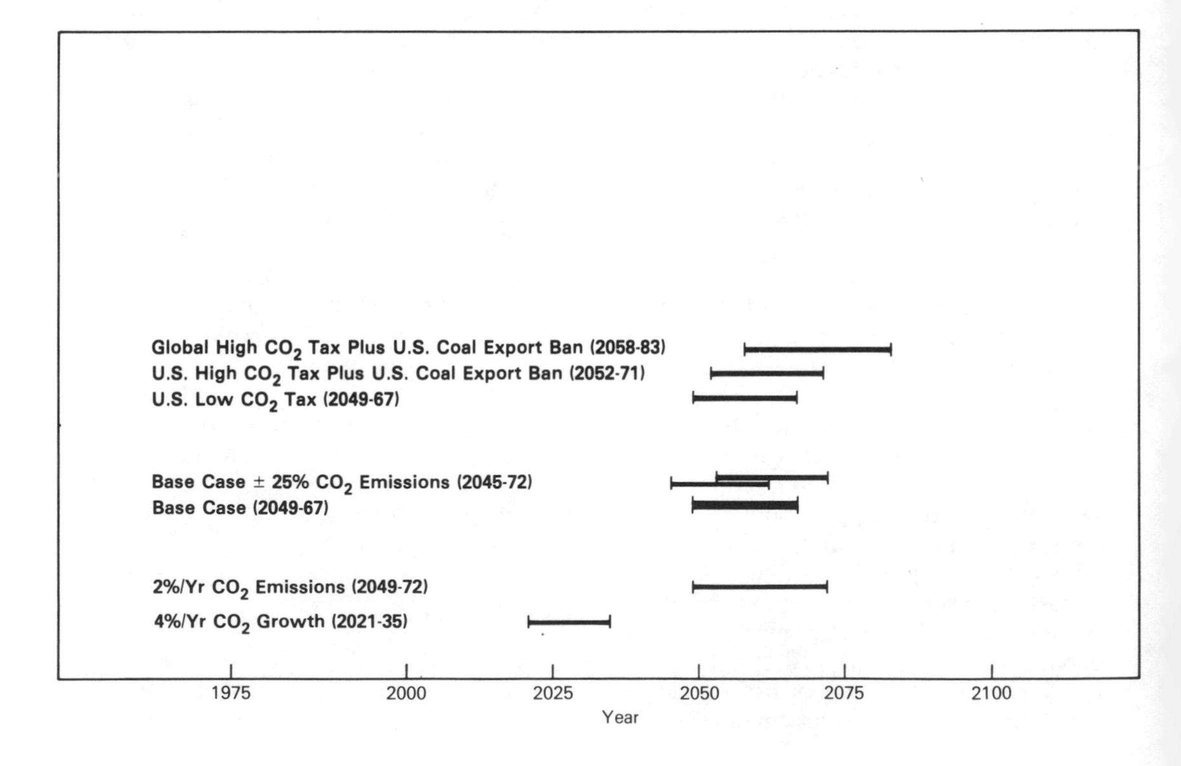

which climate change is speeded up by rapidly growing annual emissions levels. Adaptation to climate change can only be considered more difficult and costly if it is compressed into a few decades.

Because it is so far in the future, affecting our children's lives more directly than our own, the CO_2 question has an unreal aura about it. It lacks the impact of an oil embargo. Yet the question is real and while the work we present here indicates that major man-made climate changes may not be waiting just beyond the turn of the century, we do find that CO_2 concentrations will almost certainly double within 100 years; the magnitude of the temperature and climate change in response to projected increases in atmospheric CO_2 and the associated effects on human societies remain uncertain.

NOTES

1. These problems are addressed in the supply chapters on a fuel specific basis.
2. A doubling of CO_2, to 600 ppm, has become a convenient reference but is not necessarily the critical level. Noticeable climatic change from present conditions is likely at 600 ppm but critical events (i.e., disappearance of polar ice, agricultural disruption) may occur at lower or higher CO_2 levels. Carbon cycle and climate modelers disagree as to the exact temperature impact of a "doubling" of CO_2, but global average surface temperature increases in the range of 1.5 to 4.5° centigrade seem probable. See W. Kellogg, and R. Schware, *Climate Change and Society: Consequences of Increasing Atmospheric Carbon Dioxide* (Boulder, Col.: Westview Press, 1981), p. 3. In addition, the rate of change of temperature and climate may turn out to be as important to our ability to adapt as the magnitudes of the total change.
3. See A. M. Perry, "CO_2 Production Scenarios: An Assessment of Alternative Futures," in W. Clark, ed., *Carbon Dioxide Review: 1982* (New York: Oxford University Press, 1982).
4. While at first blush the focus on consequences might seem to be putting the cart before the horse, it is a natural outgrowth of the history of interest in the CO_2 question, which emerged from the natural science community. Given a historical growth rate of CO_2 production which implied a rapid buildup in CO_2, the most natural questions seemed to center on the coefficient of airborne retention, other sinks for CO_2, and the temperature and climate consequences of the buildup.
5. W. D. Nordhaus, *Strategies for the Control of Carbon Dioxide,* Cowles Foundation Discussion Paper No. 443, Yale University, 1977; W. Häfele, *Energy in a Finite World: Paths to a Substainable Future* (Cambridge, Mass.: Ballinger, 1981); and R. Rotty and G. Marland, *Constraints on Carbon Dioxide Production from Fossil Fuel Use,* Institute for Energy Analysis, Oak Ridge Associated Universities, Oak Ridge, Tenn., (ORAU/IEA-80-9), May 1980.
6. The two cases were termed the IIASA high and IIASA low cases respectively. They are not intended to bracket plausible energy-CO_2 evolutions but rather to provide two separate benchmarks.
7. See W. D. Norhaus, *Strategies for the Control of Carbon Dioxide,* p. 28. The Nordhaus numbers are about 10 percent higher for coal and 15 percent higher for oil. In addition, Nordhaus' numbers are applied at the crude stage of refinement, which tends to increase the difference somewhat due to an implicit allowance for nonoxidizing uses of fuels.
8. Since the Nordhaus scenario seems to be based on a different benchmark than the other cases, it is interesting to scale the 1980 forecast to the IIASA benchmark and then compare the three studies. In this case Nordhaus remains above this study scenario, but below the IIASA high scenario through 2020. After 2020, emissions escalate rapidly, though they level off between 2040 and 2060 at a level somewhat above this study's benchmark.
9. J. Hansen, D. Johnson, A. Lacis, S. Lebedeff, P. Lee, D. Rind, and B. Russell, "Climate Impact of Increasing Atmospheric Carbon Dioxide," *Science* (August 28, 1981) vol. 213, p. 964.
10. For example, the Hansen et al. paper, ibid., assumes in both its low and high CO_2 scenarios that emissions growth will slow after 2020, while both Nordhaus, *Strategies for the Control of Carbon Dioxide,* and this work suggest the opposite.
11. Cumulative CO_2 concentrations are calculated from the integral of emissions rates via,

$$A_t = A_0 + 0.471\, C(e^{rt} - 1)f/r$$

where A is the atmospheric concentration in period t, A_0 is the initial atmosphere concentration, C is the annual rate of emissions in the base year, r is the rate at which the rate grows, and f is the airborne fraction.
12. Even if emissions grow at a constant rate, there will be an acceleration in the rate of growth of the CO_2 concentration, though this acceleration factor diminishes to zero over long periods of time given a fixed airborne fraction.

13. By incremental emissions we mean the difference between the base year emissions and current year emissions. Thus for period t, a 25 percent increase in incremental emissions implies total emissions of

$$\hat{C}_t = C_t + (C_t - C_{1975}) * 0.25$$

where $\hat{C}$ is the revised emission rate, and C is the base case emissions rate.
14. This is a term used by economists to describe a whole class of resources or goods which are not owned by anyone, for example, air, rivers, some public lands. Because they are owned by no one, theoretically these resources are subject to abuse because there is no incentive for anyone to take care of them.
15. The policy scenarios were constructed by the U.S. Environmental Protection Agency in support of its CO_2 research program.
16. The initiation date for these taxes was assumed to be 1985.
17. These results assume an airborne fraction of 0.5. For other airborne fractions, the effects are similar with -2 ppm change if $f = 0.4$, $+3$ ppm for $f = 0.6$, and $+4$ ppm for $f = 0.7$.
18. The taxes are not optimal in the sense that the tax level reflects an assessment of the damage function.
19. Note that this applies a double CO_2 tax on coal-based synfuels, since it is taxed at its full carbon content rate as it enters the synfuel plant, but a major portion of the carbon is not released until the resulting fuel is burned.
20. Taxes were 100 percent for coal, 78 percent for oil, 56 percent for gas, and 115 percent for shale oil production.
21. Note that the initial year in the window assumes an airborne fraction of 0.8 rather than the 0.7 used elsewhere in this paper as such a rapid rate of CO_2 loading will likely be associated with a high airborne fraction. The window opens only 2 years later when 0.7 is used instead.

Global Energy Production and Use to the Year 2050 18

If the scenario we have constructed comes to pass, global energy supply and demand configurations will have undergone a dramatic evolution by the year 2050. Global reliance on conventional oil and gas is likely to fall to less than 10 percent of total energy needs, down from two-thirds in 1975. The United States, the world's largest oil importer in 1975, could well be exporting oil in 2050. Coal will exceed oil as the most important traded fuel. Coal will become the dominant primary energy source, but nearly three-quarters of all coal produced in 2050 will be transformed into liquids, gases, or electricity (35 percent will be converted to synthetic oil and gas, and approximately 40 percent will be burned by electric utilities). Despite the fact that conventional oil becomes a minor source of energy by 2050, liquids remain important, accounting for just under a quarter of demand. The continued use of liquids is made possible by shale oil production and synfuel conversions from coal, which emerge as major energy technologies after the turn of the century.

Under our base-line scenario, the rate of growth of energy demand between 1975 and 2050 slows dramatically from pre-1975 rates. Whereas global energy demand grew at an average rate of 5 percent per year between 1950 and 1975, growth rates are expected to decline to less than half that rate between 2025 and 2050. The reasons are slower rates of population and GNP growth, and higher energy prices. The share of energy demand accounted for by the developed OECD regions falls dramatically over the period, from 54 percent in 1975 to 36 percent in 2050. In contrast, the developing nations increase their relative shares of energy consumption from 11 percent to 35 percent. The global primary energy–GNP (E/GNP) ratio remains relatively stable despite major contributions by conservation and direct end-use solar applications (which save almost a third of all energy consumed by 2050). This stability results from relatively rapid economic development in third world nations where E/GNP ratios rise, counterbalanced by declining E/GNP ratios in the developed world.

Finally, projected energy price increases are expected to be dampened over time. The largest jump in price occurs between 1975 and 2000, when the price of oil increases by 75 percent. This, however, implies that between 1981 and 2000, the oil market will remain soft, with real prices of oil possibly declining from their 1981 peak. Oil prices (in real terms) increase by 36 percent between 2000 and 2025 and are nearly steady after 2025.

In this chapter we report the base-case results with all the caveats normally associated with projections—the scenario is a conditional forecast and its accuracy is dependent on the accuracy of parameter estimates, the appropriateness of the structural model, and the extent to which the implicit and explicit conditions imposed on the exercise obtain. The belief that long-term models can even approach the state of being predictors has long since died. Instead, they represent valuable tools for creating consistent scenarios.

MAJOR ASSUMPTIONS

As we have already discussed, there are major uncertainties surrounding the value of key variables which are likely to determine the energy evolution to the year 2050. To build a scenario we have had to develop our own set of judgmental values for key parameters. We have grouped a set of major assumptions into four subcategories: demographics, labor productivity, and GNP; technological change; price and income elasticities; and supply parameters.

Demographics, Labor Productivity, and GNP

Table 18-1 shows average percentage growth of population and GNP by region. The base population estimates were computed by Nathan Keyfitz (Harvard University).[1] The aggregation in table 18-1 conceals the time trends in growth rates. In general, a slowing of economic activity occurs over time, generated by declining fertility rates as populations approach stationary states.

GNP growth figures in table 18-1 are base rates of expansion. GNP is affected by changes in energy prices through a feedback mechanism with an elasticity of 0.05.

The Energy Technology Parameter

Technological change is assumed neutral across fuels. In the base-case, nonprice–nonincome-induced energy productivity gains increase over time with a constant additive factor. This leads to a slowly declining rate of increase over time. In the industrial sector of the OECD regions, the technological change coefficient rises from 1.00 in 1975 to 1.75 by 2050, adding a constant 0.1 each year.[2] In the residential/commercial and transport sectors, all technological change is assumed to occur in response to price changes. In non-OECD regions, the technological change coefficient rises from 1.00 in 1975 to 1.30 by 2050 in the single-sector economy to roughly approximate the underlying sectoral trends assumed to hold in the OECD regions.

The base assumption of no exogenous technological change in the residential/

TABLE 18-1. Population and GNP Assumptions

	Growth Rates (1975–2050) (percent per year)		*Base Year Levels*	
Region	*Pop.*	*Base GNP*	*Pop. × 10^6*	*GNP × 10^9 1975 U.S. Dollars*
1. US	0.4	2.0	214	1,520
2. CAN & WEUR	0.4	2.3	405	1,818
3. JANZ	0.4	3.0	128	586
4. EUSSR	0.4	2.0	395	966
5. ACENP	0.8	3.2	911	324
6. MIDEST	1.4	3.9	81	138
7. AFR	1.4	3.6	399	155
8. LA	1.3	3.8	313	315
9. S&E ASIA	1.3	3.5	1,130	234

commercial and transport sector is justified by our use of this variable to represent both taste changes and technological change. Taste changes appear to dominate these sectors. We assume that most technological change that occurs is in response to higher prices. Much of the energy savings which have occurred in the past in these sectors and are likely to occur take place through adoption of "known" technologies which, at present, are not in greater use because of their cost or other disadvantages relative to energy savings (e.g., insulation, passive solar design, lighter automobiles). We preserve the option of independent technological change in these sectors to represent views which hold that these types of conservation measures may become popular in their own right and in order to represent policy actions designed to legislate conservation above that induced by prices.

Price and Income Elasticities

The overall energy elasticity is constant over time. For OECD regions, the sectoral aggregate elasticities are: residential/commercial, −0.9; industrial, −0.8; transport −0.7. For single-sector regions, the overall price elasticity was assumed to be −0.8. These elasticities apply to secondary energy.[3] Interfuel substitution elasticities are controlled by a single parameter of the logit function. The share elasticities vary, depending on fuel share and, between fuels, the base weights. To give an indication of the aggregate effect of the underlying energy elasticity and interfuel substitution assumptions, we report a summary elasticity matrix (table 18-2) applicable to the demand configuration in 1975.

These elasticities are model outputs applicable to global markets in refinable oil, gas, and coal. They are generally smaller than the actual elasticities assigned to energy services because add-on costs through the transformation sectors generally tend to increase the percentage price increase in the fuel as one moves back from energy service costs, through refinery costs, to world prices. The offsetting factor (which becomes more important over time) is that primary electricity does not enter the international elasticity matrix explicitly because it is untraded. As a result, the domestic tradeoff between fossil fuels and primary electricity is counted in the conservation (own-price) fuel response in the international elasticity matrix rather than as a separate cross effect.

TABLE 18-2. Summary Demand Price Elasticities

	Oil	*Gas*	*Coal*
Oil	−0.70	0.10	0.10
Gas	0.13	−0.52	0.12
Coal	0.22	0.16	−0.56

Income elasticities are more straightforward. There are two basic income responses. The aggregate income elasticity for the residential/commercial and the transport sector in the regions is assumed to be 1.00 with respect to per capita income; in the industrial sector the elasticity is assumed to be 1.00 with respect to GNP. In the single-sector economies, the aggregate elasticity is assumed to be 1.25 in the EUSSR and 1.40 in all other regions. The second type of income response we have termed fuel preference responses. Gas and electricity have desirable attributes (clean, low labor input) which become "affordable" as per capita income increases. Small income effects have been applied to the fuel share equations to stimulate these responses.

Supply Parameters

Critical supply assumptions include primarily the resource amounts for resource-constrained technologies and ultimate breakthrough costs of essentially unconstrained technologies. Assumed resource constraints are shown in table 18-3.

TABLE 18-3. Resources of Constrained Supply Technologies (EJ)

Region	Conventional Oil	Conventional Gas	Hydro
US	1606	1583	1.83
CAN & WEUR	988	1088	3.51
JANZ	35	158	0.77
EUSSR	2047	2558	4.97
ACENP	680	403	5.76
MIDEST	3854	2089	0.61
AFR	1687	797	7.31
LA	1450	902	6.48
SEASIA	368	375	4.17

Note: These values include cumulative production or cumulative installed capacity, as applicable, and remaining resources.

Costs of nonelectric generating technologies are reported in table 18-4. These represent the minimum cost at which any of the resource will be available, except for conventional oil and gas.

Costs of electricity generation are given in table 18-5.

RESULTS

Highlights of the base-case results are presented here. We begin by discussing the price path resulting from the energy supply/demand balancing process. We then briefly examine energy use in aggregate terms, by fuel, by region, and relative to GNP. Finally, we report additional detail on supply sources for liquids, gases, solids, and electricity and detail on solids use and conversions.

TABLE 18-4. Costs of Supply Technologies (1975 U.S. dollars per gigajoule)

	Cost
Conventional oil	2.00[a]
Conventional gas	1.00[a]
Unconventional oil	3.85
Unconventional gas	3.70
Syn oil[a]	4.55
Syn gas[a]	3.30
Coal	0.26
Biomass (waste)	0.00
Biomass (energy farms)	2.10

[a]Implicitly assumed that the full resource can eventually be produced at producer costs at or less than this amount.

Prices

World oil prices (in constant dollars) are projected to fall by approximately 9 percent by 2000 from contract prices of Saudi crude as of year end 1981.[4] This price drop from 1981 prices actually represents an increase of 75 percent from the model base year (1975) oil price, and a modest increase over March 1982 spot prices.

After 2000, oil prices rise by another 36 percent before nearly leveling off after 2025. The post-2000 price rises are stimulated by gradually increasing demand coupled with reduced production of conventional oil. The combination of these factors requires a switch to somewhat more expensive synoil and shale oil. Moreover, increased demand for liquids, coupled with reduced supplies from conventional sources, will require very rapid expansion of the "new" oil industries. The rapid expansion requirements of these industries will boost oil prices 15 percent above the minimum breakthrough price by 2025. This demand pressure eases somewhat between 2025 and 2050, but pushes prices to 20 percent above the minimum breakthrough price for shale oil.[5] The actual price increase in oil between 2025 and 2050 is 6 percent.

TABLE 18-5. Electricity Generation and Transmission Costs (1975 U.S. dollars per gigajoule)

	Costs	*Conversion and Transmission Efficiency*
Oil	4.53[a]	0.27
Gas	4.51[a]	0.27
Coal w. scrubbers	6.87[a,c]	0.30
Coal w/o. scrubbers	5.86[a]	0.30
Hydro	6.12[b]	N.A.
Nuclear	9.19[b,d]	N.A.
Solar	20.00[b,d]	N.A.

N.A. = Not available.

[a]Nonfuel costs.

[b]Including costs associated with transmission losses.

[c]Scrubbers assumed to be required in OECD regions.

[d]Varies by region and over time; reported costs are low-cost regions.

Gas prices are somewhat more difficult to summarize because of the large transport cost wedge between importers and exporters. The model results shed light on this particular gas supply phenomenon. As liquefied natural gas costs become reflected in the gas prices in consuming regions, demand is restrained, tending to dampen prices. This effectively forces exporters to share the burden of transport costs by pushing the world price down. The model projects world gas prices (producer prices in exporting regions) to fall from present levels, but gas prices in consuming regions will rise. As a consequence, exporting regions have a strong tendency to consume domestic gas rather than export.

Beyond 2000, the world price of gas rises rapidly (450 percent between 2000 and 2050). The percentage rise in major consuming regions is generally less, however, because transport costs are reflected in importing regions' 2000 price.

A subject of recent interest in the United States has been the prospect for equalization of oil and gas prices in terms of price per gigajoule (or Btu). Equalization of gas and oil prices does not occur in our model. There is a tendency for the gap between them to narrow in percentage terms. Supply and demand factors interact to determine the specific gap. There are several reasons why prices do not equalize. These include different refinery and distribution costs, different nonfuel energy source costs, and specialized uses of fuels. Probably the most important factor in maintaining a gap between gases and liquids in the model is the continued exclusive use of liquids in the transport sector. The energy service costs of transportation using nonliquid fuels (electricity) remain higher than liquid-fueled transportation. Refinable liquids fractions unusable in the transport sector (e.g., residual oil) are likely to be priced competitively with fuels they compete with directly—taking into account other factors which can contribute to a price gap.

Coal prices show relatively little movement over the period. Supply factors (exhaustion of high grade–low cost coal) are not considered an important factor contributing to coal price increases on a global level. Production of coal expands at a moderately rapid (2.9 percent per year) rate but such an expansion does not generate major pressure on coal costs.

Stability of coal costs and availability of nuclear power translate to relatively stable electricity prices. A mild upward trend in prices in most regions is generated by the decreasing share of electricity produced from low-cost hydro.

Aggregate Energy Use

Primary Energy Use

Figure 18-1 shows projected global primary energy use by fuel and region for 1975, 2000, 2025, and 2050. Primary electricity is reported at its fossil fuel equivalent on the basis of average conversion and efficiency losses.[6] Considerably more detail is provided in model runs but space considerations have forced us to be selective in reporting results. Regions are aggregated from the model output for presentation.

Major results can be seen from the figure. Overall primary energy use grows at approximately 2.6 percent per year.

It is important to note that figure 18-1 refers to primary energy demand. At present, the differences between primary and secondary energy demand are largely due to conversions to electricity.[7] Beyond 2000, conversions of primary solids (coal and biomass) to liquids and gases become an important source of divergence in primary and secondary energy demand. In addition, primary electricity use grows rapidly as a share of energy supply. Together these factors widen the gap between primary and secondary energy demand.

Among the four primary fuel categories, electricity production expands most rapidly over the period—the annual average rate of expansion is nearly 6 percent. Primary solids use grows moderately (3.1 percent per year). Primary demand for oil and

FIGURE 18-1. Projected primary energy demand, by fuel and region (exajoules).

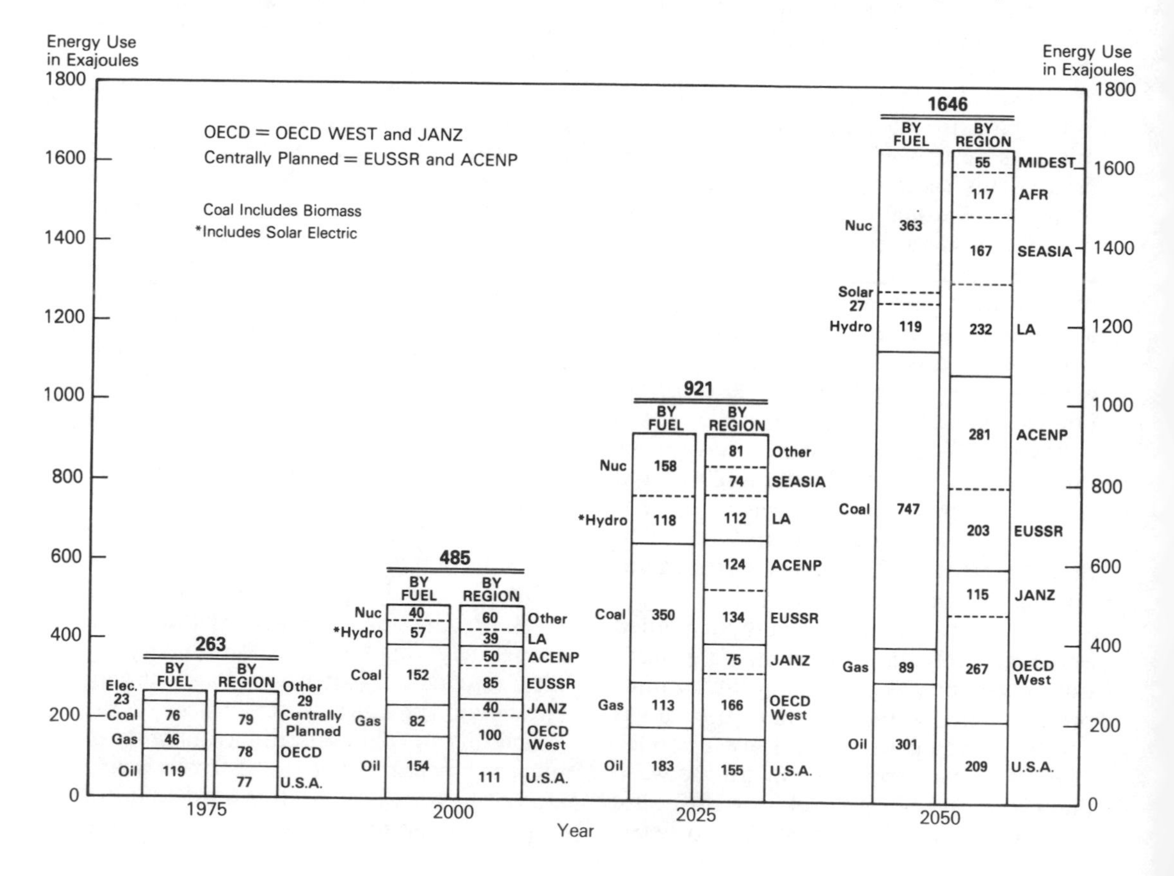

gas grows more slowly. In part, this is due to rising prices of oil and gas. However, as noted above, these categories do not include solids conversions; final demand for liquids and gases is partially supplied by such conversions.

Regional use of primary energy also changes dramatically. The clear trend over time is relatively rapid increases in energy use in the developing country regions compared with the developed regions. The U.S. share of global energy use falls from 29 percent in 1975 to 13 percent by 2050. The OECD share falls from 58 percent to 36 percent of the total. The centrally planned regions roughly maintain a 30 percent share over the period. This aggregation masks a falling share for the USSR and a rising share for China.

In 1975 our U.S. region consumed more energy than any of the other regions. By 2050 the model gives the China region that distinction, with U.S. energy consumption fourth among the nine.

The share of the developing country region (including China, South and East Asia, Africa, Latin America, and the Mideast) rises from only 19 percent in 1975 to 52 percent in 2000.

Final Energy Use

By Fuel

A significant finding in the base projection is the growing gap between final energy use and the attendant primary energy demand. The major force behind this growing gap is the increasing importance of electricity in final use. More than two-thirds of the original energy in fossil fuels and heat produced in nuclear reactors is lost in conversion to electricity and transmission of the electricity to end users. Coal synfuel conversion technologies also contribute to this gap. Overall primary energy demand is projected to increase at a 2.6 percent per year rate compared with a 1.9 percent per year increase for final energy use. (See figure 18-2.)

This finding is significant because it implies that much of the conservation in the end use of energy is thwarted by the need for higher quality fuels (i.e., liquids and electricity) as economic development continues when primarily lower quality resources (coal, biomass) are available. In many cases, solids must (for practical purposes) be converted to liquids, gases, or electricity to compete in providing a given service. The divergence in primary and final energy use is accentuated by the "fossil fuel" equivalent accounting of primary electricity production. This accounting convention does not necessarily offer the best insights into energy supply considerations.[8]

The use of all four energy carriers expands in absolute terms over the period. However, dramatic shifts in fuel shares are stimulated by price changes and fuel penetration phenomena. Liquids supplied 48 percent of end-use energy needs in 1975. By 2050 this share falls to 31 percent. As already noted, liquids are largely squeezed out of all but the transportation sector.

In contrast, overall use of electricity expands from 11 percent of final fuel use to 30 percent by 2050. Coal and gas use as a share of the total is steadier over the full period, coal remaining at about 23 percent and gas falling slightly from 17 to 15 percent.

Nonelectric Solar and Conservation

Direct use of solar energy—hot water, heating, passive solar heating or cooling through building design, active solar heating systems, etc.—is not explicitly accounted for in the model. Instead, such technologies are treated as conservation technologies in the sense that they reduce demand for commercially marketed fuels. As such, the effects of conservation and nonelectric solar can be determined by contrasting fuel use without price-induced reductions in energy use to fuel use in the base scenario. The amounts reported in figure 18-2 as non-elec solar & conservation are derived from such an exercise. As can be seen, the decrease in fuel use due to con-

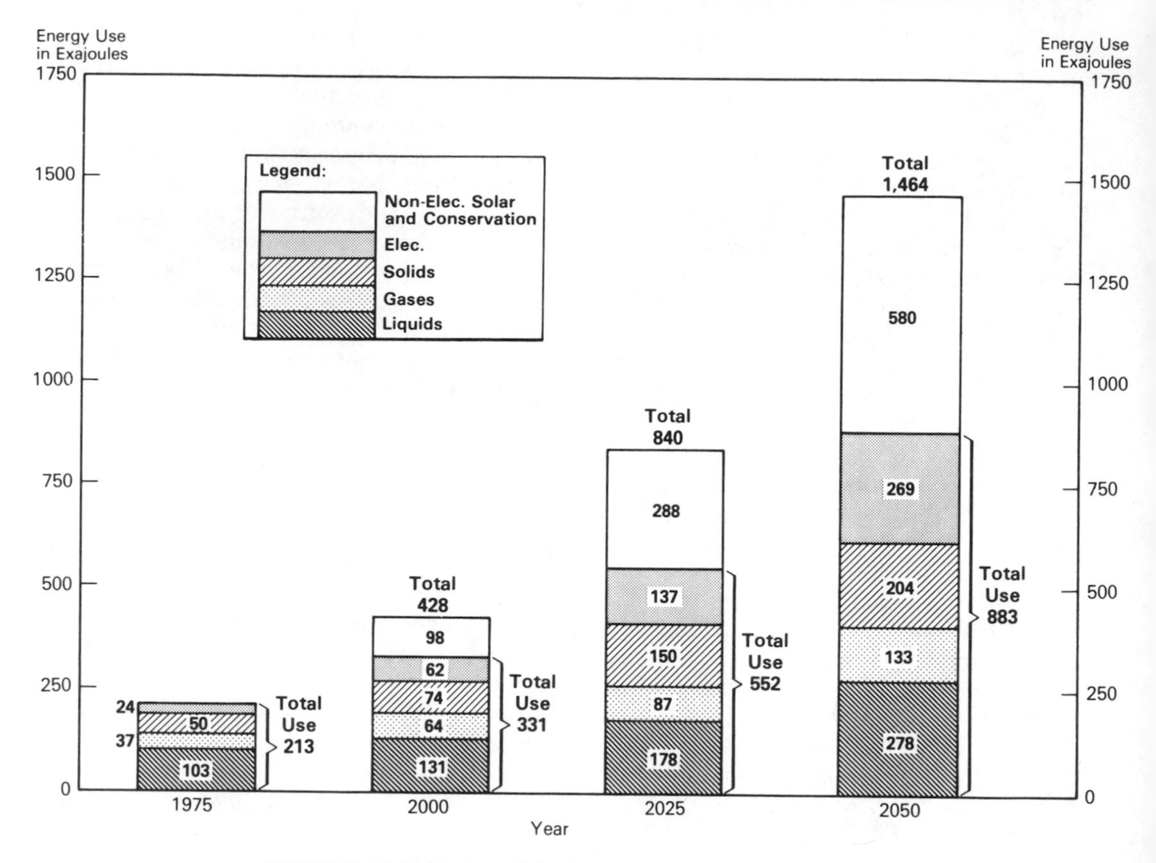

FIGURE 18-2. Projected final energy use, by fuel (exajoules).

servation and nonelectric solar is dramatic. If one views these technologies as simultaneously a "source" and "use" of energy rather than a savings, by 2000 the annual contribution is second only to liquids and by 2025 has exceeded even liquids. By 2050, nonelectric solar and conservation reaches 40 percent of total energy "use" compared with 19 percent for oil, 18 percent for electricity, 14 percent for coal, and 9 percent for gas.

E/GNP Ratios

Primary Energy Ratios

A common summary relationship used in the energy/economic literature is the E/GNP ratio. Figure 18-3 illustrates the trends in primary energy/GNP ratios as projected by the model. Income elasticities, price elasticities, and technological change assumptions underlie these results. The specific pattern of price changes over time accounts for the larger drops or moderated increases between 1975 and 2000. Care should be taken in drawing conclusions from cross-regional comparisons of the level of E/GNP ratios.[9]

We speculate that problems and differences in accounting GNP and converting it to U.S. dollars probably account for the very high E/GNP ratio for China and possibly for the USSR. In the developing countries, the E/GNP ratios are almost certainly higher than would be indicated by a comparison of real product.[10]

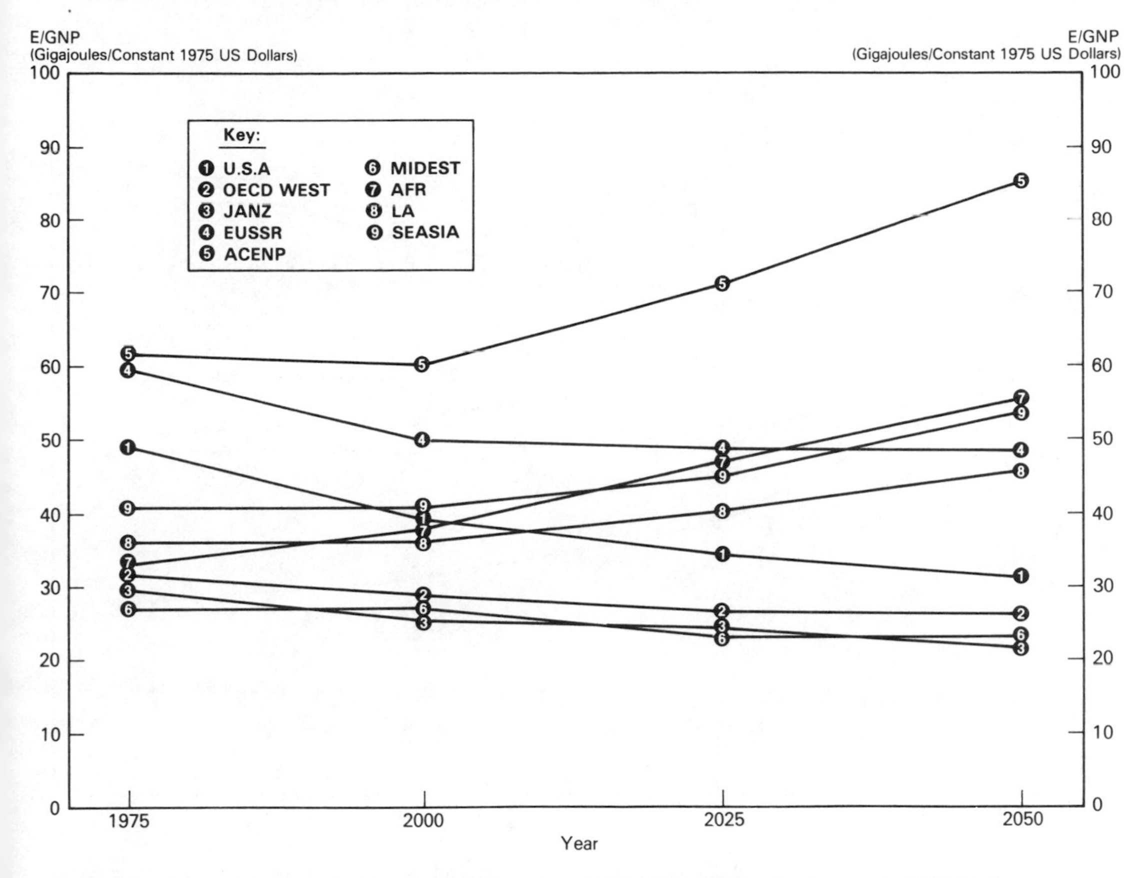

FIGURE 18-3. Projected regional E/GNP ratios, 1975–2050 (gigajoules/constant 1975 U.S. dollars).

Final Energy

The differences between primary and final energy demand that develop in the future have already been noted. Figure 18-4 depicts the E/GNP ratio where energy is measured in its final rather than its primary form. The regions have been aggregated into two groups—developed and developing. The final energy E/GNP relationship is contrasted with the primary energy E/GNP relationship.

In the developed regions, the ratio of final energy to GNP falls much more rapidly than that for primary energy—1.0 percent per year versus 0.4 percent per year. In the developing countries, the primary energy/GNP ratio rises by 0.4 percent per year compared with a drop of 0.3 percent per year in the final energy/GNP ratio.

Energy Supply Results

In general, energy supplies are implicitly reported in the preceding demand chapters since demand and supply are brought into equilibrium in the model. Several interesting elaborations of these results can be made. These elaborations become necessary because, like electricity today, secondary energy carriers of the future will be derived from several different primary supply technologies.

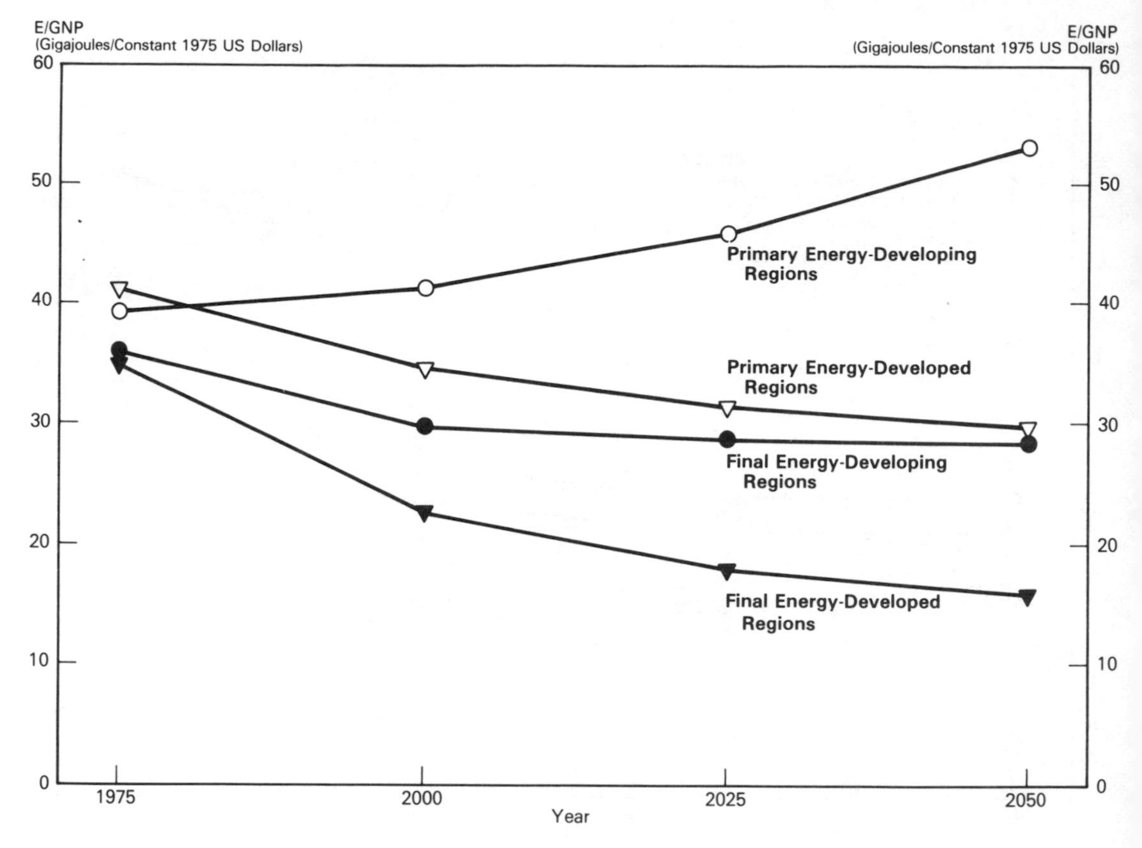

FIGURE 18-4. E/GNP ratios: developing and developed regions.

Oil

Figure 18-5 focuses on oil supply technologies. The "new" oil industries are effectively closed out of the oil supply market until after 2000 in our base-case run. Conventional oil supplies are sufficient to meet demand at prices below those necessary to make shale oil or synoil competitive. Synoil or shale oil plants in operation by 2000 are likely to be pilot plants operating at a loss or subsidized by governments—either explicitly or through some type of restrictions on imports which could make domestic production of oil using these new technologies preferable to importing oil. There is likely to be considerable private incentive to begin operating a pilot plant by 2000, even at a loss, as companies position themselves for entry into shale oil and synfuel production soon after 2000. While these results are sensitive to the level of Middle East oil production among a host of assumptions, we believe the results are robust enough to conclude that private interests are unlikely to support large synfuel or shale oil industries by 2000. Private *interest* is likely to become intense beginning around 2000 as conventional oil *prospects* become relatively poor even as then-current global oil production is peaking and oil prices are relatively depressed.

Beyond 2000, the "new oil" industries expand rapidly. Shale oil becomes and remains a significantly larger share of the total oil supply than synoil because it is estimated that the ultimate technological cost is somewhat less than that of producing liquids from coal.[11] However, the

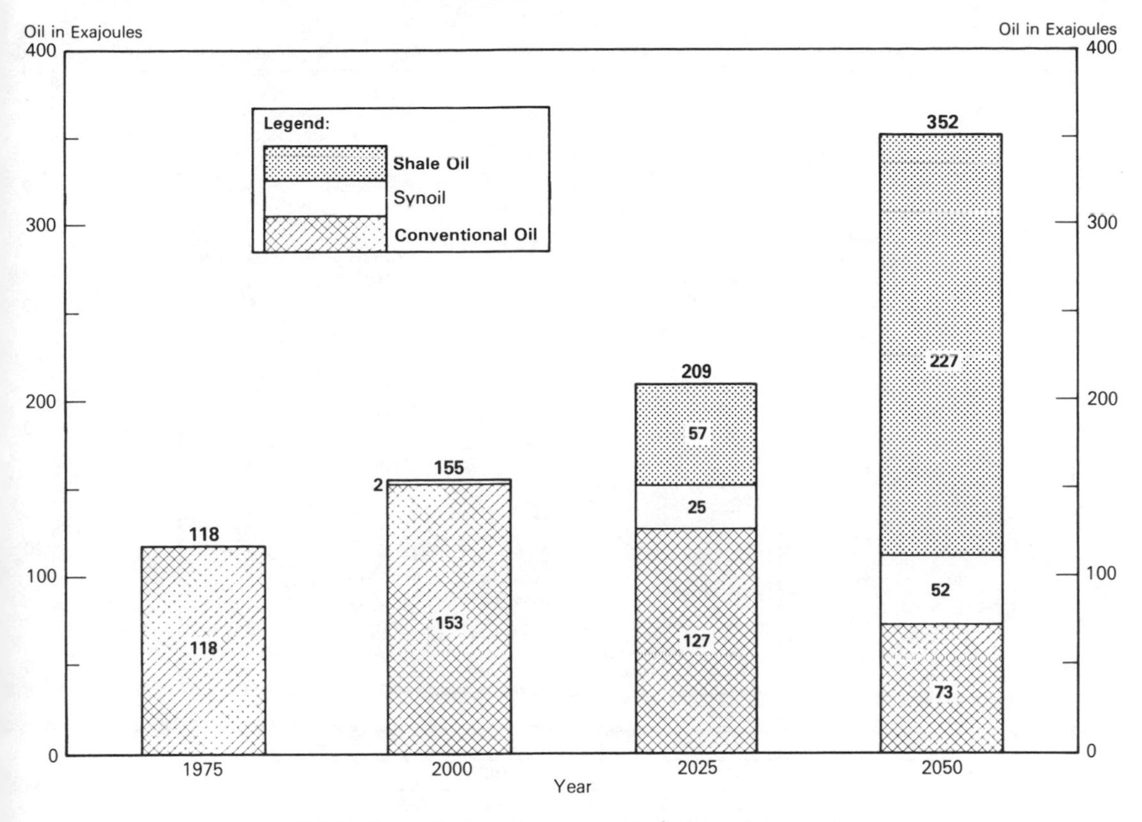

FIGURE 18-5. Projected sources of primary oil (exajoules).

model allows both to be produced if large demand pressures on the oil price in relation to oil supply cause oil prices to rise, thereby making synoil production competitive.[12]

In this base scenario, the shale oil industry is nearly twice as large in 2050 as the conventional oil industry in 1975. Even with the relatively little experience with shale oil production to date, it is clear that an industry of this magnitude could impose severe environmental costs.[13] We have not internalized environmental costs above rather minimal measures to control air pollution.

Figure 18-6 focuses on oil supply by region. The Middle East share of global oil supply drops rapidly over time. The U.S. share of the global oil supply drops through 2000 as conventional oil production falls; after 2000, production of shale oil and coal liquids expands rapidly in the United States, resulting in an expanded share of global oil production for the United States.

Coal

We have included biomass with coal because of its similar nature as a solid fuel with the capability of conversion to liquids, gases, or electricity. We also make a rather fine distinction between commercial and noncommercial biomass. Noncommercial biomass is not directly reported and is treated identically to nonelectric solar and conservation. Switches to noncommercial biomass—e.g., increased use of self-generated waste by industries or individuals—are viewed as part of the price response to rises in the aggregate price of energy. Reported amounts of biomass are

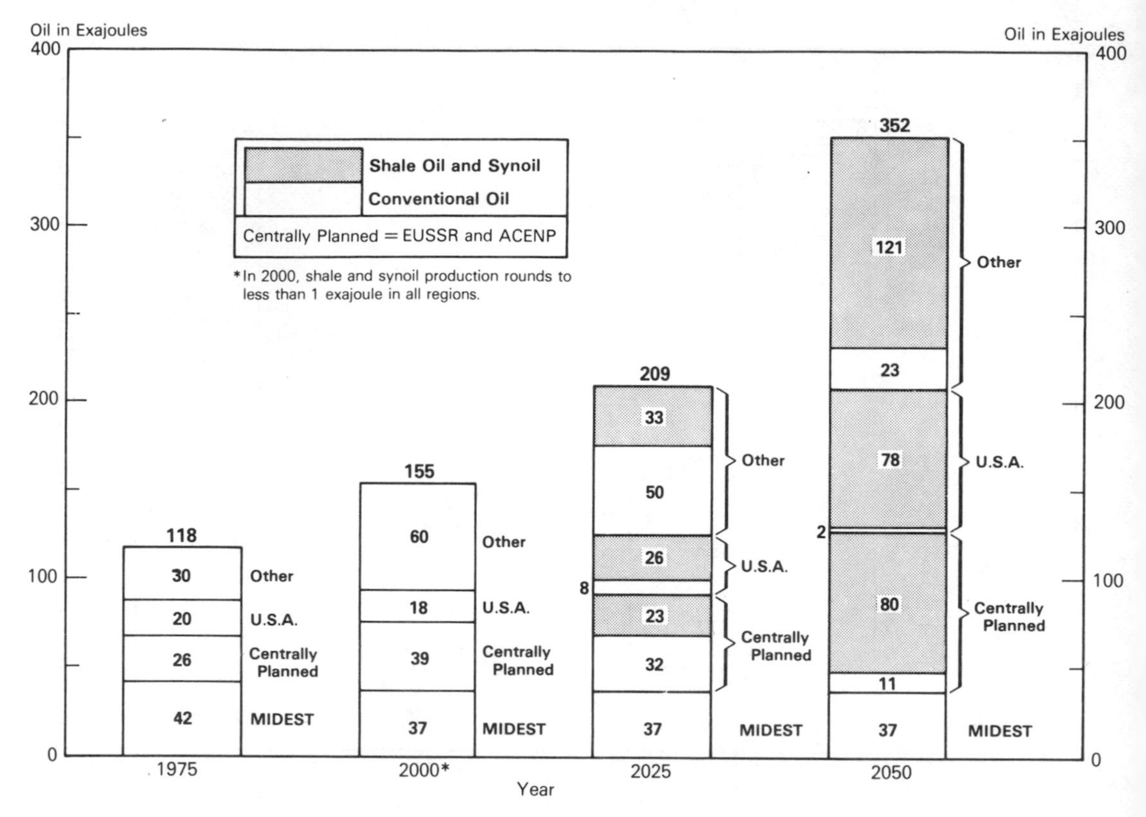

FIGURE 18-6. Projected regional supplies of oil (exajoules). For 2025 and 2050, regional totals are outside brackets; conventional oil by region is in parentheses. In 1975, no shale oil or synoil was produced. In 2000, shale and synoil production rounds to less than 1 exajoule in all regions.

commercially marketed amounts—often for the purpose of liquids or gas production but also for direct use as a solid.[14]

We generally view the use of noncommercial energy (largely biomass) to be negatively related to income per capita in the developing countries. This trend is captured by assuming the income elasticity of commercial energy use to be greater than one.[15] Price effects may in part or wholly offset this tendency.

Bearing in mind this restricted definition of biomass, its use climbs from essentially 0 in 1975 to 98 EJ by 2050. At that level of production, it is larger than total coal production in 1975. However, it represents only 13 percent of total primary solids production in 2050 and about 6 percent of total primary energy use. Most of this biomass will be available from waste material. The cost of producing biomass on "energy farms" is likely to be too expensive in most cases to allow biomass fuels (solids, liquids, or gases) from so-called "energy farms" to compete with fuels derived from coal.

Another interesting aspect of primary solids (coal and biomass) use is its conversion to other energy carriers. Figure 18-7 indicates the amounts of primary solids used in end-use sectors compared with each of the solids conversion sectors. Direct use accounted for two-thirds of solids use in 1975; electric utilities consumed the remaining one-third. By 2000, electric utilities are projected to consume 48 percent

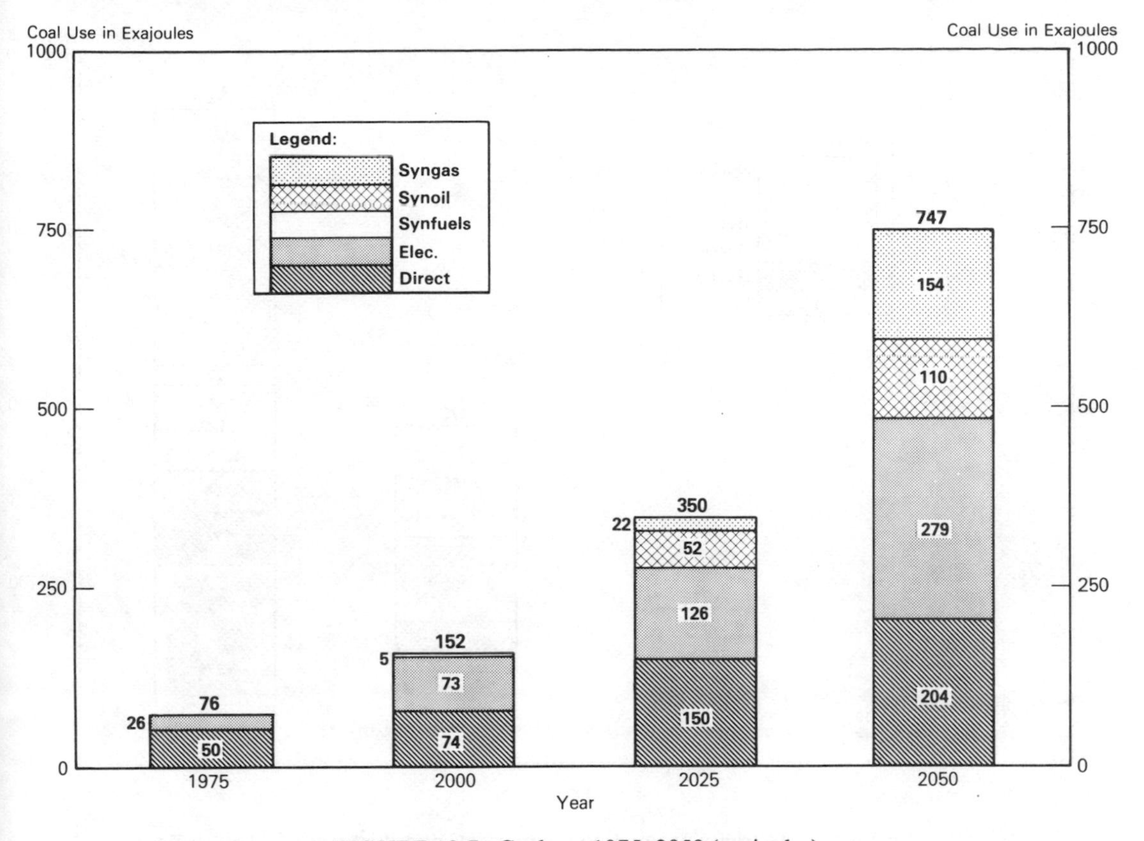

FIGURE 18-7. Coal use 1975–2050 (exajoules).

of the solids produced. Direct use of solids accounts for 49 percent and the remaining 3 percent of solids is used in synfuel conversions.

By 2025, direct use accounts for 43 percent of coal use, the share of synfuels rises to 21 percent, while electric utilities consume 36 percent of the coal produced. By 2050, direct use of solids is projected to account for only 27 percent of total solids use, synfuels production accounts for 35 percent, and electric utility consumption accounts for 38 percent.

Gas

In the base case, syngas costs remain below the cost of other unconventional gas technologies and no significant contributions from these sources are forthcoming.[16] Remaining global conventional gas resources are relatively large. Thus, syngas is only marginally competitive with conventional gas through 2025 even after including transportation charges for those regions which must import gas. By 2050, conventional gas production has begun falling worldwide, causing prices to rise to a level justifying larger amounts of syngas production. Syngas production rises from less than 1 EJ in 2000 to 14 EJ in 2025, to just over 100 EJ in 2050. Thus, syngas contributions to total refinable gas supply rise from a negligible proportion in 2000, to 22 percent in 2025, to 53 percent in 2050.

Electricity

Figure 18-8 gives projected electricity production by technology. Nuclear, solar, and

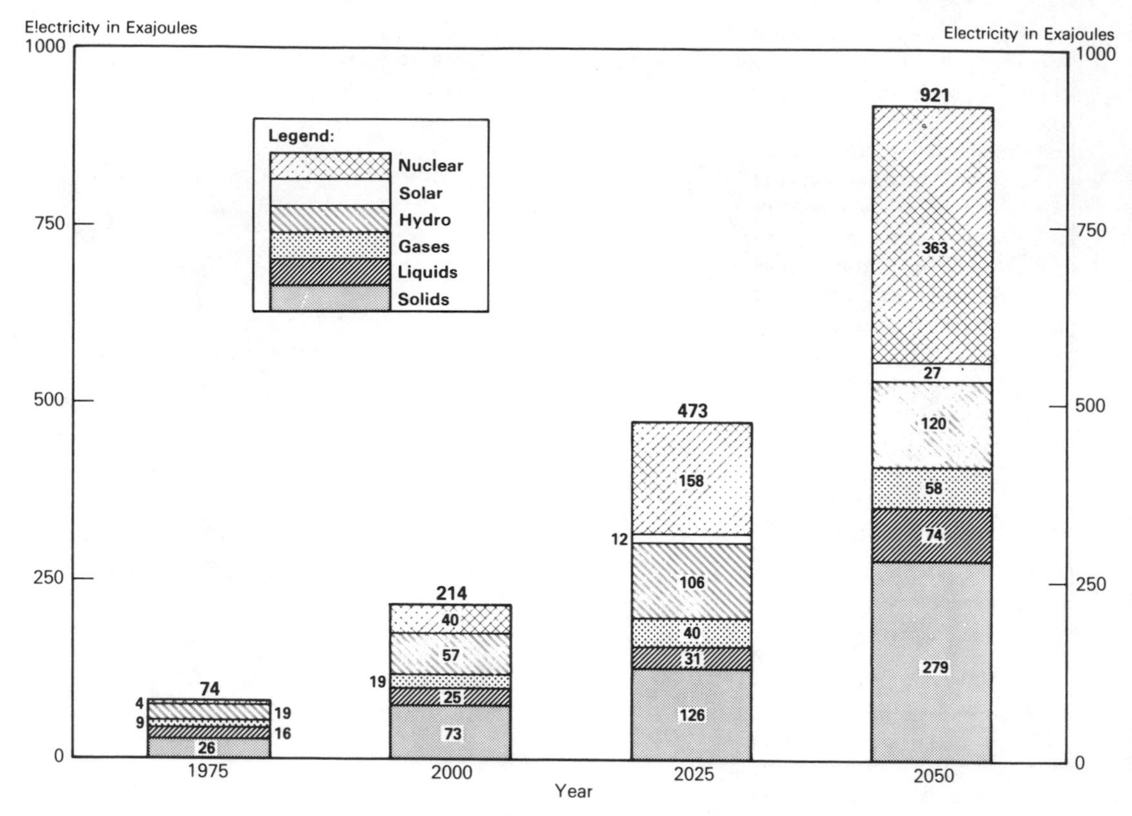

FIGURE 18-8. Fuel use in electricity production (primary electricity at fossil fuel equivalent; exajoules).

hydroelectricity are reported in fossil fuel equivalents based on the average fossil fuel efficiency for conversion to electricity.[17] The share of electricity produced from oil and gas drops from 34 to 14 percent. This drop is caused by increasing costs of oil and gas relative to other electricity-producing technologies. Some oil and gas continues to be burned in utility plants. Two considerations justify such a result: first, within the model only one refinable liquid is produced. In reality, a range of liquids is produced in the refining process. Some of the liquids, such as residual oil, are likely to be priced considerably below the average liquids price, making them potentially competitive with other technologies. In fact, residual oil has relatively few uses other than raising steam in a boiler. Second, liquids and gases have a unique position in the utility sector in supplying peaking power since the capital costs of turbines or diesel generators are relatively low compared with those of steam or pumped hydro systems.

The main electricity-producing technologies in 2050 are solids and nuclear. Solids have a 30 percent share in 2050, down from 35 in 1975. Nuclear's share rises from under 6 percent in 1978 to nearly 40 percent in 2050. The rapidly increasing share of nuclear results from its continued penetration and the relatively low cost of producing electricity from nuclear compared with other available technologies.

Hydroelectricity roughly maintains its share between 1975 and 2025 before resource constraints greatly slow expansion.

Solar technologies, including wind, solar thermal, and photovoltaics contribute just under 3 percent of total electricity production. In most areas of the world, wind appears to be the cheaper of these three technologies. In areas with high insolation, solar thermal power plants can be competitive. With present expectations about advances in photovoltaics, it is unlikely that they will contribute much if any electricity except in very specialized uses.[18] Fusion energy is dismissed as a contributor to energy supplies in the base case because it is, as yet, technologically unproven. While gains in fusion technology are being made, it is unlikely that a commercial fusion industry could be developed to a level that would have a major impact on energy supplies by 2050.

Interregional Trade

The energy trade among the nine global regions grows in our analysis of the period to 2050, albeit at a decreasing rate. The rate falls from around 1 percent per year between 1975 and 2000 to about half a percent per year between 2025 and 2050. The diminishing rate of growth of the energy trade is primarily the result of dramatic shifts in the composition of trade flows. In considering interregional trade results, it is important to keep in mind that international trade between countries within a single region is not projected in the model and is therefore not included in the results.

Oil, the dominant traded fuel in 1975, becomes decreasingly important. After the year 2000, oil trade declines absolutely, and during the period 2025 through 2050 the decline reaches an annual rate of 1 percent per year. In contrast, trade in both gas and coal increases. However, gas and coal are both hampered by high interregional transport costs and institutional constraints. While the gas trade builds rapidly to the year 2025, it stagnates thereafter at a level well below that of the oil trade.

Of all the fuels, coal shows the most rapid growth. Trade expands more rapidly than the global economy, and by 2050 coal is the most important traded fuel, surpassing the declining oil trade. Even coal, however, does not obtain the volume of trade that oil enjoyed in the year 2000.

Regional and technical changes in global energy production also influence the future course of interregional trade. This occurs most strikingly in the United States and in Latin America. Between 2000 and 2050, U.S. dependence on foreign oil steadily decreases until in 2050 the United States becomes a net exporter of liquids. This change in position is the direct result of higher oil prices, which make coal liquefaction and shale oil profitable to produce. U.S. exports are made possible by coal liquefaction and shale oil mining. In contrast, Latin America, with a smaller coal resource base, more conventional oil, and rapid economic expansion, changes its status from net exporter to net importer of liquids. Despite the rise of coal liquids and shale oil, the largest single exporting region remains the Middle East, although its exports of conventional oil decline throughout the period of analysis.

The trade in gas never develops to the point that either the oil trade or coal trade do, though it is eclipsed by coal only after 2025. Its expansion is hampered by high LNG transport costs, and by the fact that coal and biomass gasification provide domestic substitutes in most regions for natural gas imports. The most important single gas importer throughout the period of analysis remains Western Europe. Initially, European gas imports come predominantly from the Soviet Union. By 2050, however, European gas imports come increasingly from the Middle East and Africa.

The United States becomes the most important coal exporting nation in the scenario. By 2050 coal exports exceed Middle East oil exports and account for over half of all coal trade. China, Africa, and the OECD Pacific contribute the remainder of the world's coal exports in 2050. Much of the U.S. coal exports go to energy-poor Europe, which remains the largest importing region throughout the scenario. In fact, European coal imports account for 80 percent of global coal imports in 2050, with only small amounts going to the Middle East, Latin America, and South and East Asia.

NOTES

1. These estimates were also submitted to external review to Philander Claxton, who supported their use.
2. These coefficients are interpretable as the amount of energy service derived from a unit of energy.
3. In the model, elasticities are actually assigned to energy services.
4. Based on Saudi marker crude at $34 per barrel (current dollars). Softness had developed in the world markets by year end 1981. By February 1982, spot oil prices were reported as low as $28 or $30 per barrel (current dollars) for Saudi or similar grade crude.
5. The breakthrough price is used to denote strictly the supply cost of the fuel, without demand pressure on production imports. It is the lowest price that will allow breakeven production (including normal returns on capital investments).
6. The average varies by region and over time. It is around 0.29. This is somewhat lower then generally encountered for two reasons: first, it includes transmission losses [electricity *sold* divided by fossil energy inputs]. Second, it includes peaking power energy use as well as base load generation capacity. The fossil fuel equivalent measure is used as a reporting convention only. The model uses heat content of electricity generated by primary electric technologies to calculate service demands.
7. Other conversion losses are included, but, in general, are negligible.
8. If one's question is "How much oil, gas, or coal would be needed to supply the given amount of electricity?", this accounting scheme is clearly preferable. On the other hand, "fossil fuel" equivalent offers no particular insights into demands on the primary hydro, wind, solar, or nuclear resource. As these primary electricity technologies become more important, switches among these technologies may offer the more interesting questions than switches to fossil fuels. However, the conversion to "fossil fuel" equivalents does offer a measure of primary electricity on a basis more comparable with other energy carriers; electricity is 2 to 3 times more efficient in supplying mechanical drive. Its advantage is smaller in providing heating services. Because electricity is a quite different energy form than conventional energy carriers, the problem of comparability is largely unresolvable. This note is meant primarily to caution readers that the familiar problems discussed above take on heightened importance as primary electricity grows as a share of energy supply.
9. GNP figures at 1975 price levels were converted to U.S. dollars using official exchange rates. The problems inherent in such comparisons have been extensively examined by I. Kravis, R. Summers, and A. Heston in, among other publications, "International Comparison of Real Product and Its Composition: 1950–1977," *Review of Income and Wealth,* series 26, no. 1 (March 1980). An attempt to make E/GNP ratios comparable across regions by adjusting GNP and energy use (to account for fuel conversion efficiencies and noncommercial fuels) is reported in J. Dunkerley, W. Ramsay, L. Gordon, E. Cecelski, *Energy Strategies for Developing Nations* (Baltimore, Md.: Johns Hopkins University Press for Resources for the Future, 1981), pp. 10–12. An attempt to relate energy use and the more general concept of "quality of life" is reported in L. Ben-Chiech and C. Anderson, *Income, Energy Requirements, and the Quality of Life Indicators, An International Comparison, 1975* (Palo Alto, Calif.: Electric Power Research Institute, 1979).
10. Dunkerley et al., *Energy Strategies for Developing Nations,* find that the E/GNP ratio for non-Communist developed world is 1.4 times the ratio for the non-Communist developing world in 1977, after converting GDP figures to comparable U.S. dollar figures using the Kravis et al., "International Comparison of Real Product and Its Composition: 1950–1977," calculations. Dunkerley et al. go on to make calculations for noncommercial fuels and differential efficiency in end use of energy. These two efforts counterbalance one another, thereby maintaining the 1.4 relationship between developed and developing countries.
11. The technologies are very similar but shale oil is cheaper to produce because of a basic chemical difference between coal and oil shale which translates into considerably higher processing and upgrading costs for coal liquids. For a detailed discussion of relative costs, see chapter 8.
12. The methodology is discussed in chapter 16.
13. Water availability has often been cited as a limiting factor in shale oil development in the United States. We discard this as a severe constraint on the basis of recent analyses which indicate that the cost of piping water to the western United States would add well under $1.00/ barrel of oil produced. A considerably less manageable problem, at present, appears to be

groundwater pollution resulting from toxic chemicals leaching from spent shale.

14. Such use may include some processing, e.g., pelletizing wood to allow mechanical feeding and facilitate even burning in boiler operations.
15. Other factors, such as structural shifts in the economy, are also captured by this assumption and may dominate the effect.
16. Unconventional gas resources are one of the most uncertain aspects of future energy supplies. Recently a view has emerged among some geologists that some gas may be inorganic in nature; it suggests that, at somewhat deeper levels than gas from organic sources is found, an essentially inexhaustible source of combustible gas may exist.
17. Efficiency includes conversion and transmission losses, departing from the usual conventions. The conversion factor between electric output (measured at point of sale) and fossil fuel equivalent is approximately 3.5. The actual figure varies over time and by region as the share of oil, gas, and coal used to produce electricity varies.
18. These technologies are considered at length in chapter 13. Other technologies, e.g., wave, and tidal power, are considered but dismissed as potentially major contributors from various technological and cost standpoints.

Uncertainties in Long-Term Energy Projections 19

This chapter places the previous analysis of long-term global energy and CO_2 in an appropriate context. As such, it serves as a conclusion to one research effort and as an introduction to another. Thus far our research has focused on understanding the possible evolutionary paths of the global energy system. We have attempted to approach this task as analysts rather than proponents of a particular energy policy, energy source, or world view. We have been interested in describing how the world *might* look by 2050 and how policy *might* affect energy supply and use rather than how it *should* look or what policies *should* be adopted.

To this end we have concentrated on isolating the key elements which are likely to affect that evolution and on the construction of a scenario of fuel use. But, as we have observed repeatedly, one of the chief characteristics of the future is uncertainty. a nonzero probability can only be attached to a range of outcomes; a single scenario lacks such bounds. Even qualitative statements of likelihood refer to one's assessment of the actual result falling in the "neighborhood" of the predicted value. Thus, to set the context of the base-case scenario, it is useful to spend the final pages of this book exploring some attempts to define a "neighborhood" so that statements of likelihood can be addressed. A fuller examination of this aspect of energy use is the subject of current research.

UNCERTAINTY AGAIN

Models cannot predict the future with certainty for a variety of reasons which can be grouped in three broad categories: (1) the assumptions of the model were wrong, (2) even the best model is a simplification of reality and therefore technically "incorrect"; and (3) the world may have inherent random elements. One can characterize random events, i.e., describe the probability associated with various outcomes, but by definition prediction is not possible. Therefore, to the extent randomness exists in the elements which affect the forecast variable, certainty in forecasts is impossible. In the other two categories something can be done to improve forecasts and these are briefly discussed in the next two sections.

Uncertainty in Models

Of these two remaining categories, uncertainty in model structure presents the less tractable problems. Models are always a simplification of real-world phenomena and therefore cannot describe the world perfectly. Some uncertainty necessarily exists as to how much error is introduced by the simplifying process as well as the extent to which one's understanding of the system is correct. That is, does the model capture the key features of the global energy system? The oil price spikes of 1974 and 1980 illustrate several important aspects of modeling uncertainty. The failure of short-term models to predict these increases illustrates the limited knowledge of the system used to construct these models. While knowledge of energy system structure probably increased as a result of the 1970s experience, an adequate description remains elusive in ways only the future will make apparent. Having the benefit of hindsight, model structures which predated 1973 did not include a cartel model for OPEC behavior, did not include a model of political stability in the Middle East and how oil production might be affected by instability, and did not accurately describe the disequilibrating role oil stock build-up played in increasing prices. The cartel model issue illustrates that appropriate model structures can change over time or (depending on whether one views the event as random or causally based) our inability to model the formation and disintegration of cartels. The other two examples indicate more clearly a failure to fully understand the important aspects of the system and how it works. Again, in the case of political disruption, there is almost certainly a mixture of randomness and causality.

The above discussion highlights the problems forecasters face: Are events random or causal? We would maintain that how one *treats* events in a modeling exercise depends on one's time horizon. If the forecast horizon is one year, the forecaster would hope to evaluate the impact of political events on the price of oil (perhaps in a relatively subjective manner). For example, one might decide that continued fighting in Iran/Iraq is expected to reduce crude oil reaching the world market by *"X"* million barrels per day. Behind such a statement is some assessment of the current situation and how it is likely to proceed. However, analysis of the current political situation is a very poor guide to the political situation 10, 20, or 50 years hence. Thus, even if in the short-term war and political disruption can be traced to causal factors, it is useful to treat them as random events in long-term forecasting.

Two general kinds of randomness are important. First is the "white noise" variety, that is, nonsystematic noise around a trend. There is little hope of predicting yearly ups and downs over a 75-year period; in terms of predictive forecasting (statements of likelihood) it is possible only to forecast a smooth trend, and for our purposes that is sufficient.* The second type of randomness is of the "random walk" variety. For example, one may consider it, at least partly, a random event as to which one of two new competing technologies is proved first; once the outcome is known, it has a lasting impact on the course of events. The success of one technology may slow down testing and research on the other. The "random walk" element cannot be ignored, but it can be explored with scenario analysis.

In general, there is no magic formula for anticipating the unforeseen (except hindsight) or for formally modeling the elements of society on which there is little agreement about causal structure. While we recognize that this adds to the overall uncertainty surrounding the reference case, we make no attempt here to quantify it.

Uncertainty in Assumptions

The more conventional and tractable problem is that of addressing uncertainty associated with alternative exogenous assumptions. There are various ways of

*One may still, for some purposes, wish to construct "what if" scenarios; e.g., what if oil prices suddenly doubled in 1995? The model developed here has not been designed with such questions in mind.

approaching the problem. These include sensitivity and uncertainty analyses using alternative methodologies to generate the underlying probability distributions for the set of exogenous assumptions. Less formal approaches such as scenario analysis are also possible. The remainder of this chapter attempts to explore in an informal way how our reference scenario might change as a result of alternative assumptions.

The general problem of uncertainty is approached in three ways. First, we address the potentially large environmental impacts likely under the energy use estimates given here. If environmental problems are large and the costs of these problems are internalized in fuel choice and energy use decisions, how will energy and CO_2 emissions be different than in the reference case? Second, the consequences of major simultaneous changes in important variables are explored. The example used represents a radically different "world view" about how the energy system is likely to evolve. It is obtained by altering the input assumptions rather than the model structure. Thus, the "orthodoxy" of the reference case is seen largely as a result of input assumptions rather than constraints imposed by model structure. Finally, we attempt an initial quantification of the uncertainty associated with future projections of energy use and CO_2 emissions as it relates to uncertainty in model inputs. While the quantification is subjective, it provides an indication of the degree of uncertainty inherent in long-term forecasts. In all of these exercises, we stress the need for additional supporting research.

ENERGY POLICY AND UNCERTAINTY

Uncertainty in future global energy production and use stems from numerous sources, including technological, demographic, economic, political, and institutional factors. Energy policy, including direct actions such as taxing or subsidizing various energy sources and uses and indirect actions such as creating a climate in which various energy options either flourish or are discouraged, represents a major source of uncertainty. In addition, energy policy covers an interesting set of issues because it is at once a source of uncertainty to the analyst and a tool for dealing with the world for the policymaker.

Model assumptions can be adjusted to describe energy paths consistent with various policies explicitly (e.g., taxes, quantity limits), or in some cases, implicitly (e.g., a favorable climate for a given technology can be represented as lower or falling costs). It is important to note, however, that the resulting scenarios are the result of both policy and nonpolicy assumptions. To measure the impact of policy alone, it is customary to compare scenarios which are identical except for the implementation of a particular policy. The differential results are then attributed to the policy. In such exercises, one should not lose sight of the fact that the particular result is contingent on the underlying base assumptions as well as on the policy in question. Some general policy results remain invariant over a broad range of plausible base cases. We call such results "robust." Others may be highly dependent on the base case environment. As an obvious example, limitations on shale oil production would have no impact on CO_2 emission if shale oil were not produced in the base scenario. This base case effect might be called a *ceteris paribus* problem since it demonstrates the importance of the set of *ceteris paribus* conditions under which a policy experiment is conducted.

The model can be used to assess the effectiveness and robustness of alternative energy policies. For example, in our analysis of policies designed to reduce global CO_2 emissions from fossil fuel use, we found that the United States, acting alone to reduce the demand for fossil fuels, can do little to reduce global fossil fuel emissions. We have found this result to be robust across a broad range of plausible scenarios.

In addition to the recognition that the

choice of base-case conditions is important, there is another *ceteris paribus* problem: Where does one draw the line between the base and the policy? If, for example, it is CO_2 emissions that one is concerned with, one would generally wish to remove from the base case any assumptions in which the policy actor of interest took explicit recognition of the CO_2 problem. Others actors who explicitly recognize CO_2 as a policy problem and act on it and other policies which affect CO_2 emissions, but which are directed primarily at other targets, should be taken into account in such a study. For example, an assessment of policy responses to acid rain which inhibit coal production and consumption and therefore affect CO_2 emissions should be included in the base case.

While the model described in chapter 16 can answer many of the "what if" kinds of questions about the impacts of a specific policy or set of policies under specific "ceteris paribus" (base case) conditions, the model is not designed to address the "What is the optimal policy?" question. Like the "what if" policy questions just discussed, optimal policy models yield answers which are conditional on the base case conditions embedded in the model's data base. In addition, however, optimal policies also depend upon the definition which one chooses for the term "optimal." But optimality is not a uniquely defined term. Its definition depends on who is asking the question. All of the definitions which are currently invoked to make optimal policy models operational (usually maximum GNP) are derived from an arbitrary value system.

Occasionally such models are used as descriptive rather than as normative tools. In these cases one is forced to argue that either policies are always invoked which lead to an optimal state, or that the system evolves as if it were an approximately optimal state. The "as if" position is clearly stronger since the optimal policy position is belied by the complicated and idiosyncratic processes societies evolve for regulating themselves. The "as if" merely asserts that the optimal evolution need not be literally true, but that as an approximation it is workable. This is certainly a much weaker condition. Whether or not the approximation is in fact workable enough for the particular question under scrutiny is another matter altogether. We have chosen to follow the descriptive rather than the normative approach to modeling.

Some Examples

The above discussions outline the philosophy of how we believe policy analysis should proceed. This section presents some examples of the types of policy analysis discussed.

Non-CO_2 Policy Impacts on CO_2 Emissions

Table 19-1 gives primary energy and CO_2 production resulting from parameter changes which can be interpreted as policy-induced impacts. The policy actions implied by the parameter changes are at least imaginable. It is easy to produce these cases with a computerized model. The aspect of this research that requires considerable additional effort is relating particular environmental hazards of energy use to particular restrictions or safeguards and then determining the equivalent quantity restrictions or cost add-ons to the basic technology. In general, environmental research has not looked as far into the future as is necessary for the CO_2 issue. That is, the impacts of a greatly increased scale of activities are not addressed. The parameter changes reported in table 19-1 were arrived at fairly arbitrarily and therefore we stress the illustrative nature of the exercise. We leave it to the reader to judge whether the various policy actions are particularly plausible or likely.

Again, as an illustrative exercise, general observations can be made from the results of table 19-1. The various policies tend to have a major impact on CO_2 emissions.

TABLE 19-1. Impacts of Various Energy Policies

	Primary Energy (EJ)				*CO_2 Emissions (10^9 tons carbon)*			
Case No.	*1975*	*2000*	*2025*	*2050*	*1975*	*2000*	*2025*	*2050*
(1) Base case	263	485	921	1646	4.68	6.98	12.26	26.38
			Year				Year	
Policy case		2000	2025	2050		2000	2025	2050
(2) Nuclear moratorium[a]		452	823	1455		6.93	12.98	29.01
(% change from base)		(−6.8)	(−10.6)	(−11.6)		(+0.9)	(+5.9)	(+10.4)
(3) Coal environmental[b]		472	869	1533		6.38	9.83	21.25
(% change from base)		(−2.7)	(−5.6)	(−6.9)		(−7.1)	(−19.8)	(−19.1)
(4) Shale oil environmental[c]		485	888	1549		6.87	11.02	19.22
(% change from base)		(0.0)	(−3.6)	(−5.9)		(0.0)	(−10.1)	(−26.86)
(5) Solar electric subsidy[d]		490	934	1674		6.82	12.05	25.66
(% change from base)		(+1.0)	(+1.4)	(+1.7)		(−0.7)	(−1.7)	(−2.4)
(6) All of above policies together		453	789	1366		6.44	9.34	16.52
(% change from base)		(−6.6)	(−14.3)	(−17.0)		(−6.3)	(−23.8)	(−37.1)

[a]All nuclear plants gradually phased out.
[b]Environmental considerations force a quadrupling of coal use costs by 2025.
[c]Environmental considerations double shale oil breakthrough costs.
[d]Research support or direct production subsidy reduces solar electric costs to one-half those in the base case.

And, as was argued earlier in the chapter, if such policies are instituted in their own right, they have a considerable effect on dimensions of the CO_2 issue. Because of the illustrative nature of the exercise, we have not proceeded to the next step, i.e., accepting one of the policy cases at a best guess and therefore as an appropriate base from which to evaluate CO_2 policy. However, obviously, controls on carbon emissions could be considerably less restrictive if one viewed case (6) as the best-guess case rather than case (1).

A METHODOLOGY FOR COMPARISON OF ALTERNATIVE WORLD VIEWS

There are a variety of "world views" concerning future energy use. These often correspond to an entire philosophy of how society responds to the problems and potentials presented by new and changing sources of energy. The debate over "hard" versus "soft" energy paths is illustrative. These views have been arrived at using a variety of analytic techniques, including various formal models and informal mental models or back-of-the-envelope calculations. An exercise which has been pursued with some success is the comparison of different models.[1] This is one-half of an exercise in determining comparability and points of difference in future forecasts but it tends to include only computerized models. Nevertheless, it is an analysis which is valuable because it gives insights into differences in the way models behave. The question asked in such an exercise is: Given that basic exogenous factors are the same, what differences result in forecasts because of structural differences in the model?

This question can be turned around. That is, one can ask: Given a common model structure, how must a set of exogenous input values differ to produce scenarios representing a variety of views? This exercise becomes meaningful when one accepts the fact that most parameters and exogenous assumptions are uncertain. One can then examine the necessary differences in input assumptions to determine where scenarios differ. Presumably the input assumptions, while uncertain, are values on which direct evidence exists or on which further research can determine a reasonable value range. Differences among views of the future are thus reduced to a level where discussions can be meaningful. The particular value of such an exercise is that all input assumptions must be considered explicitly and the parameters and exogenous variables are strictly comparable among the different scenarios. Like the model comparison exercise, it requires a suspension of disbelief during the course of the exercise. In the modeling exercise, participants accept the exogenous variable assumptions. In the proposed exercise, participants accept the model structure for the course of the exercise.[2]

It, of course, is the case that any conclusions are sensitive to the particular model structure chosen. One could imagine a full exercise where both model structure and world views are compared in a matrix of participants with various views and several different model structures. This could become a very large exercise. However, some guidelines in selecting models can reduce the number of models which might be selected or, alternatively, make the exercise more valuable even if a single model is chosen. The basic guideline is to choose (a) model structures which do (does) not impose restrictions which the participants cannot accept. Thus, if one participant believed that price was not an important determinant of energy demand while another believed it was, an acceptable model structure would provide the ability to vary the price elasticity. This criteria would exclude two kinds of models. It would exclude models which fail to explicitly consider price and it would exclude models where the price response is an endogenously (econometrically) determined quantity. Similarly, if some participants would posit that all conservation is price driven while others would argue that advancing technology leads to either improving energy efficiency or a more energy-intensive mix of products apart from any price changes, then a more general model should have the capability to represent both of these views.

Of course, no model is perfectly general—the very purpose of modeling is to structure the problem—and various models are more general in some representations and less general in others. Having said this, we propose that the model developed in chapter 16 is a good candidate for this exercise.

An Example

The main requirement of a model used in such an exercise is that it be flexible enough to represent diverse scenarios and assumptions. We have chosen to replicate a low energy use, slow growth, high conservation and decentralized power scenario.[3] This "soft" energy path represents a radically different future energy use path than the base case.

In order to replicate the scenarios, we designate a set of target outputs and a set of instrumental variables. We assume that other variables remain unchanged. The process of hitting target outputs is an iterative procedure, which is not mechanized. There is not, in general, a unique set of changes in assumptions which generate a given set of target outputs. This arises because there are more instrumental variables than target outputs and, in addition, any given instrument affects more than one output. Eventually one needs a single "best" set of input values. Here "best" ought to be defined by the individual holding the view. Thus, the exercise is seen as one where perhaps a week is spent in a workshop atmosphere where each participant is able to create his "best" scenario. The example reported in this section was not produced in such a manner and, therefore, is meant to be primarily illustrative.

Table 19-2 reports some of the results. GNP growth was assumed to be lower as part of the scenario, commercial energy use was a target, and the other outputs are incidental results. One can, perhaps, discuss the likelihood of each of the outputs of table 19-2. However, the value of the exercise is in comparing input assumptions since these are, in general, more easily related to specific, identifiable relationships which are subject to some form of empirical testing. The illustrative nature of the exercise precludes any firm conclusions. For this reason, we will basically report the process of developing the soft energy path reported in the table. In some respects the process remains incomplete since additional iterations might be proposed as making a better case for a "soft" energy path.

TABLE 19-2. A Soft Energy Path and the Base Case

	1975	*Base Case in 2050*	*Soft Energy Path in 2050*
Primary commercial energy use (EJ)	263	1647	116
Nonelectric solar and conservation[a]	0	174	857
E/GNP ratio[b] (gigajoules/1975 constant U.S. dollars)	42.8	38.7	5.1
Energy use per capita[b] (gigajoules)	65	201	15
GNP per capita (1975 constant U.S. dollars)	1,523	5,198	3,034
GNP growth rate (1975–2050, %/yr)	N.A.	2.6	1.8

[a]See chapter 18 for a description of how nonelectric solar and conservation are calculated.
[b]Commercial primary energy use.

We began by introducing slower GNP growth. The rates were those used by Lovins.[4] Beyond 2025, GNP growth fell to 0.0 in the developed regions and 1.1 in the developing regions. This drop in GNP growth was insufficient, by itself, to drop energy use to its target range.

We next observed that a major theme of soft energy path proponents is the current economic viability or desirability of energy-saving technologies and life-styles. We characterized this as merely requiring time for acceptance of these technologies and life-styles. The model parameter of interest was a time trend change in energy use per unit of output. For the aggregate economy this rate was changed from 0.3 percent per year to 2 percent per year.

This led to some interesting results. Basically, fuel prices declined precipitously because of the tremendous drop in demand. At this iteration, energy consumption fell more but remained above target levels because of increased demand in response to the falling price of fuels.

In the next iteration, it became necessary to adjust supply specifications to maintain prices. As a benchmark, we assumed that oil and gas resources (the amount producible over all time) consisted of currently proved reserves plus historically produced amounts, thus allowing for no additional proving of oil in known fields or discoveries of other relatively low-cost fields.

By 2050, low-cost oil is still being produced in the Middle East, with very little being produced elsewhere. Some conversions of solids to liquids and gases occur but, in general, very little of either is demanded, and most electricity is generated by hydro power. A fair amount of energy is supplied by solids (which includes biomass).

This is generally consistent with the soft path view of the world, which sees a good share of energy demand being met by direct solar (nonelectric solar), doing without (conservation), and renewables (solids such as biomass and hydropower). Thus, we settled on this iteration. The parameter changes required to generate the scenario pose several questions; among them is whether a 2 percent per year reduction in energy use per unit output without price incentives is plausible. And is it reasonable to believe that fossil fuel resources are as limited as indicated by proved reserves? The research reviewed in chapters 2 through 15 has some bearing on the answers to these questions. However, we leave the questions posed as a stimulus to discussion and as a direction for further research.

Quantifying Uncertainty

This section reports results of model experiments which seek to define subjective probability limits on global energy use and CO_2 emissions. As discussed above, such an exercise is necessary if any statement of likelihood is to be associated with a set of results. While the formal treatment of uncertainty was beyond the scope of our current research program, the process of formulating and implementing the assumptions in an informal manner is instructive.

What we have chosen to do is to take key factors, which influence model outcomes, and determine standard deviation bounds for the value of each variable. These bounds have been determined subjectively. Extreme high values (in relation to their impact on CO_2 emissions) for parameters are used jointly to create a high CO_2 emission scenario. Low values are similarly used to construct a low emissions scenario. For the most part, the high CO_2 emissions scenario is also a high energy use scenario. The high CO_2 scenario has rapid economic expansion, low rates of exogenous energy productivity advances, cheap shale oil, and rapid exploitation of coal resources. On the other hand, the high CO_2 emissions scenario has prohibitively expensive nuclear and solar electric power generation costs and no electric vehicles, both of which might accompany a high energy use scenario (see table 19-3).

TABLE 19-3. Key Assumptions for the Reference and Extreme Scenarios

	Low	*Reference*	*High*
Base GNP growth (%/yr)			
Developed economies	1.8	2.3	3.1
Developing economies	3.1	4.0	4.4
World	2.1	2.9	3.5
Exogenous energy productivity (%/yr)	1.01	0.35	0.00
Power generation costs ($/GJ)			
Nuclear	9	9	Prohibitive
Solar	9	20	Prohibitive
Shale oil costs at breakthrough ($/GJ)	19.25	3.85	3.85
Base rate of coal expansion (%/yr)	1.4	2.9	3.3
Electric car	Yes	No	No

In most cases, we view the assumptions used to create the reference scenario as moderate; however, in some cases the reference scenario appears overly optimistic about the world's ability to achieve technological fixes to the problems of environmental degradation associated with energy extraction and use. Further, it seems likely that actions will be taken to limit these impacts. Thus, the reference case probably represents a case toward the high side of the probability distribution. Nevertheless, a best-guess case probably remains substantially above the low case. Results are given in table 19-4.

The high and low CO_2 cases encompass a wide range of possible futures. The high case is seven times the low CO_2 emissions case (compared to a factor of four for energy). Subjectively, we would assign a 90 percent confidence to the interval described by the high and low scenarios. While these boundaries are very wide by the year 2050, encompassing most other studies (see figure 17-2), including Nordhaus, Rotty–Marland, Häfele, and our own reference case, they do exclude our own soft path scenario in which global CO_2 emissions fall to 2 gigatons of carbon per year (GTC/yr) by the year 2050, and the extrapolations of historical (pre-1973) time trends which yield 125 GT/yr rates by the year 2050.

This impressionistic view of the bounds on likely futures of energy and CO_2 is only a place to start. In future research it will be necessary to explore more formally the uncertainty which pervades energy and CO_2 futures.

TABLE 19-4. Energy and CO_2 in the Low, Reference, and High Cases

	Energy (EJ/yr)			*CO_2 (10^9 tons carbon/yr)*		
Year	*Low*	*REF*	*High*	*Low*	*REF*	*High*
1975	259	259	259	4.7	4.7	4.7
2000	413	485	525	6.1	6.9	8.4
2025	533	922	1071	6.6	12.2	19.1
2050	560	1647	1998	6.5	26.3	47.4

NOTES

1. See the Energy Modeling Forum series of reports: Energy Modeling Forum, Terman Engineering Center, Stanford, California.
2. This approach has been successfully applied to the policymaking process in the Federal Republic of Germany. See J. Conrad, "Future Nuclear Energy Policy—The West German Enquete Commission," *Energy Policy* (September 1982), pp. 244–249.
3. The particular scenario is one put forth by Amory Lovins [see Lovins, A. B., L. H. Lovins, D. Krause, and W. Bach, *Energy Strategies for Low Climate Risks,* prepared for the German Federal Environmental Agency, San Francisco International Project for Soft-Energy Paths (June 1981)]; however, we take it as representative of a more general view and do not make any claim that it necessarily accurately represents any individual's particular view of the world's energy future.
4. Ibid.

Conclusions 20

EVOLUTION, DIVERSITY, AND UNCERTAINTY

Energy Evolution

A popular conception of the future path of energy use portrays the world entering a period of transition during which the world economy either builds a sound foundation for a sustainable future or fails to adjust its appetite for dwindling fossil fuels and faces a day of reckoning when the last oil and gas run out. The perception of potantially catastrophic resource constraints is not new. It dates almost to the beginnings of the economics discipline. It appeared first in Malthus' *Essay on the Principle of Population* in 1789 and was later revived as a theory of natural resource constraint by Jevons, who forecast the near-term decline of British economic power as a consequence of limited coal reserves in the United Kingdom. Its most recent incarnation began with *The Limits to Growth* which portrayed a world rapidly running out of the basic mineral resources upon which modern economic growth was built.[1] *Limits to Growth* posed the problem as a period of 20 years or so during which the world must adjust. Later studies took issue with the conclusions but seemed to maintain the transition paradigm as a reference.

The perception of a period of energy transition from a steady-state solution existing prior to 1973 to an implied steady state some time after the turn of the century is reflected in the titles of what may be a majority of important energy studies of the past several years. Included among these are *Energy in Transition, 1985 to 2010* (CONAES, 1979),[2] *Energy in a Finite World: Paths to a Sustainable Future* (IIASA, 1981),[3] *Rays of Hope: The Transition to a Post-Petroleum World* (Hays, 1977),[4] *Energy Transitions: Long-Term Perspectives* (AAAS, 1981),[5] *Coal: Bridge to the Future* (WOCOL, 1980).[6]

The eventual steady-state solution has appeal because it offers a final simple answer. The study is finished, policy is completed, when the final fuel mix is indefinitely sustainable. One lists those energy technologies which can provide a flow of energy over an indefinite period of time (solar, the breeder, wind, fusion, biomass) and then chooses among them through an implicit or explicit weighing of their desirable characteristics (cost, environmental impacts, societal impacts). The energy projection problem then becomes one of mapping out the transition period or finding some combination of exhaustible resource supplies which can serve as a bridge to the future while governments vigorously develop the foundations for a sustainable society built on renewable sources of energy. This simplistic caricature of the energy transition is certainly unfairly applied to any of the above studies. More often it would seem that the study title and design reflect the basic transition view of the world but the research has shown, even when designed with the belief that a transition to renewables is around the corner, that forces governing the timing of such a transition were not nearly as powerful as first believed.

In light of the findings of these studies and our own work, we have come to understand the energy developments which will likely transpire in the next three-quarters of a century in a new light—as what we call an "energy evolution." We use the term evolution to denote a continuous gradual change over the period, with changes likely to continue beyond the time horizon of the model run. In view of the historical evidence, a paradigm of continuous energy evolution appears much more accurate than one of transition between steady states. The historical plots of the IIASA market penetration model show little in the way of a steady state at any time in the 100-year historical period examined in the study. The record shows a continuous shift of fuels as new technologies are developed. The forces behind the changes have been both necessity (e.g., local exhaustion of wood supplies in Europe can be identified with the introduction of coal) and technological advances (e.g., the development of nuclear power).

Unfortunately the ability to project continuous evolution is limited by the ability to foresee technological breakthroughs that bring about change. If we attempted to run the present version of our model well beyond the 2050 terminal date, it would very likely show eventual convergence to something like a steady state solution (few changes occurring from one period to the next). After examining the historical evolution of energy systems as well as the results of other projection efforts, it is our conclusion that such a result would indicate only that the model was pushed beyond the horizon of the underlying analysis.

A fair question to ask is: How does a revision in the analytical paradigm change or add to energy projections and their policy implications? The differences are subtle but important. In terms of projections, it means that a steady-state solution indicates a failure of the model and an attempt to extend it further than it should be. In terms of policy implications, a model of energy evolution directs policy toward aiding or easing the problems of change rather than searching for the ultimate solution. Policies must continuously adapt as energy markets and systems evolve as well as when policy goals change.

Energy Diversity

The concept of an ultimate "backstop" technology has been almost as popular as the concept of an energy system which either is transformed into a steady state or collapses as a consequence of resource depletion. In fact, the two concepts are not far removed from one another. If the world does achieve the transition to a new energy age, based on nonresource-constrained energy sources, there must be a technology available which is equal to the task. The technological backstop varies, depending upon the particular advocate's position, but nuclear and solar/conservation energy are the current favorites, though fusion energy has its supporters. Yet we see no support for the view that a single technology can fill all or most of societies' energy needs, either in the modern historical record or in our analysis of key factors influencing energy supply and demand over the period leading to 2050. Diversity of energy supply technologies follows from the interaction of supply and demand considerations. The nonenergy cost of providing energy services in different applications varies enough to allow supply technologies to compete in some markets even though one fuel may clearly supply energy at a lower cost per joule. Even though coal has been a cheaper source of energy, nonenergy costs and considerations have caused it to lose its market share continuously over a long period.

The historical record shows an increasingly diverse pattern of energy supply over the modern period. Earliest man used his own muscle power, animal power, and firewood. Later he developed the capability to use wind and water power, then coal, oil, and gas. Most recently, nuclear power has been developed as a commercial energy source.

Modern economic development has led to greater sophistication in the ways in which societies use energy. Energy in low grade forms has been transformed into higher, more concentrated and more usable forms. Crude oil is refined, and heat from various lower grade sources is transformed into electricity. Diversified applications of energy have led to similarly diversified energy using capital stocks and associated technologies. Personal transportation has been most successful in utilizing liquids; electricity is clearly preferred in lighting; natural gas has proved to be a clean, convenient heating fuel; and solids have proved an inexpensive source of low-grade energy.

Our analysis suggests a continued diversification of the global energy system. Coal will again become an important primary energy source. However, the uses of coal are likely to change dramatically, because some coal uses require higher grades of energy. Coal will be used increasingly to produce electric power or be converted to synthetic liquids and gases. The direct use of coal, as an energy source already confined to the industrial sector in the developed economies, will continue to shrink as a share of total energy use. The share of energy supply in the form of conventional liquids and gases also will shrink. The reduced importance of these fuels is brought on by the necessary shift toward diverse new supply technologies, including shale oil and synthetics from coal and biomass. Nuclear power and coal compete to provide a large share of electricity but solar (including wind) and hydro contribute significant shares by 2050. The more important contribution of renewable technologies results from the use of direct solar (nonelectric) and from conservation. By 2050 these energy "sources" contribute 35 percent of final energy use.

While uncertainty must dominate any projection for as distant a period as 2050, we have every reason to believe that the geographically and technologically diverse nature of future energy systems is a robust finding. If anything, we see increasing diversity, with numerous solar, biomass, nuclear, and other energy sources and technologies used to provide energy in solid, liquid, gaseous, and electrical forms.

Uncertainty

There are many ways to look at the future. One way is to offer predictions, which by their nature can only be tenuous. One must not only understand the way in which independent variables of the system interact to determine the values of the dependent (predicted) variables, but one must be able to predict the values of the independent variables as well.

An alternative way of formulating the problem is to offer conditional predictions. That is, the prediction takes the values of important independent variables to be exogenously determined. This places the responsibility for determining the values of key independent variables outside the purview of the modeler. The modeler is only responsible for the determination of the model structure and the values used in the model.

Parametric models take a still more agnostic view. Here neither model parameters nor independent variables are predicted. Only the model structure is held sacred. As one moves across this spectrum of modeling metaphors, the model results are held with decreasing subjective certainty by the modeler.

In trying to look ahead three quarters of a century, very little seems fixed. Many of the reference points that anchor energy forecasts fade with time. After 75 years, almost none of the industrial capital stock will remain intact. The personal transport fleet will have been built and retired several times over. Residential uses of energy will likely have evolved beyond recognition. There is less and less with which to orient one's forecasts, making it ever more difficult to develop a "best guess" scenario that has any validity.

In view of the great uncertainty surrounding the forecast of global energy production and use to the year 2050, we present our base case not as the most likely future, but rather as a plausible, consistent, and reproducible scenario of the future. As such, it provides a mechanism for conducting controlled experiments. Those experiments involve a systematic examination of responses to policy actions and alternative underlying assumptions.

Resource amounts, costs of new technologies, economic growth, and price and income responses are all highly uncertain. While we have explored alternative assumptions within the framework of the model, a systematic documentation of the impact of these uncertainties on the evolution of energy systems is left for future work. Our base scenario is based on reasonable, plausible assumptions set in a conservative, business-as-usual environment. This does not necessarily make the projections highly probable. We urge the reader to view our results as a plausible rather than a most likely case. As such, views of the world which are much different than our base case are not necessarily much less likely than a scenario very close to the base case; not only is the distribution of plausible scenarios wide, but an accurate subjective assessment suggests the distribution is nearly flat over fairly wide ranges.[7] We view the existence of alternative world views as evidence of the uncertainty surrounding the future course of man's energy system, but not necessarily a measure of it. Amory Lovin's notable time trend of U.S. energy demand forecasts indicates that the entire spectrum of forecasts may shift dramatically. What was once thought a likely future state may be demoted to "unthinkable," while forecasts regarded as extremely unlikely may come to be regarded as the common wisdom. The relationship between forecasts and the underlying probability of future events is not as solid as forecasters would wish. The conventional wisdom surrounding energy forecasts at any given time may understate the true uncertainty since the conventional wisdom is built up by a consensual process and is based on earlier forecasts and projections. And, of course, there is the uncertainty of human behavior, which may not fulfill the expectations of the model.

Perhaps the most important uncertainties associated with the energy future are the interaction of energy and the environment. Major new technologies will continue to evolve, including coal liquefaction, coal gasification, shale oil production, solar electricity generation, and fast-breeder reactors. It is not clear what the environmental impacts of these technologies will be if and when they become major contributors to global energy needs. Even less certain are the policy responses to the foreseen and unforeseen environmental impacts of existing and new technologies. Other uncertainties include the capital costs of energy technologies and the requisite global savings rates, the international trade environment, the OPEC policy on Middle East oil and gas production, and global population and economic growth rates.

A FINAL COMMENT

We have attempted to document in as complete a manner as possible what we have uncovered in our work to date. In doing so we hope it offers guidance to those who must plan for the future, but more important, we hope that it serves as a basis upon which we and others can build in future research. In many ways the distillation of our findings inadequately conveys the richness of what we have found. As with so many endeavors, the rewards, in terms of understanding, associated with actually doing the work—making the mistakes, sorting through the varying qualities of information, and piecing together a coherent whole—are greater than can be gained by simply following the well-marked trail of final results. There is much to be said for the process of analysis and modeling as teacher. It is often as useful to know the "wrong" answers and why they are "wrong" as it is to know the "right" answers and why they are "right." All of this cannot be packed between two covers.

We, of course, bring varying degrees of prejudice and a full set of preconceptions to our work. Some of these are the legitimate distillations of research performed in the service of past work and from formal training. Many are unrecognized. All are by-products of a much longer process of study and work and by their nature remain undocumented.

NOTES

1. D. H. Meadows et al., *The Limits to Growth* (New York: Universe Books, 1972).
2. Committee on Nuclear and Alternative Energy Systems, *Energy in Transition, 1985–2010* (San Francisco: W. H. Freeman, 1979).
3. Wolf Häfele, *Energy in a Finite World: Paths to a Sustainable Future.* Report of the International Institute of Applied Systems Analysis (IIASA) (Cambridge, Mass.: Ballinger, 1981).
4. Denis Hayes, *Rays of Hope: The Transition to a Post-Petroleum World* (New York: W. W. Norton, 1971).
5. Lewis J. Perelman, A. W. Gubelhaus, and M. D. Yohell (eds.), *Energy Transitions: Long-Term Perspectives.* American Association for the Advancement of Science Selected Symposium (Boulder, Colo.: Westview Press, 1981).
6. Carroll L. Wilson, *Coal—Bridge to the Future.* Report of the World Coal Study (Cambridge, Mass.: Ballinger, 1980).
7. See W. Nordhaus and G. Yohe, "Modeling Uncertainty About CO_2," preliminary draft paper prepared for the International Institute for Applied Systems Analysis Conference, July 1982 for further evidence of large uncertainty in forecasting long-term energy and CO_2 emissions.

Index